McCook(

Basic Studies in the Field of
High-temperature Engineering

Second Information Exchange Meeting
Paris, France
10-12 October 2001

Co-organised by the
Japan Atomic Energy Research Institute (JAERI)

NUCLEAR ENERGY AGENCY
ORGANISATION FOR ECONOMIC CO-OPERATION AND DEVELOPMENT

ORGANISATION FOR ECONOMIC CO-OPERATION AND DEVELOPMENT

Pursuant to Article 1 of the Convention signed in Paris on 14th December 1960, and which came into force on 30th September 1961, the Organisation for Economic Co-operation and Development (OECD) shall promote policies designed:

- to achieve the highest sustainable economic growth and employment and a rising standard of living in Member countries, while maintaining financial stability, and thus to contribute to the development of the world economy;
- to contribute to sound economic expansion in Member as well as non-member countries in the process of economic development; and
- to contribute to the expansion of world trade on a multilateral, non-discriminatory basis in accordance with international obligations.

The original Member countries of the OECD are Austria, Belgium, Canada, Denmark, France, Germany, Greece, Iceland, Ireland, Italy, Luxembourg, the Netherlands, Norway, Portugal, Spain, Sweden, Switzerland, Turkey, the United Kingdom and the United States. The following countries became Members subsequently through accession at the dates indicated hereafter: Japan (28th April 1964), Finland (28th January 1969), Australia (7th June 1971), New Zealand (29th May 1973), Mexico (18th May 1994), the Czech Republic (21st December 1995), Hungary (7th May 1996), Poland (22nd November 1996) and the Republic of Korea (12th December 1996). The Commission of the European Communities takes part in the work of the OECD (Article 13 of the OECD Convention).

NUCLEAR ENERGY AGENCY

The OECD Nuclear Energy Agency (NEA) was established on 1st February 1958 under the name of the OEEC European Nuclear Energy Agency. It received its present designation on 20th April 1972, when Japan became its first non-European full Member. NEA membership today consists of 27 OECD Member countries: Australia, Austria, Belgium, Canada, Czech Republic, Denmark, Finland, France, Germany, Greece, Hungary, Iceland, Ireland, Italy, Japan, Luxembourg, Mexico, the Netherlands, Norway, Portugal, Republic of Korea, Spain, Sweden, Switzerland, Turkey, the United Kingdom and the United States. The Commission of the European Communities also takes part in the work of the Agency.

The mission of the NEA is:

- to assist its Member countries in maintaining and further developing, through international co-operation, the scientific, technological and legal bases required for a safe, environmentally friendly and economical use of nuclear energy for peaceful purposes, as well as
- to provide authoritative assessments and to forge common understandings on key issues, as input to government decisions on nuclear energy policy and to broader OECD policy analyses in areas such as energy and sustainable development.

Specific areas of competence of the NEA include safety and regulation of nuclear activities, radioactive waste management, radiological protection, nuclear science, economic and technical analyses of the nuclear fuel cycle, nuclear law and liability, and public information. The NEA Data Bank provides nuclear data and computer program services for participating countries.

In these and related tasks, the NEA works in close collaboration with the International Atomic Energy Agency in Vienna, with which it has a Co-operation Agreement, as well as with other international organisations in the nuclear field.

FOREWORD

In the field of basic research on materials pertinent to high-temperature and neutron irradiation tests, great international interest exists concerning the study of irradiation effects, notably on advanced ceramic and metallic materials and on the development of in-core instrumentation.

For this reason, the Nuclear Science Committee (NSC) of the OECD Nuclear Energy Agency (NEA), along with the Japan Atomic Energy Research Institute (JAERI), organised the Second Information Exchange Meeting on Basic Studies in the Field of High-temperature Engineering. The meeting was held in Paris on 10-12 October 2001, and attended by 43 participants from eight countries and two international organisations. A total of 28 papers were presented, including 26 orally, dealing with recent progress in the following research domains:

- overviews of high-temperature engineering research in each country and organisation;

- improvement in material properties by high-temperature irradiation;

- development of in-core material characterisation methods and irradiation facilities;

- basic studies on the behaviour of irradiated graphite/carbon and ceramic materials including their composites under both operation and storage conditions;

- basic studies on high-temperature, gas-cooled reactor (HTGR) fuel fabrication and performance.

The final session was dedicated to discussing possible co-operative works to be undertaken within the framework of the OECD/NEA Nuclear Science Committee. It was recommended that a further information exchange meeting, dealing with possible international collaboration as well as some of the topics addressed here, should be planned to be held in Oarai, Japan in 2003.

TABLE OF CONTENTS

Session Summaries

Chair: Y. Sudo

EXECUTIVE SUMMARY

Co-organised by the OECD Nuclear Energy Agency and Japan Atomic Energy Research Institute, the Second Information Exchange Meeting on Basic Studies in the Field of High-temperature Engineering was a great success. It was held in Paris on 10-12 October, 2001, and its main topics were:

- The exchange of information with regard to the latest activities for studying high-temperature irradiation-induced damage/effects in advanced materials and fuels.

- The discussion of the possibility of co-operative studies in the field of high-temperature engineering using the HTTR and other research reactors for irradiation research programmes in the framework of international co-operation organised by the OECD/NEA/NSC.

- The development of a framework of possible international co-operative activities concerning topics of interest.

- The proposal/recommendation of possible co-operative works to be undertaken within a framework or under the auspices of the OECD/NEA/NSC.

Forty-three participants gathered together from eight countries and two international organisations. Twenty-eight papers were submitted, of which 26 were oral presentations dealing with recent progress and national/international programmes in the field of high-temperature engineering. The papers were presented in five sessions as follows:

- *Session I*: Overviews of high-temperature engineering research in each country and organisation (six papers).

- *Session II*: Improvement in material properties by high-temperature irradiation(four papers).

- *Session III*: Development of in-core material characterisation methods and irradiation facilities (six papers).

- *Session IV*: Basic studies on the behaviour of irradiated graphite/carbon and ceramic materials including their composites under both operation and storage conditions(eight papers).

- *Session V*: Basic studies on HTGR fuel fabrication and performance(two papers).

In Session I, the current status of high-temperature engineering research in the EC, Japan, the Netherlands, France, Germany and the United Kingdom were summarised.

In Session II, examples of improvement in material properties through high-temperature irradiation, the effects of superplastic deformation on thermal and mechanical properties of 3Y-TZP ceramics and the development of an innovative carbon-based ceramic material and carbon dioxide partial condensation cycle were presented.

Session III was dedicated to the development of in-core material characterisation methods and irradiation facilities. New irradiation equipment for the HTTR, optical diagnostic system for the high-temperature gas-cooled reactor, in-core neutron and gamma-ray distributions measurement with scintillator optical fibre detector and self-powered detector, in-core test capabilities for material characterisation in the OECD Halden reactor, work function measuring technique to monitoring/characterisation of material surfaces under irradiation, and materials irradiation tests at high temperatures in JMTR were some of the topics presented.

In Session IV, presentations were made on the subjects of finite-element analyses of the mechanical behaviour of graphite cores, mesoscopic analyses for mechanical properties of graphite, graphite oxidation models, irradiation creep in graphite, and planning for disposal of irradiated graphite and database on irradiated graphite properties are presented.

Session V is related to the basic studies on HTGR fuel fabrication and performance. Fission product release from ceramic materials and reactor test technique of HTGR fuel elements were also discussed.

In the last session of the meeting, the floor was open for discussion on possible activities in the framework of OECD/NEA/Nuclear Science Committee for each session. In addition to the topics discussed in this meeting, it was recommended to include topics relevant to fission product behaviour and safety issues of HTGR in next meeting. It was proposed that a further information exchange meeting, dealing with possible international collaboration using HTTR and other irradiation facilities, be held in autumn of 2003 at Oarai, Japan.

OPENING SESSION

WELCOME ADDRESS

Carol Kessler
Deputy Director-General, OECD/NEA

Good morning, ladies and gentlemen.

It is a great pleasure for me to welcome you, on behalf of OECD Nuclear Energy Agency (NEA), to this Second Information Exchange Meeting on Basic Studies in the Field of High-temperature Engineering. I would like to thank the members of the Organising Committee and of the Scientific Advisory Board for all their efforts in preparing this meeting and I am sure it will be a successful one. This meeting was also co-organised by the Japan Atomic Energy Research Institute.

The first such information exchange meeting was held September 1999 in Paris and proved quite successful. At this meeting some topics were recommended for further development and we will be focusing today on those recommendations. I understand that the subjects of these recommendations have been revised slightly, but the point is we now know where we want to head and this meeting will take us along the next steps. I understand there is a desire to meet again in two years time based on the progress made in this meeting.

My personal experience is that these kinds of information exchange meetings are very useful for building an overall picture of ongoing and planned research activities in a specific area. They also provide the opportunity to meet and discuss issues with others with similar interests.

The meeting's focus on high-temperature gas-cooled reactors (HTGR) meets the growing interest of many OECD and non-OECD member countries. I am sure many of you are aware that while we are meeting, another part of the international community is meeting in Florida in the US to discuss the next steps in the Generation IV International Forum. This forum is moving towards decisions in the late 2002 timeframe on possible research projects needed to confirm the viability of truly advanced reactor designs. They will have an explicit need for an improved knowledge of materials that can be used in different high-temperature nuclear applications. For example, I understand this meeting will have a discussion about further international collaboration using irradiation facilities. This collaboration can play an important role in future study in this research field.

This meeting is also timely for the NEA and especially the Nuclear Science Committee. The NEA has a biennial programme of work and will soon start to prepare the programme for 2003-2004. This new programme will be discussed for the first time by the NEA Steering Committee in their spring 2002 meeting. Thus any recommendations from this meeting today can be taken into account as we plan for the next two years' programme of work.

In fact, there are a number of ongoing projects within the Nuclear Science Committee that could benefit from your discussion and recommendations. I am thinking primarily about two possible activities, one within the field of radiation-induced material damage and the other, the setting up of an expert group on the nuclear production of hydrogen.

I would like to thank the Japan Atomic Energy Research Institute (JAERI) for co-organising this meeting and especially Dr. Toshiyuki Tanaka and Dr. Yukio Sudo, the former and present directors general of the Oarai Establishment for their support of this information exchange meeting.

I wish you all a very interesting and productive meeting.

Thank you for your attention.

OPENING ADDRESS

Shori Ishino
Tokai University, Japan

Good morning, ladies and gentleman.

On behalf of the Organising Committee, I cordially welcome all of you to the Second Information Exchange Meeting on Basic Studies in the Field of High-temperature Engineering.

It was in November 1997 at the workshop on High-temperature Engineering Research Facilities and Experiments held in Petten, the Netherlands, that the first recommendation for this meeting was made. The proposal for the meeting was made after consultation with experts from Japan, the Netherlands and Germany, and was then discussed and approved at the OECD/NEA/NSC meeting in June 1998.

The information exchange meetings' background was briefly summarised by Mrs. Kessler in her welcome address, so I would like to concentrate my speech on its purpose and meaning.

This is a follow-up to the First Information Exchange Meeting on Basic Studies in the Field of High-temperature Engineering, which was held in Paris, on 27-29 September 1999. We are here, as all of you know, to exchange information on the latest methods of understanding high-temperature irradiation-induced phenomena in the field of high-temperature engineering and for improvement of properties in advanced materials and fuels using high-temperature irradiation.

Possible international collaboration, using HTTR and other research reactors for irradiation research programmes in the field of high-temperature engineering, the framework of co-operation activities organised by OECD/NEA/NSC and some related topics, recommended for further development at the first meeting, will be discussed at this meeting. For this reason, a final summary session will be held on the last day for discussion and preparation of recommendations.

It is my great pleasure that more than 40 participants have gathered together here today to take part in these six plenary sessions, during which 28 papers will be presented. These numbers are more than expected, and have greatly encouraged the Organising Committee. This level of interest is, I believe, due to the importance of basic studies in high-temperature engineering.

In the present situation, HTGRs are developed as a new generation-type gas-cooled reactor including inherent safety. Some of the papers submitted concern feasibility studies on HTGR; all of these have been included in the program. These current topics concerning the development of HTGR have received great interest among researchers, and this will be a driving force for us to undergo basic studies and to promote international collaboration within the framework of the NEA/NSC.

The Organising Committee hopes that this information exchange will present a good opportunity to elucidate the basic research subject of high-temperature engineering, and to develop international co-operative works under the auspices of the OECD/NEA/NSC.

Finally, I would like to express my sincere gratitude to the Scientific Advisory Members for their support, the NEA Secretariat for the preparation of this meeting and the Japan Atomic Energy Research Institute as a co-organiser for the technical support to this meeting. Also, I would like to thank all of you, the participants, especially those who have come from countries outside Europe.

SESSION I

Overviews of High-temperature Engineering Research in Each Country and Organisation

Chair: W. von Lensa

RESEARCH ACTIVITIES ON HIGH-TEMPERATURE GAS-COOLED REACTORS (HTRs) IN THE 5TH EURATOM RTD FRAMEWORK PROGRAMME

Joaquín Martín-Bermejo
European Commission, DG RESEARCH, Brussels, Belgium

Michel Hugon, Georges Van Goethem
European Commission, DG RESEARCH, Brussels, Belgium

Abstract

One of the areas of research of the "nuclear fission" key action of the 5th EURATOM RTD Framework Programme (FP5) is the safety and efficiency of future systems. The main objective of this area is to investigate and evaluate new or revisited concepts (both reactors and alternative fuels) for nuclear energy that offer potential longer term benefits in terms of cost, safety, waste management, use of fissile material, less risk of diversion and sustainability. Several projects related to high-temperature gas-cooled reactors (HTRs) were retained by the European Commission (EC) services. They address important issues such as HTR fuel technology, HTR fuel cycle, HTR materials, power conversion systems and licensing. Most of these projects have already started and are progressing according to the schedule. They are the initial core of activities of a European Network on "High-temperature Reactor Technology" (HTR-TN) recently set up by 18 EU organisations.

Introduction

The 5th Framework Programme sets out the priorities for the European Union's research, technological development and demonstration activities for the period 1998-2002 [1]. These priorities have been identified on the basis of a set of common criteria reflecting the major concerns of increasing industrial competitiveness and the quality of life for European citizens.

The 5th Framework Programme differs considerably from its predecessors. It has been conceived to help solve problems and to respond to major socio-economic challenges facing Europe. To maximise its impact, it focuses on a limited number of research areas combining technological, industrial, economic, social and cultural aspects. Management procedures have been streamlined with an emphasis on simplifying the process and systematically involving key players in research.

The 5th Framework Programme has two distinct parts: the European Community framework programme covering research, technological development and demonstration activities, and the EURATOM framework programme covering research and training (RT) activities in the nuclear sector. A budget of euros14 960 million has been agreed for the period up to the year 2002 of which euros 13 700 million is foreseen for the implementation of the European Community section of the 5th Framework Programme and euros1 260 million have been allocated to the EURATOM programme.

In practice, this Programme is implemented either via indirect actions co-ordinated by Directorate General RESEARCH of the European Commission (EC) or via direct actions under the responsibility of the Joint Research Centre (JRC) of the EC. The indirect actions consist mainly of research co-sponsored by DG RESEARCH but carried out by external public and private organisations as multi-partner projects. The direct RTD actions normally comprise research of an institutional character that is carried out directly by the JRC using their unique expertise and facilities.

The 5th EURATOM Framework Programme (1998-2002)

Overview

The strategic goal of the 5th EURATOM Framework Programme (FP5) for research and training activities is to help exploit the full potential of nuclear energy in a sustainable manner by making current technologies even safer and more economical and by exploring promising new concepts. The specific programme (indirect actions) comprises two "key actions", controlled thermonuclear fusion and nuclear fission, generic research on radiological sciences and support for research infrastructure.

The budget breakdown of the 5th EURATOM specific programme is as follows:

Key action 1: Controlled thermonuclear fusion	euros 788 million
Key action 2: Nuclear fission	euros 142 million
Generic R&D activities on radiological sciences	euros 39 million
Support for research infrastructures	euros 10 million
Joint Research Centre (direct actions)	euros 281 million

The key action "nuclear fission"

The main objectives of this key action are to enhance the safety of Europe's nuclear installations and improve the competitiveness of Europe's industry. Within these broader objectives, the more detailed aims are to protect workers and the public from radiation and ensure safe and effective management and final disposal of radioactive waste, to explore more innovative concepts that are sustainable and have potential longer term economic, safety, health and environmental benefits and to contribute towards maintaining a high level of expertise and competence on nuclear technology and safety.

Research focuses on issues that currently hinder the fuller exploitation of nuclear energy (e.g. cost, safety, waste disposal, public attitudes) and aim to demonstrate the availability of practical solutions to the outstanding scientific and technical problems and public concerns. It should bring together at the research stage different stakeholders sharing common strategic objectives. It covers four principal areas of research:

- Operational safety of existing installations.

- Safety of the fuel cycle.

- Safety and efficiency of future systems.

- Radiation protection.

Their objectives and main priorities are briefly described below.

Operational safety of existing installations

The objectives are to provide improved and innovative tools and methods for maintaining and enhancing the safety of existing installations, for achieving evolutionary improvements in their design and operation and for improving the competitiveness of Europe's nuclear industry. Three top research priorities have been identified:

- Plant life extension and management.

- Severe accident management.

- Evolutionary concepts.

Safety of the fuel cycle

The objective is to develop a sound basis for policy choices on the management and disposal of spent fuel and high-level and long-lived radioactive wastes and on decommissioning and to build a common understanding and consensus on the key issues. Three top research priorities have been identified:

- Waste and spent fuel management and disposal.

- Partitioning and transmutation.

- Decommissioning of nuclear installations.

Safety and efficiency of future systems

The objective is to investigate and evaluate new and updated concepts for nuclear energy that offer potential longer term benefits from the points of view of cost, safety, waste management, use of fissile material, less risk of diversion and sustainability. Innovative or revisited reactor concepts, and innovative fuels and fuel cycles have been identified as the top research priorities.

Radiation protection

The objectives are to help operators and safety authorities to protect workers, the public and the environment during operations in the nuclear fuel cycle, to manage nuclear accidents and radiological emergencies and to restore contaminated environments. Four top research priorities have been identified:

- Governance (assessment and management) of risk.

- Monitoring and assessment of occupational exposure.

- Off-site emergency management.

- Restoration and long-term management of contaminated environments.

Selection of projects in FP5

Proposals for "indirect" actions under the Framework Programmes are normally invited by the EC via calls for proposals with fixed deadlines. The eligible proposals are evaluated by independent experts who gave their advice to the Commission services, which takes it into account when selecting the portfolio to be negotiated. Up to date there have been three of these calls for proposals with deadlines 17 June 1999, 4 October 1999 and 22 January 2001 with a total funding of euros 153 million. The details of the research areas and priorities are given in the Work Programme, which is updated annually and is available on the CORDIS website (www.cordis.lu/fp5-euratom).

After the evaluation of the proposals received, a number of research contracts covering the objectives of the above-mentioned areas has been successfully negotiated by the EC and different EU organisations (see Table 1).

Table 1. Results of the first three calls for proposals of EURATOM FP5

Research area	Eligible proposals	Projects accepted	Commission funding (million euros)
Operational safety of existing installations	152	66	39
Safety of fuel cycle	102	57	57
Safety and efficiency of future systems	22	15	12
Radiation protection	49	19	9.5
Radiological sciences	97	44	29
Support for research infrastructures	24	19	6.5
Total	**446**	**220**	**153**

The research is implemented through: shared-cost actions (the main mechanism for the key action), support for networks and databases, training fellowships, concerted actions and accompanying measures. These indirect actions are managed and co-ordinated by Unit J-4 (Nuclear fission and radiation protection) of Directorate J (Preserving the ecosystem II) within DG *RESEARCH*. The actual work is carried out by laboratories, research centres, universities, regulatory authorities, utilities, engineering companies and consulting firms. The principal end-users of the results are nuclear power plants, national regulatory authorities and the industry sector in general.

Safety and efficiency of future systems

This is a newly identified research area for FP5 with the objectives of assessing new and updated concepts for nuclear energy that offer potential longer term benefits in terms of safety, economy, waste management, use of fissile material, less risk of diversion and sustainability. It covers two sub-areas, namely "Innovative or revisited reactor concepts" and "Innovative fuels and fuel cycles". Relevant work had been previously undertaken, in particular in the cluster of projects "INNO" of Area A ("Exploring innovative approaches") of the specific programme on Nuclear Fission Safety in the EURATOM 4th Framework Programme (1994-1998).

The majority of research projects selected for funding in this area are related to high-temperature reactors (HTR) and are grouped in a cluster. They address the main technical issues to be solved before using the HTRs at the industrial scale to produce energy: fuel technology, fuel cycles, waste, reactor physics, materials, power conversion systems and licensing. The rest of the projects in this area are assessing the sate of the art and R&D needs of other reactor concepts and of other applications of nuclear energy, such as supercritical pressure light water reactors, gas-cooled fast reactors, molten salt reactors and sea water desalination. Finally, a thematic network is addressing the competitiveness and sustainability of nuclear energy in the European Union.

The list of HTR-related projects co-sponsored by the EC is shown in Table 2. Four of them (HTR-F, HTR-M, HTR-N and HTR-C) were accepted in the call of proposals of 4 October 1999 whereas the rest (HTR-F1, HTR-M1, HTR-N1, HTR-E and HTR-L) were accepted in the call of 22 January 2001.

Research on HTRs in EURATOM FP5

A number of HTRs were developed throughout the 1960s and 1970s (i.e. Peach Bottom and Fort St. Vrain in the US, AVR and THTR in Germany, and Dragon in the UK) but then abandoned. However, its inherent safety features, its potential for use in high-temperature industrial processes and the possibility of using direct cycle gas turbines has kept the concept alive. In fact, there is renewed interest in other parts of the world (Japan, China, South Africa, Russia) and their potential for deployment later on in Europe is now being re-considered. However, a full development programme will require a large, fairly long-term effort for which a more proactive public R&D strategy might be necessary.

Current EC-sponsored projects

The nine HTR-related projects selected by the EC form a consistent and structured cluster covering both fundamental research and technological aspects. They were selected after two calls for

Table 2. Ongoing HTR-related research projects in EURATOM FP5

Acronym	Subject of research	Co-ordinator (country)	Number of partners	Duration (months)	EC funding (million euros)
HTR-F	HTR fuel technology	CEA (F)	7	48[*]	1.7
HTR-F1					0.8
HTR-N	HTR reactor physics and fuel cycle	FZJ (D)	14	54[**]	1.0
HTR-N1					0.55
HTR-M	HTR materials	NNC Ltd. (UK)	8	54[***]	1.1
HTR-M1		(UK)			0.7
HTR-E	Innovative components and systems in direct cycles of HTRs	FRAMATOME (F)	14	48	1.9
HTR-L	HTRs licensing safety approach and main licensing issues	Tractebel (B)	8	36	0.5
HTR-C	HTR programme co-ordination	FRAMATOME (F)	6	48	0.2

[*]   Duration of combined projects HTR-F and HTR-F1.
[**] Duration of combined projects HTR-N and HTR-N1.
[***] Duration of combined projects HTR-M and HTR-M1.

proposals with deadlines 4 October 1999 and 22 January 2001. The latter targeted on complementary R&D activities on HTRs with emphasis on issues which it was not possible to address in the former due to budget and scheduling constraints.

This section describes the objectives as well as the main experimental and analytical activities foreseen within the above-mentioned projects. Around 25 different organisations, representing research centres, universities, regulators, utilities and vendors from nine EU member states and Switzerland are involved.

Projects HTR-F and HTR-F1

These projects are "shared-cost" actions to be carried out by a consortium of seven organisations (CEA, FZJ, JRC-IAM, JRC-ITU, BNFL, FRAMATOME and NRG) under the co-ordination of CEA. The duration foreseen for the combined projects is 48 months.

The objectives of HTR-F are: (i) to restore (and improve) the fuel fabrication capability in Europe, (ii) to qualify the fuel at high burn up with a high reliability and (iii) to study innovative fuels that can be used for applications different from former HTR designs. The project started in October 2000 and its work programme includes the following activities:

- To collect data from the various types of fuels tested in the past in European reactors (e.g. HFR, THTR, DRAGON, OSIRIS, SILOE, etc.) and to analyse them in order to better understand the fuel behaviour and performance under irradiation.

- To define experimental programmes (in-pile and out-of-pile) in order to qualify the fuel particle behaviour under irradiation and high temperatures. A first irradiation test is planned in the HFR reactor on pebbles from the last German high quality fuel production with the objective to reach a burn-up of 200 000 MWd/t. Concerning the heat-up tests, the Cold Finger Furnace (KÜFA) facility, in which temperatures can reach up to 1 800°C, was transferred from Jülich (FZJ) to Karlsruhe (JRC/ITU) where it will be commissioned after having tested one irradiated pebble.

- To model the thermal and mechanical behaviour of coated fuel under irradiation and to validate it against the experimental results available. The models in existing codes (e.g. PANAMA, FRESCO, COCONUT, etc.) will be used to develop a common European code.

- To review the existing technologies for fabrication of kernels and coated particles, to fabricate first batches of U-bearing kernels and coated particles, to characterise them and to study alternative coating materials (e.g. ZrC and TiN). Kernels and particles will be fabricated in different laboratories (two at CEA and one at JRC/ITU) and the first coatings tests will be performed on simulated and depleted uranium kernels.

The programme of HTR-F1, which should start in November 2001, is fully complementary to HTR-F. It will enable to complete the irradiation of the German pebbles in the HFR in Petten, to carry out their post-irradiation examination (PIE) and to perform heat-up tests under accident conditions in the modified KÜFA facility at JRC/ITU. Also, the code developed in HTR-F for modelling the thermal and mechanical behaviour of the coated fuel particles should be validated. Finally, the production of coated particles and kernels should start at CEA and JRC/ITU.

Projects HTR-N and HTR-N1

These projects are "shared-cost" actions to be carried out by a consortium of 14 organisations (FZJ, Ansaldo, BNFL, CEA, COGEMA, FRAMATOME ANP SAS and GmbH, NNC Ltd., NRG, JRC-ITU, Subatech, and the Universities of Delft, Pisa and Stuttgart) under the co-ordination of FZJ. The duration foreseen for the combined projects is 54 months.

The main objectives of HTR-N are: to provide numerical nuclear physics tools (and check the availability of nuclear data) for the analysis and design of innovative HTR cores, to investigate different fuel cycles that can minimise the generation of long-lived actinides and optimise the Pu-burning capabilities, and to analyse the HTR-specific waste and the disposal behaviour of spent fuel. The project started in September 2000 and its work programme includes the following activities:

- To validate present core physics code packages for innovative HTR concepts (of both prismatic block and pebble bed types) against tests of Japan's high-temperature test reactor (HTTR) and to use these codes to predict the first criticality of China's HTR-10 experimental reactor.

- To evaluate the impact of nuclear data uncertainties on the calculation of reactor reactivity and mass balances (particularly for high burn-up). Sensitivity analyses will be performed by different methods based on today's available data sets (ENDF/B-VI, JEFF-3, JENDL 3.2/3).

- To study selected variations of the two main reactor concepts (i.e. hexagonal block type and pebble-bed) and their associated loading schemes and fuel cycles (i.e. the static batch-loaded cores and continuously loaded cores) in order to assess burn-up increase, waste minimisation capabilities, economics and safety.

- To analyse the HTR operational and decommissioning waste streams for both prismatic block and pebble bed types and to compare them with the waste stream of LWR.

- To perform different tests (e.g. corrosion, leaching, dissolution) with fuel kernels such as UO_2 and $(Th,U)O_2$ and coating materials of different compositions (e.g. SiC, PyC) in order to evaluate and generate the data needed to model the geo-chemical behaviour of the spent fuel under different final disposal conditions, i.e. salt brines, clay water and granite.

The HTR-N1 project proposes to: extend the nuclear physics analysis of HTR-N to the hot conditions of low-enriched uranium (LEU) cores with data from HTTR and HTR-10; to investigate the potential to treat or purify specific HTR decommissioning waste (e.g. structural graphite) on the basis of samples taken from the AVR side reflector and to continue the leaching experiments for disposed spent fuel with irradiated fuel (instead of dummies) for initial commissioning of the test rigs. The project is due to start in October 2001.

Projects HTR-M and HTR-M1

These projects are "shared-cost" action to be carried out by a consortium of eight organisations (NNC Ltd., FRAMATOME, CEA, NRG, FZJ, Siemens, Empresarios Agrupados and JRC-IAM) under the co-ordination of NNC Ltd. The duration foreseen for the combined projects is 54 months.

The objectives of HTR-M are to provide materials data for key components of the development of HTR technology in Europe including: reactor pressure vessel (RPV), high-temperature areas (internal structures and turbine) and graphite structures. The project started in November 2000 and its work programme consists of the following basic activities:

- Review of RPV materials, focusing on previous HTRs in order to set up a materials property database on design properties. Specific mechanical tests will be performed on RPV welded joints (FRAMATOME facilities) and irradiated specimens (Petten HFR) covering tensile, creep and/or compact tension fracture.

- Compilation of existing data about materials for reactor internals having a high potential interest, selection of the most promising grades for further R&D efforts, and development and testing of available alloys. Mechanical and creep tests will be performed at CEA on candidate materials at temperatures up to 1 100°C with focus on the control rod cladding.

- Compilation of existing data about turbine disk and blade materials, selection of the most promising grades for further R&D efforts, and development and testing of available alloys. Tensile and creep tests (in air and vacuum) from 850°C up to 1 300°C and fatigue testing at 1 000°C will be performed at facilities at CEA while creep and creep/fatigue tests in helium will be performed at JRC.

- Review the state of the art on graphite properties in order to set up a suitable database and perform oxidation tests at high temperatures on: (i) a fuel matrix graphite to obtain kinetic data for advanced oxidation (THERA facility at FZJ) and (ii) advanced carbon-based materials to obtain oxidation resistance in steam and in air respectively (INDEX facility at FZJ).

The HTR-M1 project complements HTR-M, as it concentrates on the long-term testing of the materials for the turbine and irradiation tests for the HTR graphite components. Special attention is placed on the fact that previous graphites are no longer available because the coke used as the raw material has either run out and the manufacturer's experience lost, or production techniques and equipment no longer exist. The work programme includes verification of models describing the graphite behaviour under irradiation and screening tests of recent graphite qualities. The project should start in November 2001.

Project HTR-E

This project is a "shared-cost" action to be carried out by a consortium of 14 organisations (FRAMATOME ANP SAS, Ansaldo, Balcke Dürr, CEA, Empresarios Agrupados, FRAMATOME ANP GmbH, FZJ, Heatric, Jeumont Industrie, NRG, NNC Ltd., S2M, University of Zittau and Von Karman Institute) under the co-ordination of FRAMATOME ANP SAS. The duration foreseen for this project is 48 months and the expected commencement date is December 2001.

This project addresses the innovative key components, systems and equipment related to the direct cycle of modern HTRs. These include turbine, recuperator heat exchanger, active and permanent magnetic bearings, rotating seals, sliding parts (tribology) and the helium purification system. The programme contains both design studies (e.g. computer fluid dynamics and finite element analyses) and also experiments (e.g. magnetic bearing tests at Zittau facility, validation tests of the recuperator at CEA's CLAIRE loop or tribological investigations at FRAMATOME's Technical Centre).

Project HTR-L

This project is a "shared-cost" action to be carried out by a consortium of eight organisations (Tractebel, Ansaldo, Empresarios Agrupados, FRAMATOME ANP SAS, FRAMATOME ANP GmbH, FZJ, NRG, and NNC Ltd.) under the co-ordination of Tractebel. The duration foreseen for this project is 36 months and the commencement date is October 2001.

The project proposes a safety approach for a licensing framework specific to modular high-temperature reactors and a classification for the design basis operating conditions and associated acceptance criteria. Special attention will be put on the confinement requirements and the rules for system, structure and component classification as well as a component qualification level being compatible with economical targets.

Project HTR-C

This is a "concerted action" to be carried out by a consortium of six organisations (FRAMATOME, FZJ, CEA, NNC Ltd., NRG, and JRC) under the co-ordination of FRAMATOME. The duration foreseen is 48 months.

This project, which started in October 2000, is devoted to the co-ordination and the integration of the work to be performed in all the above-mentioned projects. Moreover, HTR-C should organise a world-wide "technological watch" and develop international co-operation, with first priority to China and Japan, which have now the only research HTRs in the world. In order to promote and disseminate the achievements of the EC-sponsored projects, HTR-C will organise presentations in international conferences.

The "High-temperature Reactor Technology Network" (HTR-TN)

In the beginning of 2000, 15 EU organisations signed a multi-partner collaboration agreement to set up a European Network on "High-temperature Reactor Technology" hereinafter referred to as the "HTR-TN". The agreement does not involve cash flow between the members and all contributions are made in kind. The operating agent and the manager of this network is the JRC-IAM (Petten) and the rest of the partners are: Ansaldo (I), Belgatom (B), BNFL (UK), CEA (F), Empresarios Agrupados (E), FRAMATOME (F), FZJ (D), FZR (D), IKE (D), University of Zittau (D), Delft University (NL), NNC (UK), NRG (NL) and Siemens (D). Many of these organisations had already been working together in the "INNOHTR" Concerted Action of the EURATOM FP4 (contract FI4I-CT97-0015).

The general objective of this network is to co-ordinate and manage the expertise and resources of the participant organisations in developing advanced technologies for modern HTRs, in order to support the design of these reactors. The primary focus will be to recover and make available to the European nuclear industry the data and the know-how accumulated in the past in Europe and possibly in other parts of the world. The network should also work on the consolidation of the unique safety approach and of the specific spent fuel disposal characteristics of HTR, providing data, tools and methodologies which could be available for the safety assessment of European safety authorities. The EC-sponsored projects mentioned in the previous section are the initial "kernel" from which the HTR-TN has departed.

The activities of this network started officially in April 2000 at the kick-off meeting held in Petten (The Netherlands). During this meeting the Steering Committee of the network was constituted and different task groups were set up in order to implement the agreement. Six technical task groups were created to address the following areas: components technology, system and applications studies, material performance evaluation, safety and licensing, fuel testing, physics and fuel cycle including waste. In addition to these technical task groups some "horizontal" task groups were also formed to cover aspects such as strategies for future common projects, internal and external communications, and international relationships.

At the second Steering Committee meeting of the HTR-TN held in Brussels on November 2000 three new organisations, Balcke-Dürr (D), COGEMA (F) and VTT (FI) joined HTR-TN. The network remains open for further partners or associates from Europe and elsewhere. An HTR-TN web page has been set up by the network members using the CIRCA server of the JRC (http:/www.jrc.nl/htr-tn).

Conclusions

At present, there is renewed interest in different countries of the world (Japan, China, South Africa, Russia) in high-temperature gas-cooled reactors (HTRs). Their potential for deployment later on in Europe is also being re-considered. Its inherent safety features, its potential for use in high-temperature industrial processes and the possibility of using direct cycle gas turbines has made of HTRs a very attractive concept in the current socio-economic and political environment. However, a full development programme will require a large, fairly long-term effort for which a more proactive public R&D strategy might be necessary.

The EC is supporting a number of research projects on HTRs in its 5th EURATOM Framework Programme. They are aimed at investigating and evaluating the potential of this type of reactor in terms of safety, economy, waste management, use of fissile material, less risk of diversion and sustainability. These projects address important issues such as HTR fuel technology, HTR fuel cycle, HTR materials, power conversion systems and licensing.

The research activities, implemented mainly through shared-cost actions, concerted actions and networks, are managed by Unit J-4 (Nuclear fission and radiation protection) of Directorate J (Preserving the ecosystem II) within DG RESEARCH. The actual work is carried out by laboratories and research centres, utilities, engineering companies, regulatory authorities and universities. This enforced collaboration of organisations of different EU member states contributes to improve the consistency of the research fabric within the EU so that optimal use can be made of the available resources in the present and the future EURATOM Framework Programmes.

A European Network on "High-temperature Reactor Technology ("HTR-TN") has been set up by 18 EU organisations with the JRC-IAM (Petten) as the operating agent. The primary objective will be to recover and make available to the European nuclear industry the data and the know-how accumulated in the past in Europe and possibly in other parts of the world. The EC-sponsored projects in EURATOM FP5 mentioned above are the initial core of activities from which the HTR-TN departed.

REFERENCE

[1]   EUR 18764 "The Fifth Framework Programme – The Research Programmes of the European Union 1998-2002", European Communities, Luxembourg, 1999.

UPDATE ON IAEA HIGH-TEMPERATURE GAS-COOLED REACTOR ACTIVITIES

J. Kendall, M. Methnani
International Atomic Energy Agency
P.O. Box 100, A-1400, Vienna, Austria

Abstract

IAEA activities on high-temperature gas-cooled reactors are conducted with the review and support of Member States, primarily through the Technical Working Group on Gas-cooled Reactors (TWG-GCR). This paper provides an update on related IAEA activities in recent years with a focus on Co-ordinated Research Projects (CRPs). In particular, the status of the ongoing CRP on the evaluation of high-temperature gas-cooled reactor performance is presented. Information on IAEA gas-cooled reactor activities and relevant technical documents are also available on the Internet at the following address: http://www.iaea.org/inis/aws/htgr/index.html.

Introduction

International interest in HTGR technology has increased in recent years due to its potential design characteristics featuring inherent safety and high efficiency. Industry-supported concept development and future deployment feasibility studies have been initiated in South Africa with its PBMR project, Japan with its HTTR, China with its HTR and the collaboration of Russia, the US and France on their GT-MHR project. The European Commission has also increased its support for modular HTGR technology development through the 5th Framework Research and Development Programme.

In line with this surge of global activity, the IAEA has played a leading role as a catalyst, facilitating international collaboration and information exchange among Member States on issues related to this technology. The activities have mainly focused on organising Technical Working Group (TWG-GCR) meetings and Co-ordinated Research Projects (CRPs). The TWG-GCR currently includes participants from 12 countries: China, France, Germany, Indonesia, Japan, the Netherlands, Poland, the Russian Federation, South Africa, Switzerland, the United Kingdom and the United States. It is currently set to meet every 18 months to exchange information related to gas-cooled reactor technologies in the Member States and to advise the IAEA regarding future plans. CRPs, on the other hand, are developed in relation to a well-defined research topic on which a number of institutions agree to collaborate, and represent an effective means of bringing together researchers to solve a problem of common interest.

A CRP is typically 3 to 5 years in duration and involves 5 to 15 participating Member States. The scope and schedule of a CRP is established by IAEA staff based on expressed interest and recommendations from Member States. The participants provide a substantial majority of the resources required to conduct the CRP, organising and performing the work and documenting the results. Typically, the IAEA contributes staff time for facilitating the CRP, support for travel to periodic Research Co-ordination Meetings (RCMs), and in some cases a small funding contribution to participants from developing countries.

Recently-completed CRPs

CRP on Validation of Safety-related Physics Calculations for Low-enriched HTGRs

This CRP was formed to address core physics aspects for advanced gas-cooled reactor designs. It focused primarily on development of validation data for physics methods used for core the design of HTGRs fuelled with low-enriched uranium. Experiments were conducted for graphite-moderated LEU systems over a range of experimental parameters, including carbon-to-uranium ratio, core height-to-diameter ratio and simulated moisture ingress conditions, which were defined by the participating countries as validation data needs. Key measurements performed during the CRP provide validation data relevant to current advanced HTGR designs including measurements of shutdown rod worth in both the core and side reflector, effects of moisture on reactivity and shutdown rod worth, critical loadings, neutron flux distribution and reaction rate ratios. Countries participating in this CRP included China, France, Germany, Japan, the Netherlands, Poland, the Russian Federation, Switzerland and the United States of America. Work under the CRP has been completed and documented in the form of an IAEA TECDOC [1].

CRP on Heat Transport and Afterheat Removal for Gas-cooled Reactors Under Accident Conditions

Within this CRP, the participants addressed the inherent mechanisms for removal of decay heat from GCRs under accident conditions. The objective was to establish sufficient experimental data at realistic conditions, and validated analytical tools to confirm the predicted safe thermal response of advanced gas-cooled reactors during accidents. The scope included experimental and analytical investigations of heat transport by natural convection, conduction and thermal radiation within the core and reactor vessel, and afterheat removal from the reactor vessel. Code-to-code comparisons and code-to-experiment benchmarks were performed for verification and validation of the analytical methods. Countries participating in this CRP included China, France, Germany, Japan, the Netherlands, the Russian Federation and the United States of America. Work under the CRP has been completed and documented in the form of an IAEA TECDOC [2].

Ongoing CRP on Evaluation of HTGR Performance

CRP objective

Modular HTGRs rely on the integrity of ceramic fuel elements of the core as a main barrier for confining fission products. A combination of a strongly negative temperature coefficient, a low power density, a large core thermal capacity and a passive decay heat removal path promise to assure fuel integrity under all postulated severe accidents. The CRP is intended to validate the calculation codes used by the Member States for this type of analysis by comparing results of several benchmark problems to available experimental results.

CRP scope

The CRP, initiated in October 1997 and scheduled to be completed in October 2004, has the following Member States as participants: China, France, Germany, Indonesia, Japan, the Netherlands, the Russian Federation, South Africa and the United States The following scope has been defined for the project:

- Reactor physics benchmark analysis.

- Thermal hydraulic transient benchmark analysis.

- Demonstration of HTGR safety characteristics.

Benchmark problems have been defined for the purpose of comparing analytical results with experimental data from HTR-10, HTTR and PBMR projects. In addition, a code comparison problem has been defined by the GT-MHR project. The primary activities of the CRP include the specification of experimental benchmark problems and the exchange of data necessary to conduct analyses of the problems and compare results. The scope of the CRP centres around problems identified for research reactors and power plant designs as summarised below. In each case the participant specifying the problem was responsible for providing the necessary facility data, defining the problem in detail and compiling and comparing the analysis results.

Reactor physics benchmark problems

Four sets of reactor physics benchmark problems have been defined for the HTTR, HTR-10, the GT-MHR and the PBMR (via the ASTRA critical facility).

HTTR

The following six benchmark problems were defined for the HTTR facility:

- *HTTR-FC (Phase 1, first criticality).* The first core loading of HTTR was conducted by initially loading the reactor with graphite blocks, then replacing the graphite blocks with fuel blocks one column at a time. The outer ring of fuel was loaded first, then successive inner rings. The problem involved prediction of the number of fuel columns required to achieve the first criticality, including the excess reactivity when the column resulting in criticality was fully loaded.

- *HTTR-FC (Phase 2, first criticality).* The initial results generally underpredicted the number of fuel columns required, thus a second phase of calculations was conducted. This involved the following adjustments to the data used in the original calculation:

  - Allowance for air in the graphite voids.

  - Revised impurity contents in the initially loaded graphite blocks.

  - Aluminium in the temporary neutron detector holders.

- *HTTR-CR (control rod worth).* Evaluation of the control rod insertion depths at the critical condition for the three cases listed below. All control rod insertion levels to be adjusted on the same level except three pairs of control rods in the outermost region in the side reflectors, which were assumed to be fully withdrawn.

  - 18 columns (thin annular core).

  - 24 columns (thick annular core).

  - 30 columns (fully loaded core).

- *HTTR-EX (excess reactivity).* Determination of the excess reactivity for the three cases mentioned above, assuming moderator and fuel temperatures of 300°C and one atmospheric pressure of helium as the primary coolant condition.

- *HTTR-SC (scram reactivity).* Scram reactivity for a 30 column core fully loaded with fresh fuel for the following two cases:

  - All reflector CRs are inserted at the critical condition.

  - All CRs in reflector and core are inserted at the critical condition.

- *HTTR-TC (temperature coefficient of reactivity).* Isothermal temperature coefficients for a fully loaded core should be evaluated from 280 to 480°K, assuming a fixed control rod position based on critical conditions at 480°K.

HTR-10

The following benchmark problems have been defined for the HTR-10 facility:

- *HTR-10FC (first criticality).* Amount of loading (given in loading height, starting from the upper surface of the conus region) for the first criticality: $K_{eff} = 1.0$ under the atmosphere of helium and core temperature of 20°C, without any control rod being inserted.

- *HTR-10TC (temperature coefficient of reactivity).* Effective multiplication factor K_{eff} of full core (5 m^3) under helium atmosphere and core temperature of 20°C, 120°C and 250°C respectively, without any control rod being inserted.

- *HTR-10CR (control rod worth).* The following sets of conditions are defined:

  - Reactivity worth of the ten fully inserted control rods under helium atmosphere and core temperature of 20°C for full core.

  - Reactivity worth of one fully inserted control rod (the other rods are in withdrawn position) under helium atmosphere and core temperature of 20°C for full core.

  - Reactivity worth of the ten fully inserted control rods under helium atmosphere and core temperature of 20°C for a loading height of 126 cm.

  - Differential reactivity worth of one control rod (the other rods are in withdrawn position) under helium atmosphere and core temperature of 20°C for a loading height of 126 cm, when the lower end of the control rod is at the following axial positions: 394.2 cm, 383.618 cm, 334.918 cm, 331.318 cm, 282.618 cm, 279.018 cm, 230.318 cm.

GT-MHR

The following code comparison problems are defined for the GT-MHR design with plutonium fuel:

- Cell calculations:

  - Dependence of K_{inf} for unit cell versus burn-up.

  - Content of the main isotopes vs. burn-up.

- Reactor calculations:

  - Isothermal reactivity coefficients versus temperature at the beginning and at the end of fuel cycle.

  - Control rod worth in the active core.

  - Control rod worth in the side reflector.

PBMR-ASTRA

The PBMR benchmarks are defined with respect to the ASTRA critical facility, which is being used to provide reactor physics for the PBMR design.

- *Core height for criticality*. The requirement for this task is to determine the height at which the ASTRA facility will achieve criticality, assuming that no control rods or shutdown rods are inserted.

- *Control rod worth*. This task involves the determination of the worth of control rods with the pebble bed raised to a height of $H_{pb} = 268.9$ cm.

    - *The worth of control rods depending on their position in the side reflector*. This task requires the determination of the control rod worth for six distances from the core boundary to the axis of the control rod.

    - *Interference of a system of control rods*. Control rods worth are to be determined for single control rods as well as for combinations of two and three rods. For these combinations it is necessary to determine the worth as well as the interference coefficient.

- *Control rod differential reactivity*. This task requires the evaluation of the differential reactivity of the CR5 and MR1 control rods as a function of depth of insertion.

- *Investigation of critical parameters with varying height*. For this task it is required to determine the change in reactivity due to an increase in the height of the ASTRA critical assembly pebble bed.

Thermal hydraulics benchmark problems

Thermal hydraulics problems have been defined for the HTTR facility as follows:

- *HTTR-VC (vessel cooling)*. Participants should predict heat removal by the vessel cooling system (VCS) and temperature profile on the surface of the VCS side panel at 30 MW power operation. The analytical results will be compared with measured heat removal and temperature profile. Conditions for this benchmark problem such as power distribution, mass flow rate of coolant, reactor inlet temperature, etc., are given by JAERI.

- *HTTR-LP (loss of off-site electric power)*. This benchmark problem requires evaluation of the transient response of the HTTR to a loss of off-site electric power from initial steady-state operation at 15 and 30 MWt. During normal operation, called parallel loaded operation, the intermediate heat exchanger and the primary and secondary pressurised water coolers are operated simultaneously, removing heat from the primary system in parallel. The following problems are defined:

    - *HTTR-LP (15 MW)*. Analytical simulation of the transient behaviour of the reactor and plant during a loss of off-site electric power from normal operation at 15 MW thermal power.

    - *HTTR-LP (30 MW)*. Analytical simulation of the transient behaviour of the reactor and plant during a loss of off-site electric power from normal operation at 30 MW thermal power.

In both simulation cases, the items to be estimated on a time dependent basis are as follows: (1) hot plenum block temperature, (2) reactor inlet coolant temperature, (3) reactor outlet coolant temperature, (4) primary coolant pressure, (5) reactor power, (6) heat removal of the auxiliary heat exchanger. Estimation duration is for 10 hr from the beginning of the loss of off-site electric power.

Demonstration of HTGR safety characteristics

Participants in the CRP will review the definition, conduct and data gathering plans for safety demonstration tests to be conducted with the HTTR and HTR-10 facilities. Representatives from each facility will provide information on planned tests for review by the CRP. Specific aspects of some of the tests may be identified as benchmark problems for the CRP (e.g. HTTR-LP).

CRP schedule

The CRP schedule is centred around the RCMs, which serve as forums to present and discuss information benchmark problems, including problem definition as well as experimental and analytical results. Sufficient contractual agreements with participating Member States were in place to initiate the CRP by October 1997, with a five-year duration ending in October 2002. In 2000, the International Working Group on Gas-cooled Reactors recommended that the CRP be extended two additional years. This recommendation was reviewed and approved by the IAEA, resulting in extension of the CRP to October 2004. The results of the first three RCMs are briefly summarised below.

First Research Co-ordination Meeting

The first RCM was initially scheduled to be held in Japan in March 1998, but administrative difficulties required rescheduling, and the meeting was held in Vienna in August 1998. CRP participants at the time of the first RCM included representatives from China, Germany, Indonesia, Japan, the Netherlands, the Russian Federation and the United States. Information exchanged in advance of the RCM allowed for presentation and discussion of results on the HTTR-FC, HTTR-CR and HTTR-EX benchmarks. Also, initial discussions were held regarding the definition of the HTTR-SC, HTTR-TC, HTR-10 SC, HTR-10 TC and HTR-10 CR benchmark problems. Participants gave indications of intent regarding participation in the active benchmarks, and the date and location of the second RCM were tentatively agreed upon.

Second Research Co-ordination Meeting

The second RCM was held in Beijing, China in October 1999. Participants included the Member States attending the first RCM plus representatives from South Africa and France. Discussions were held regarding HTTR-FC, HTTR-CR and HTTR-EX, including calculational results completed after the first RCM and consideration of data from the HTTR initial criticality, which had been achieved in November 1998. Results were also presented and discussed for HTTR-SC, HTTR-TC, HTR-10 SC, HTR-10 TC and HTR-10 CR benchmarks. Additional benchmarks tentatively defined included HTTR-FC (Phase 2), HTTR-VC, HTTR-LP, GT-MHR code comparisons and PBMR/ASTRA benchmarks. Initial indications of intent to participate in the additional benchmark problems were provided by the meeting participants, and the third RCM was tentatively scheduled for March 2001 in Oarai, Japan.

Third Research Co-ordination Meeting

The third RCM was held in Oarai, Japan in March 2001. Participants included China, France, Germany, Indonesia, Japan, the Netherlands, the Russian Federation, South Africa and the United States. Presentations and discussions at the meeting addressed analytical and experimental results for the core physics and thermal hydraulic benchmarks previously defined. Additional benchmarks were also proposed by the Chief Scientific Investigators (CSIs).

Future work

A fourth RCM is planned in Vienna in October 2002 to evaluate the status of the CRP and review new results. Two TECDOCS are also planned to document these results.

REFERENCES

[1]   IAEA-TECDOC-pc4232, "Critical Experiments and Reactor Physics Calculations for Low-enriched HTGRs" (2001).

[2]   IAEA-TECDOC-1163, "Heat Transport and Afterheat Removal for Gas-cooled Reactors Under Accident Conditions" (2000).

PRESENT STATUS OF THE INNOVATIVE BASIC RESEARCH ON HIGH-TEMPERATURE ENGINEERING USING THE HTTR

Y. Sudo[1], T. Hoshiya[1], M. Ishihara[1], T. Shibata[1], S. Ishino[2], T. Terai[3],
T. Oku[4], Y. Motohashi[5], S. Tagawa[6], Y. Katsumura[3],
M. Yamawaki[3], T. Shikama[7], C. Mori[8], H. Itoh[1], S. Shiozawa[1]

[1]Japan Atomic Energy Research Institute
[2]Tokai University
[3]University of Tokyo
[4]University of the Air
[5]Ibaraki University
[6]Osaka University
[7]Tohoku University
[8]Aichi Institute of Technology

Abstract

The high-temperature engineering test reactor (HTTR) of JAERI aims at establishment and upgrading the HTGR technologies as well as the innovative basic research in the field of high-temperature engineering. It is now under commissioning tests and will soon achieve the rated power of 30 MW. This paper presents a brief overview concerning the innovative basic research in the following fields:

- New materials development of high-temperature oxide superconductors (HTSC), high performance silicon carbide (SiC) semiconductor, and heat-resistant ceramic composite materials.

- High-temperature radiation chemistry research on polysilane decomposition into SiC fibres, and high-temperature radiolysis of heavy oils and plastics, as well as *in situ* measurement of property changes of solid tritium breeding materials under irradiation.

- High-temperature in-core instrumentation development of a heat- and radiation-resistant optical fibre system and of devices for monitoring neutron and gamma-ray spectra.

JAERI is preparing for the first HTTR irradiation in 2003 and will then proceed to the international collaboration phase on high-temperature irradiation test for development of the new materials in the field of high-temperature engineering.

Introduction

The HTTR is the first high-temperature gas-cooled reactor in Japan. It is currently in the commissioning stage at the Oarai Research Establishment of the Japan Atomic Energy Research Institute (JAERI). The construction of the HTTR was decided to establish and upgrade HTGR technology and to develop an irradiation test for innovative basic research in the field of high-temperature engineering (Figure 1). For the establishment of HTGR technology, basic data concerning HTTR performance is to be accumulated on the focus of fuel integrity, helium gas leakage and so on during the long-term rated power operation. For the upgrading of HTGR technology, several research and development activities are to be performed in the field of evaluation of reactor performance on core physics and thermal analysis, etc., safety demonstration tests under simulated accident conditions, development of a process heat application system and so on. For the innovative basic research preliminary studies have been started since 1994 under the direction of the Utilisation Committee chaired by Prof. S. Ishino.

The construction of the HTTR was initiated in 1991. The reactor achieved first criticality on 10 November 1998 and the full core loading in December 1998. It is now at the stage for commissioning tests through a medium power of 20 MW (spring 2001) up to the full power of 30 MW (autumn 2001) with an outlet coolant temperature of 850°C. The maximum outlet coolant temperature of 950°C will be attained in 2002. The HTTR operational schedule is shown in Figure 2.

Irradiation fields in HTTR [1,2,8]

Major components such as a reactor pressure vessel, primary cooling system component, etc., are all installed inside a steel containment vessel. A vertical section of HTTR is shown in Figure 3, and major specifications of the HTTR are listed in Table 1. Special types of the prismatic fuel elements using hexagonal graphite blocks are used for HTTR irradiation. The blocks are 580 mm in height, and their facing surfaces are 360 mm apart. The maximum irradiation space corresponds to a cylinder which is about 300 mm in diameter and about 500 mm in height.

Figure 4 indicates the locations of the HTTR irradiation regions. The rig for the first HTTR irradiation, after completion, will be loaded into one of the three holes, 123 mm in diameter, of a graphite block of the control-rod insertion type, which is located in the replaceable reflector region A. In these regions, maximum fast neutron flux (E > 0.18 MeV) and thermal neutron flux (< 2.38 eV) are 2×10^{16} and 2×10^{17} m^{-2} s^{-1}, respectively, and gamma ray intensity is 30 Gy s^{-1}. Additionally, the attainable temperature in the He gaseous environment is below 850°C.

Recent activities of the innovative basic research [1,2,5,9]

HTTR utilisation research committee

The innovative basic research concerns basic science and technology to be associated with high-temperature irradiation environments. The research directions involve novel and bold challenges both in nuclear and non-nuclear fields. This innovative basic research on high-temperature engineering is one of the key subjects using HTTR and research reactors and concerns with basic science and technology to be associated with the field of high-temperature irradiation.

JAERI organised the HTTR Utilisation Research Committee in 1993 to promote basic research. The HTTR Utilisation Committee consists of three subcommittees in the field of new materials development, high-temperature radiation chemistry and high-temperature in-core instrumentation.

Proposed research themes spanning a wide variety of scientific and technological interests were discussed from the viewpoint of the evaluation on functional characteristics, electrical, magnetic, dynamical and mechanical, chemical and physical properties as shown in Figure 5.

At the beginning of the programme, a maximum of 66 subjects were proposed over a wide range of research fields. Only eight tasks have been finally selected for preliminary tests using out-of-pile facilities and currently available irradiation facilities such as accelerators, research and testing reactors. The selection was made through helpful and critical comments and discussion by the Utilisation Committee. The purpose of the preliminary testing is to examine and demonstrate the scientific and technical feasibility and effectiveness of the tasks selected.

Outlines of preliminary tests of three ongoing research subjects are summarised in Table 2 [11]. These are briefly described below.

Recent results of innovative basic research

New materials development

Scientific and technological significance and method have been investigated for three research subjects in the field of new materials development. This research field is in a technical level deserving concrete investigations for the HTTR irradiation. Major results obtained are summarised in Table 3.

Neutron transmutation doping (NTD) of high-temperature semiconductor

The decreases in impurities of silicon carbide (SiC), which can be applied to the field of high-temperature, high power and high speed switching devices, is a very important issue to be solved with regard to the practical application of SiC. SiC is not readily permeable to impurity elements by conventional diffuse method. On the other hand, phosphorus (P) transmutation doping (NTD) of P using thermal neutrons is recognised as a promising method. The ^{30}Si (natural abundance 3.1%) is transmuted into ^{31}Si (half life 2.62 h) by a thermal neutron capture reaction, followed by β decay to ^{31}P (stable) which acts as a shallow electron donor in SiC.

As a preliminary test, after implantation with N, P and Al ions at temperatures ranging from 1 073 K to 1 473 K and subsequently annealing at higher temperatures ranging from 1 773 K to 1 873 K, residual defects of 3C-SiC (cubic) and 6H-SiC (hexagonal) films have been annealed out by increasing the implantation temperature up to 1 473 K. This anneal-out is explained as the formation of an amorphous island-like structure. Amorphous island-like structures in SiC after room implantation contain impurity P atoms as well as point defects. The island structures are not readily recovered during annealing at high temperatures and even after the annealing. In contrast, such amorphous island structures are not formed in high-temperature implantation, and therefore radiation-induced defects are almost perfectly recovered by the post-implantation annealing at higher temperatures. Almost 100% activation of the impurity atoms, therefore, can be achieved by the high-temperature implantation or neutron irradiation with subsequent annealing.

Several kinds of apparatuses for post-irradiation examinations (PIEs) of SiC semiconductors are under preparation in JAERI's JMTR Hot Laboratory. The Hall effect apparatus and annealing furnace up to 1 973 K is under installation in a hot experimental room. After preparation, the first post-annealing data will be obtained on SiC samples to be irradiated in JMTR as a preliminary neutron irradiation test.

Irradiation modification of high-Tc superconductor

As for the irradiation-induced modification of an oxide high-temperature superconductor, Bi-2212 ($Bi_2Sr_2CaCu_2O_x$) with a high critical temperature (Tc), its main purpose is to increase the critical current density (Jc). In this field, neutron irradiation is utilised as an important device to introduce and effectively control the size and the space distribution of lattice defects, which act as trap sites of external magnetic fluxes that are going to invade the superconductor and lead to degradation of the superconductivity [13]. After irradiation with a neutron fluence of $5 \times 10^{22} m^{-2}$ (E > 1 MeV) in JMTR, the critical current density of irradiated Bi-2212 sample revealed the drastic increase in Jc value (measured at 40 K) to be a tenth as large as that of an irradiated specimen by post annealing at 673 K for 86.4 ks (1 day) as reported elsewhere [13] and as shown in Figure 6.

Research on radiation damage in ceramic composite materials

This research subject is aimed at a mechanistic understanding of the radiation damage of heat-resistant ceramic composites, carbon- and SiC-composite materials, and is focused on the mezzoscopic approach in order to clarify the relationship between microstructures and dynamical and mechanical properties [14].

High-temperature radiation chemistry

Elaborate studies have been undertaken to utilise radiation-induced chemical reactions at high temperatures, in production of SiC fibres from a macromolecule compound of polysilane as well as in decomposition of heavy oils and plastics.

Radiation—enhanced thermal deposition of heavy oils and plastics at high temperatures

The purpose of this research is to clarify the mechanism of the high-temperature radiation-induced decomposition reactions of heavy oils and plastics and to seek their application. From a viewpoint of reformation of heavy oils and decomposition and recovery of wasted plastics by means of high efficiency thermal decomposition in radiation environments, model compounds of normal hexadecane (n-$C_{16}H_{34}$) and polyethylene were selected for gamma-ray exposure experiments and recent studies are focused on the usage of supercritical water as a powerful tool for decomposition of toxic wastes.

Elucidation of singular chemical reactions in high-temperature radiation fields and their application to process development

This subject deals with radiation-assisted precursor curing of carbon and SiC fibres at high temperatures. Some radiation-induced chemical reactions accelerate much higher decomposition rates even at temperatures around 473 to 573 K than at room temperature. It is also recognised that some reactions can take place effectively at high temperatures. If we elucidate the mechanisms of these complicated reactions, we can develop new chemical processes as applications of these reactions. In the preliminary studies, gamma-ray irradiation experiments have been performed to clarify the effect of the exposure on the decomposition behaviour of SiC-based polymers including polycarbosilane to form SiC fibres.

Preliminary test on surface and functional properties of solid tritium breeder under irradiation

Elaborate studies have been made to enable *in situ* measurement of irradiation-induced property changes of solid tritium breeding materials under irradiation [15]. This subject is concerned with physicochemical properties of lithium oxide ceramics for the fusion reactor blanket. In the present test, *in situ* measurement of a contact potential difference (CPD) on the surface and of the bulk electrical conductivity of oxide ceramic materials can first be measured by using experimental apparatus. Reasonable results are obtained by *in situ* measurement of work function on surface reaction. This sophisticated system is useful in quantifying the irradiation-induced surface and bulk damages in lithium oxides as candidates of the tritium breeding materials and can be applied to nuclear and non-nuclear field by using the effective cut of irradiation-induced noise.

High-temperature in-core instrumentation

Efforts have been made to develop optical fibres with the heat- and radiation-resistance system as well as other devices for detection of neutron and gamma-ray spectra at high temperatures.

Development of a heat- and radiation-resistant optical fibre system

The optical fibres to be doped fluorine (F) or hydroxyl group (OH) were newly developed with radiation resistance. In these fibres, optical loss reveals quite a lower value through the media rather than that of the conventional type of SiO_2 fibre. From the recent irradiation experiments, the effect of neutron irradiation on optical fibre system with the light emitting source made of high-purity alumina (sapphire) was investigated in JMTR. The tests reveal a high potential for this method to be a sensitive and effective tool for real-time detecting of optical transmission at temperatures up to 1 273 K and intensity of gamma ray of $5 \times 10^3 Gys^{-1}$. In the present stage, radioluminescence phenomena of several kinds of ceramics including the rare earth elements are systematically investigated by the JMTR irradiation test.

Development of detection methods of neutron and gamma-ray spectra in high-temperature reactor core

New types of neutron and gamma-ray detectors are being developed for use at temperatures up to 1 273 K on the high-temperature in-core instrumentation. A polycrystalline boron nitride (BN) sensor is a first candidate for a neutron detector, which is based on a thermal neutron absorption, $^{10}B(n,\alpha)^7Li$. Several neutron detectors with different configurations and dimensions have been tested for measuring the magnitude of electric currents induced either by a thermal mechanism or by thermal neutron reactions. A new measurement system composed of high quality BN ceramics as sensor materials and optical fibre as transmitting materials, is investigated on the in-core response of gamma-ray and neutron spectra.

JAERI is also investigating and preparing the PIE apparatuses [5], as described above. In the NTD research for the high-temperature SiC semiconductor, somewhat advanced plans have been made for installation of a set of special apparatuses for the specific resistance/Hall effect measurement, together with a post-irradiation annealing furnace. As for the research on high-temperature superconductor materials and ceramic composite materials, programmes have been tentatively set up to install some general apparatus such as mechanical and thermal property test devices for the post-irradiation examination in the Hot Laboratory.

Future plans for innovative basic research

JAERI is now making great efforts to prepare for the realisation of HTTR irradiation in the field of innovative basic research.

The first task is selection of candidate research subjects for the initial stage of the HTTR irradiation. The above-described research subjects in the new materials development field are expected to provide favourable and definitive results through further preliminary tests for an additional few years.

The second task is to clarify the HTTR irradiation methods including requirements for the HTTR irradiation capsule and reveal the steady progress for practical use.

Recent efforts [11] to carry out these tasks have led to a plan for the first HTTR irradiations to be started in 2003 (Figure 2). JAERI is now preparing the HTTR irradiation and the activities in this field will shift to the international research collaboration phase, including Japan materials testing reactor (JMTR). The HTTR is aimed at the COE for HTGR development through the international contribution to HTGR development in the world.

Concluding remarks

The HTTR is now in the stage of commissioning tests up to 30 MW. Preliminary tests of the innovative basic research are ongoing in the following fields:

- New materials development.

- High-temperature radiation chemistry research.

- In-core measurement of property changes of solid tritium breeders.

- High-temperature in-core instrumentation development.

Interesting results which reveal the effectiveness of high-temperature irradiation has been obtained on R&D status of the high-performance SiC semiconductor and high-temperature oxide superconductor.

JAERI will proceed to an international collaboration phase on the HTTR irradiation for development of new ceramic materials and for development of the basic upgrade HTGR technology, and is now preparing for the HTTR irradiations (2003).

Acknowledgements

The authors are deeply indebted to university members for the innovative basic research, and to members of the HTTR Utilisation Research Committee of JAERI, Prof. Hideki Matsui of Tohoku University, Prof. M. Nakazawa of the University of Tokyo and Prof. Takao Hayashi of Tokai University, and to those of the Special Committee of the Atomic Energy Society of Japan.

Thanks are also due to JAERI members, Dr. T. Tanaka, Dr. K. Hayashi, Dr. T. Arai, Mr. T. Kikuchi, Dr. K. Sawa, Mr. M. Niimi, Mr. S. Baba, Mr. T. Takahashi and Ms. J. Aihara, for their co-operation and encouragement.

REFERENCES

[1]   Japan Atomic Energy Research Institute, "Present Status of HTGR Research & Development", 1996.

[2]   Japan Atomic Energy Research Institute, "Present Status of HTGR Research & Development", 2000.

[3]   S. Saito, *et al.*, "Design of the High-temperature Engineering Test Reactor", JAERI 1332 (1994).

[4]   O. Baba and K. Kaieda, "Present Status and Future Prospective of Research and Test Reactors in JAERI", Proc. 6th Asian Symposium on Research Reactors, JAERI-Conf. 99-006 (1999) pp. 49-55.

[5]   K. Sawa, M. Ishihara, T. Tobita, J. Sumita, K. Hayashi, T. Hoshiya, H. Sekino and E. Ooeda, "R&D Status and Requirements for PIE in the Field of the HTGR Fuel and the Innovative Basic Research on High-temperature Engineering", Proc. of the 3rd JAERI-KAERI Joint Seminar on the PIE Technology, JAERI Oarai, Japan, 25-26 March 1999, pp. 335-355, JAERI-Conf. 99-009 (1999), pp. 341-361.

[6]   Japan Industrial Forum, Inc., "Japan's First HTGR – JAERI's HTTR Reaches Criticality", *Atoms in Japan*, 42 (1998) pp. 4-5.

[7]   M. Ishihara, S. Baba, J. Aihara, T. Arai and K. Hayashi, "A Preliminary Study on Radiation Damage Effect in Ceramics Composite Materials as Innovative Basic Research Using the HTTR", Proc. 6th Asian Symposium on Research Reactors, Mito, Japan, 29-31 March 1999, Session 9-4, JAERI-Conf. 99-006 (1999), pp. 207-212.

[8]   K. Sanokawa, T. Fujishiro, T. Arai, Y. Miyamoto, T. Tanaka and S. Shiozawa, Proc. of the OECD/NEA/NSC Workshop on High-temperature Engineering Research Facilities and Experiments, 12-14 Nov. 1997, Petten, The Netherlands, ECN-R-98-005 (1998), pp. 145-156.

[9]   T. Arai, H. Itoh, T. Terai and S. Ishino, Proceedings of the OECD/NEA/NSC Workshop on High-temperature Engineering Research Facilities and Experiments, 12-14 November 1997, Petten, The Netherlands, ECN-R-98-005 (1998), pp. 113-124.

[10]   M. Ishihara, S. Baba, J. Aihara, T.Arai, K, Hayashi and S. Ishino, "A Preliminary Study on Radiation Damage Effect in Ceramics Composite Materials as Innovative Basic Research Using the HTTR", Proc. 6th Asian Symposium on Research Reactors, JAERI-Conf. 99-006 (1999), pp. 207-212.

[11]   S. Ishino, *et al.*, "An Investigation of High-temperature Irradiation Test Program of New Ceramic Materials", JAERI-Review 99-019 (1999) [in Japanese].

[12] K. Abe, T. Ohshima, H. Itoh, Y. Aoki, M. Yoshikawa, I. Nishiyama and M. Iwami, *Materials Science Forum*, 264-268 (1998), pp. 721-724.

[13] T. Terai, "Modification of HTSC by High-temperature Neutron Irradiation in HTTR", Proceedings of the 1st Information Exchange Meeting on Basic Studies on High-temperature Engineering, 27-29 September 1999, Paris, France (1999), pp. 319-329.

[14] M. Ishihara, S. Baba, J. Aihara, T. Arai, K. Hayashi and S. Ishino, "Radiation Damage on Ceramics Composites as Structural Materials", Proceedings of the 1st Information Exchange Meeting on Basic Studies on High-temperature Engineering, 27-29 September 1999, Paris, France (1999), pp. 299-308.

[15] M. Yamawaki, K. Yamaguchi, A. Suzuki, T. Yokota, G. N. Luo and K. Hayashi, "Development of a New Method for High-temperature In-core Characterization of Solid Surfaces", Proceedings of the 1st Information Exchange Meeting on Basic Studies on High-temperature Engineering, 27-29 September 1999, Paris, France (1999), pp. 357-364.

[16] T. Shikama, T. Kakuta, M. Narui, M. Ishihara, T. Sagawa and T. Arai, "Application of Optical Fibres for Optical Diagnostics in High-temperature Gas-cooled Reactor", Proc. OECD/NEA Workshop on High-temperature Research Facilities and Experiments, 12-14 November 1997, ECN Petten, The Netherlands, pp. 125-144.

[17] T. Shikama, M. Ishihara, T. Kakuta, N. Shamoto, K. Hayashi and S. Ishino, "Development of Heat- and Radiation-resistant Optical Fibres", Proceedings of the 1st Information Exchange Meeting on Basic Studies on High-temperature Engineering, 27-29 September 1999, Paris, France (1999), pp. 379-386.

[18] The HTTR Utilisation Research Committee, "Results and Future Plans for the Innovative Basic Research", JAERI-Review 2001-016, 2001 [in Japanese].

Table 1. Major specifications of HTTR

Thermal power	30 MW
Outlet coolant temperature	850°C/950°C
Inlet coolant temperature	395°C
Fuel	Low enriched UO_2 (ave. 6%EU)
Fuel element type	Prismatic block
Direction of coolant flow	Downward
Pressure vessel	Steel
Number of cooling loop	1
Heat removal	IHX and PWC (parallel loaded)
Primary coolant pressure	4 MPa
Containment type	Steel containment

Table 2. Summary of preliminary tests

Research Field and Title	Experimental Method	Measurement
< New Materials Development > - High temperature SiC semiconductor by neutron transmutation doping	- Sample: SiC single crystal - Hot implantation of P ions and annealing	- Residual defects by ESR - Electron concentration (Hall effect)
- Improvement of high-Tc superconductor by neutron irradiation	- Sample: Bi-2212 single crystal - Fast neutron irradiation in JMTR/ annealing	- Critical current density at 4-60K
- Mechanistic radiation damage studies on ceramic-based structural composites	- Sample: SiC/SiC and C/C composites etc. - Fast neutron irradiation in JMTR	- Microstructural features by SEM - Thermal & mech. properties
< High Temperature Radiation Chemistry > - Radiation-assisted precursor curing for SiC and carbon fibers	- Sample: SiC containing polymers etc. - Gamma-ray irradiation at RT - 400°C	- Mass spectrum by EPR spectrometry
- Radiation enhanced thermal decomposition of heavy oils and plastics	- Sample: n-hexadecane & polyethylene etc. - Gamma-ray irradiation at RT - 400°C	- Gas chromatography - Photo-ionization detector
< Fusion-related Research > - Measurement of physicochemical properties of lithium oxides under irradiation	- Sample: LiO_2, Li_4SiO_4, Li_2TiO_3, Li_2ZrO_3 etc.	- Contact potential difference and electric conductivity
<High Temperature In-core Instrumentation> - Development of a heat- and radiation-resistant optical fiber system	- Preparation of OH and F doped silica - performance tests under irradiation in JMTR	- Optical transmissivity by optical power and spectrum analyses
- Development of ceramic-based neutron/gamma detectors including SPNDs	- Sensor: polycrystalline BN etc. - Performance tests under irradiation in KUR	- Electrical current induced by neutrons and gamma-rays

Table 3. Major results obtained on new material development

Subject	Objective	Countermeasure	Results
High Temperature Semiconductor SiC	Enhancement of carrier concentration	Post annealing 1873K>T>1773K Hot implantation 1473K>T>1073K Decrease in residual defects	Enhancement 100%
High Temperature Superconductor $Bi_2Sr_2CaCu_2O_x$	Enhancement of Jc Jc :critical current density	Post annealing at 673K Φ_f: $10^{22} m^{-2}$ Rearrangement of residual defects	Enhancement > Ten times (at 40K)
Ceramics Composite C-, SiC-	**Mechanistic approach** (Mesoscopic)	Under way	Under way

Figure 1. Research and development using HTTR

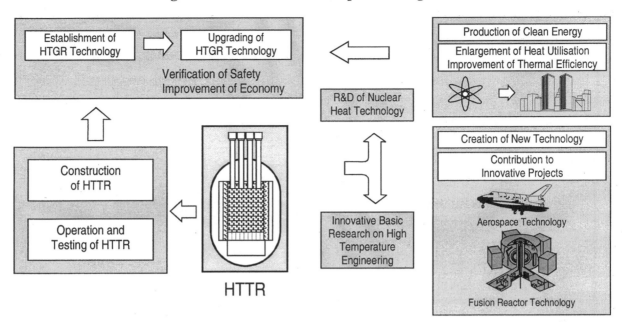

HTTR

Figure 2. Operational schedule of HTTR

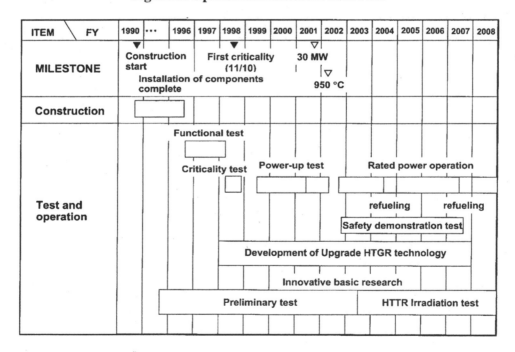

Figure 3. Vertical section of HTTR reactor pressure vessel

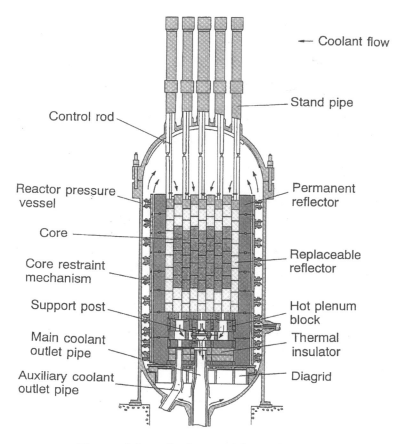

Figure 4. Irradiation positions of HTTR

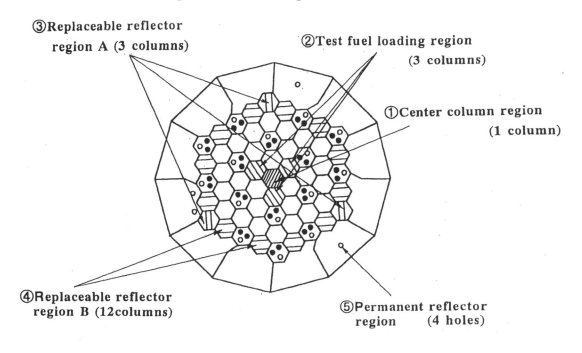

Figure 5. Organised HTTR Utilisation Research Committee

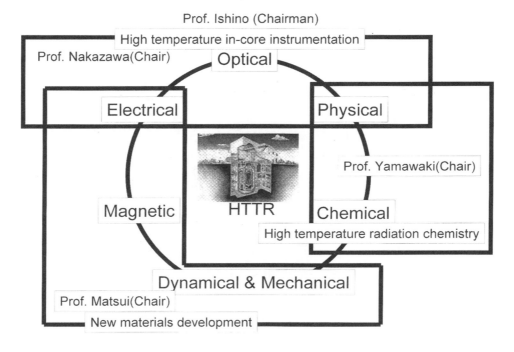

Figure 6. Changes in Jc due to neutron irradiation and thermal annealing

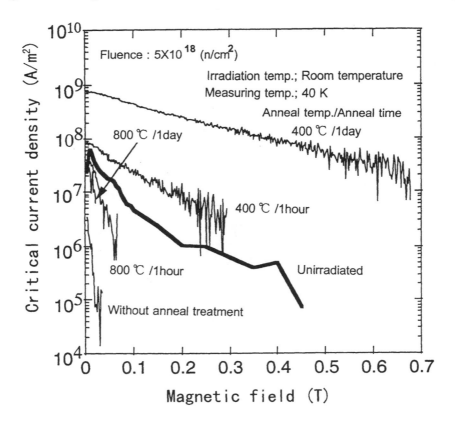

PRESENT STATUS IN THE NETHERLANDS OF RESEARCH RELEVANT TO HIGH-TEMPERATURE GAS-COOLED REACTOR DESIGN

J.C. Kuijper, A.I. van Heek, J.B.M. de Haas
NRG, P.O. Box 25, NL-1755 ZG Petten, The Netherlands

R. Conrad, M. Burghartz
EC-JRC-IE, P.O. Box 2, NL-1755 ZG Petten, The Netherlands

J.L. Kloosterman
IRI, Delft University of Technology, Mekelweg 15, NL-2629 JB Delft, The Netherlands

Abstract

Research relevant to the design of high-temperature gas-cooled reactors (HTR), is performed in the Netherlands at NRG Petten/Arnhem, JRC-IE Petten (previously JRC-IAM) and IRI–Delft University of Technology, Delft. These present as well as past HTR activities in these organisations fit into an international – European and world-wide – framework.

An important role in these activities is played by the JRC's high flux reactor (HFR) in Petten. Because of its favourable design and operational characteristics and the availability of dedicated experimental equipment, the HFR has been used extensively as a test bed for HTR fuel and graphite irradiations for more than 30 years. Since the First Information Exchange Meeting in 1999 new HTR-related irradiation experiments are being planned in the HFR irradiation programme, e.g. HFR-EU1 (irradiation of fuel pebbles as part of the "HTR-F" project in the European 5th Framework Programme) and HFR-EU2 (irradiation of General Atomics fuel compacts).

Besides computational support for the irradiation experiments mentioned above, e.g. design calculations and prediction of sample composition after irradiation, more general HTR-related computational analyses and auxiliary investigations have been and still are being carried out by the organisations in the Netherlands. Examples include the INCOGEN study, investigations on the ACACIA combined heat and power concept, the South African PBMR, the Chinese HTR-10 and the Japanese HTTR and also the implications of Pu incineration in HTR systems and innovative burnable poison concepts. These analyses comprise computational HTR core physics, thermal hydraulics and shielding analysis, HTR system safety related transient analysis and also the development, improvement, verification and validation of software for performing these types of analyses.

Auxiliary investigations in the INCOGEN framework covered e.g. plant layout, design of the energy conversion system, control philosophy, inspection and maintenance, licensing, economics and market potential. Presently, auxiliary supporting experiments and studies are being carried out on chemical aspects of HTR fuel under final repository conditions and the compilation of a database of properties of potential HTR vessel materials, as part of the "HTR-N" and "HTR-M" projects in the European 5th Framework Programme, respectively.

Introduction

The high-temperature gas-cooled reactor (HTR) has been a topic of interest for research institutes in the Netherlands for a number of years. Research relevant to the design of such reactors has been and still is being performed in the Netherlands at NRG Petten/Arnhem (a merger of the former nuclear departments of KEMA Arnhem and ECN Petten), the JRC Institute for Energy (JRC-IE) Petten (previously JRC-IAM) and IRI, Delft University of Technology, Delft. The HTR activities in these organisations comprise irradiation of HTR fuel and structural materials in the HFR, including calculational support, auxiliary supporting experiments, model calculations and software development and auxiliary studies.

This paper briefly describes the past and ongoing HTR activities in these organisations, also indicating the international – European and world-wide – framework, in which irradiation experiments in JRC's high flux reactor (HFR) at Petten play an important role.

HFR characteristics

The HFR is a multi-purpose materials testing reactor of the tank-in-pool type. It is light-water cooled and moderated, and operated at a power of 45 MW with highly enriched uranium (~93% in ^{235}U) as a fuel (Al + U alloy). The reactor tank being part of the closed primary coolant circuit is immersed into a water-filled pool. The core lattice is a 9×9 array containing 33 fuel assemblies, six control members, 19 experimental positions and 23 beryllium reflector elements (Figure 1). The useful height of the core is 600 mm. With the control rods inserted from below the tank, the top lid provides easy access for the in-vessel irradiation positions. In addition to the 19 in-vessel irradiation positions, there is one pool side facility (PSF). This PSF is used for materials and fuel irradiation experiments and for the production of radioisotopes. Larger objects such as graphite blocs can be tested as well.

The HFR has a present annual operating schedule composed of 11 cycles and 25 operating days; two prolonged stops of several weeks each are used for maintenance and related activities. Reactor start-up from 0 to 45 MW takes one to a few hours, depending upon on the experimental requirements. An important feature of the HFR is its high reliability and annual availability. More than 270 operation days have consistently been achieved since the start-up of the reactor in 1961. The maximum achievable annual fluence level is 10 dpa for graphite. Comprehensive information on the HFR and on local radiation field characteristics and its irradiation facilities is given in [1].

Presently, the HFR is in the process of changing over from highly-enriched uranium (HEU) to low-enriched uranium (~20% in ^{235}U) (LEU) fuel. The impact on the reactor behaviour and on the attainable neutron flux in the experimental positions is expected to be limited.

HFR irradiation experiments and calculational support

One argument for the continuous involvement of HFR in R&D for HTR is the fact that it is embedded into the infrastructure of a large nuclear research centre (JRC-IE Petten and NRG). With its high and consistent annual availability of more than 270 operating days, the HFR has an outstanding performance record. High variety in irradiation positions covers the entire required range of nuclear characteristics for HTR fuel testing. The infrastructure at the Petten site provides the entire range of high-level services, which is required for cost effective irradiation testing. These services are provided by different on-site organisations:

- Computational studies are provided by NRG; thermal-hydraulic and mechanical computations and modelling by means of finite element (FE) computer codes such as ANSYS 5.4 and MARC; nuclear modelling by means of computer package WIMS and MNCP codes are used. Predictive pre-irradiation calculations are verified by follow-up calculations and either on-line or PIE measured data sets.

- Design of entire irradiation facilities is provided by JRC-IE and NRG.

- Fabrication and quality control (ISO 9001 certificate) of facilities are provided at ECN.

- NDE laboratory for neutron radiography and X-rays are available at JRC-IE and NRG.

- Neutron metrology services are provided by NRG.

- Post-irradiation examinations in dedicated hot cell laboratories are provided by NRG. The hot cells have extensive facilities for non-destructive and destructive PIE. The PIE includes visual inspection, dimensional measurements, gamma spectrometry, ceramography, EPMA, SEM, TEM, mass spectrometry, X-ray with image analysis, chemical analysis of HTR fuel components, etc.

- Waste disposal is taken care of by NRG.

The joint management structure between JRC-IE and NRG and the entirety of on-site services makes the HFR an effective instrument in HTR fuel testing.

Past experiences with HTR fuel and graphite irradiation experiments (JRC-IE, NRG)

The HFR has been used extensively for fuel and graphite irradiation experiments to support R&D for the German modular HTR and the US HTGR. A more extensive description can be found in [2].

Fuel and fuel element structures are the first barriers against fission product release. The assessment of fission product retention under normal and off-normal operating conditions is therefore the primary goal of HTR fuel irradiation experiments. Starting from screening tests of experimental coated particle fuel embedded in coupons or compacts, large irradiation programmes concentrated in the 80s and 90s on performance testing of reference coated fuel particles and reference fuel elements. In particular, the UO_2 low-enriched uranium (LEU) fuel cycle for the German concepts of HTR-Modul and HTR-500 [7] and the LEU fissile/ThO_2 fertile fuel system for former US HTGR have been extensively investigated at the HFR, see Table 1. The programmes on fuel testing were terminated in 1994.

The protection of spherical HTR fuel elements by means of a corrosion-resistant SiC-layer has been proposed for concepts of inherently safe HTRs. On behalf of FZJ and within their R&D Programme on Innovative Nuclear Technology, CVD-based SiC coatings on graphite spheres have been irradiated at HFR Petten under specific operating conditions of a modular HTR [3]. Three in-pile tests were performed between 1995 and 2000 (see Table 1).

The major test objectives of the in-pile tests that are mentioned in Table 1 were the demonstration of the integrity and the retention capability of the particle coating against fission product and fission gas release and the irradiation stability of full-size spherical fuel elements fabricated on a production scale under power plant conditions with respect to temperature history, correlation of fast fluence and burn-up and power history. The irradiations were not only conducted at normal, but also at off-normal operating conditions, e.g. simulating core water ingress and temperature transients.

Unique irradiation facilities have been developed by JRC-IAM, and the design has continuously evolved to meet the extensive range of requirements for HTR fuel testing [1,2,5]. Typical for all HTR fuel testing rigs at HFR are the multi-capsule designs, which allow loading of independent experiments into one core position. Small samples with diameters up to 30 mm can be loaded in rigs (type TRIO or QUATTRO) that provide three or four parallel and independent channels. The inserts of each channel can independently be loaded and unloaded. Larger samples with diameters of up to 64 mm can be loaded in multi-capsule rigs (type BEST or REFA) that can accommodate up to four independent superposed capsules. Each individual capsule is connected with an independent sweep loop that enables automatic control of temperature, gas pressure and purge gas mass flow. Surveillance of gas quality and the release of volatile fission products are provided as well. Furthermore, the capsules are fully instrumented with thermocouples, fluence detector sets, gamma scan wires and self-powered neutron detectors. Those containments, which contain fuelled specimens, are continuously purged with pure helium.

The irradiation temperature can be controlled between 870 and 1 770 K with an accuracy of less than 10 K by means of gas mixture technique and vertical adjustment of capsules in the flux profile. Modelling by means of FE computer codes enable the design of required temperature fields by tailoring the appropriate dimensions of capsule components. The downstream of the purge gas of each capsule is continuously monitored on-line on the release of gaseous fission products. The fractional fission gas release R/B (release rate to birth rate) can be determined by means of intermittently taken gas samples. The release rate is measured and the birth rate is calculated.

The multi-capsule design features the possibility to perform parametric studies at identical nuclear conditions with a large variety of specimens. The required burn-up-fluence correlation can be adjusted by flux tailoring, by which the material of the direct environment of the in-pile sections can be adequately selected.

Present activities in HTR fuel irradiation experiments (JRC-IE, NRG)

Presently, two HTR fuel-related irradiation experiments in the HFR are being set up: HFR-EU1, concerning the irradiation of fuel pebbles, and HFR-EU2, concerning the irradiation of fuel compacts.

The HFR-EU1 irradiation at HFR Petten is the first fuel test that will be conducted within the new European network HTR-TN as the Shared Cost Action project "HTR-F" of the EU 5th Framework Programme [6]. A consortium of six contractors participates in work package 2 of the HTR-F project, i.e. CEA as co-ordinator and FRAMATOME, NRG, FZJ, JRC-ITU and JRC-IE.

The HFR-EU1 irradiation experiment will be carried out at the high flux reactor (HFR) Petten in close co-operation between FZJ, NRG and JRC-IE. Post-irradiation examinations (PIE) will be performed both at NRG Petten and at JRC-ITU. Accident simulation tests will be done at JRC-ITU.

The main objective of the HFR-EU1 irradiation test is the demonstration of the feasibility of high burn-up for the existing German LEU fuel with TRISO coated particles. It will include in particular:

- The irradiation up to a burn-up of 20% FIMA, to be achieved within two years.

- The evaluation of fuel performance at such ultra-high burn-up to explore the real limits of the existing CP that have formerly been designed for operational conditions of the HTR-Modul.

- The extension of the existing database for the EOL metallic fission product release, particularly the silver isotope [110m]Ag for an improved assessment of the particle choice for the gas turbine HTGR concept.

- The demonstration of the ability of the LEU-TRISO coated particle for fission product retention at accident scenarios, e.g. post-irradiation heating beyond 1 600°C. These PIE tests, which will be performed in the KÜFA facility at JRC-ITU, are the main goal of the HFR-EU1.

The irradiation samples of the HFR-EU1 test will be four spherical fuel elements (FEs) of German production with 60 mm outer diameter. These spheres exist and are of the type AVR GLE-4 (AVR reload 21-2) with 16.7% [235]U enrichment. These type of FEs have been irradiated earlier in the AVR. They have reached a calculated average burn-up of 8.6% FIMA with a maximum burn-up of about 20% FIMA. No particle defects were observed during the on-line gas release measurements. This excellent result indicates a certain potential of the LEU-TRISO particles for the application in FE's with a prospective extreme high burn-up.

Excellent results have also been obtained from a large variety of tests on LEU-TRISO fuel at the HFR Petten. As mentioned before, these tests (Table 1) have been conducted under conditions beyond nominal conditions of HTR concepts. These tests comprised spherical fuel elements and a large variety of compacts and loose coated particles. The experience of three decades of fuel testing resulted in optimised design and instrumentation of the facilities and irradiation under well-controlled conditions.

The proposed facility for the HFR-EU1 test will be designed for the simultaneous irradiation of five spherical fuel elements. The in-pile section will consist of a sample holder with two independent and superposed capsules, which will be inserted into a thimble that forms the second controlled containment. The thimble is a standard facility, code-named REFA-172. The useful diameter of the REFA is 72 mm. The upper capsule will contain three spheres and the lower capsule will contain two spheres. They are held in position by cylindrical half-shells. The capsules will be instrumented with thermocouples, fluence monitors and flux detectors. Each capsule and the second containment will be connected with an independent sweep loop circuit that allows continuous purging with an inert carrier gas under controlled conditions. The purge gas of the second containment will be composed of an adjustable helium-neon gas mixture and serves for automatic temperature control. The purge-gas of the two capsules will basically be pure helium and serves for continuous surveillance of the fission gas release rate of the specimens. The fractional release of the main Kr and Xe isotopes will be determined quasi on-line. Further features of the HFR-EU1 test are the feasibility to tailor the neutron spectrum for a proper burn-up/fluence correlation, and a vertical displacement unit, which allows fine tuning of fluence/burn-up and temperature, and the cycle-wise turning by 180° to realise a homogeneous azimuth fluence/burn-up distribution in the large specimens. Design of the structural part of the capsules ensures that the total amount of solid fission product released from the fuel elements can be measured by leaching these parts after irradiation. The required burn-up level should be reached after irradiation of about two years in a peripheral in-core position of the HFR (see Figure 1).

The HFR-EU1 test shall be conducted such that the central temperatures of all fuel specimens be held constant at about 1 100 C. This will require a gradual increase of the initial surface temperature of 800°C to compensate for the reduction of the fuel central temperature with increasing burn-up. The procedure involves for the accompanying calculation inherent uncertainties such as power profile and material properties as a function of temperature and fluence. Temperature transients will not be performed. The fission power of a single particle should remain < 250 mW. The irradiation shall be conducted until a maximum burn-up of the highest loaded FE of approximately 20% FIMA is achieved. Fast neutron fluence shall not exceed $6.5*10^{25}$ m^{-2} (E > 0.1 MeV).

A similar test programme, the HFR-EU2 project, is being set up to irradiate General Atomics HTR fuel compacts in a TRIO facility.

Present activities regarding irradiation of HTR structural materials (JRC-IE, NRG)

In the EU 5th Framework Programme project "HTR-M", an experimental programme will be established and performed to verify selected materials. This will include the procurement of representative welded test pieces and the machining of tensile, creep and/or compact tension fracture specimens to determine representative properties. The irradiation experiment will concentrate on the properties of a relevant weld material, both in the as-fabricated state and after irradiation to end-of-life neutron fluence conditions. NRG will perform preparatory work on the test facilities and test specimens. After test weld procurement, creep tests (including cross weld creep and fatigue) and toughness properties of a welded joint representative of HTR vessel welds will be carried out. The irradiation tests are performed on selected plate and weld materials under relevant HTR conditions using the LYRA facility in the pool side of the HFR with post irradiation examination carried out in the NRG hot cell facilities. Results from this work will be reviewed when completed and included in the materials database. The synthesis for design and defect assessment applicability will be performed towards the end of the project taking account of information available at the time. A further review and synthesis is planned on completion of the irradiation test work and examination.

The work on vessel materials will yield the following results:

- Database on pressure vessel steel plate and weld material relevant for HTR RPV applications.

- Recommendations for HTR RPV feasibility, choice of material, defect assessment and leak-before-break applicability.

- Evaluation of factors for defect tolerance assessment for HTR relevant welds.

- Irradiation data for HTR RPV relevant irradiation conditions (temperature and fluence) on selected properties of the HTR steel and weld.

Graphite irradiation data for new graphites is needed for HTR feasibility assessments. It is also an important consideration for future decommissioning activities This work will look at suitable graphites for future European HTRs and initiate an irradiation programme on a European test facility as a basis for continuation and extrapolation to higher doses. Numerous grades of reactor graphite have been produced in different countries as part of the development of HTR and gas-cooled reactors, however much of this specialist information and experience has disappeared. The most recent European graphite reactors went into operation in 1989 (Torness, UK) and 1990 (Smolensk 3, Russia). Many graphites used in previous core designs are no longer manufactured commercially. New data is needed on present day candidate grades for comparative assessments with existing graphites to provide a provisional assessment and identification of candidate graphite grades that may offer potential long-term benefits in terms of cost, safety, waste management, use of fissile materials and safeguards. The NRG and JRG-IE activities on this subject are part of the EU 5th Framework Programme project "HTR-M1". An HFR irradiation experiment will be set up and performed in the period 2001-2005.

Auxiliary experiments

HTR-PROTEUS reactivity effects (IRI)

Experimental and calculational studies have been performed on various reactivity effects in the HTR-Proteus facility of the Paul Scherrer Institute (Switzerland) [7]. Several methods to measure (changes in) the reactivity were reviewed including the pulsed-neutron source method, the inverse kinetics technique and noise analysis techniques. New expressions were derived that better represent the expectation values of the experimental quantities. Inserting CH2 rods into the core and/or the reflector simulated the local ingress of water. The measurement of the safety and shutdown rods worth showed significant spatial effects, which made necessary the use of calculated correction factors in the analyses. Both Monte Carlo and deterministic codes have been used to analyse the experiments and to calculate the kinetics parameters and the reactivity worth of the rods.

HTTR start-up core physics (NRG, IRI)

Both NRG and IRI take part in the benchmark of start-up core physics of the high-temperature engineering test reactor (HTTR) [8], which is part of the IAEA Co-ordinated Programme "Evaluation of HTGR Performance". To compare the performance of the SCALE-based IRI code package with that of the WIMS/PANTHER code package of NRG, a calculational intercomparison has been performed. In Phase 1 of the HTTR benchmark NRG and IRI used the Monte Carlo code KENO Va (3-D) and the diffusion theory codes BOLD VENTURE (2-D) and PANTHER (3-D) to calculate k_{eff} for different configurations and also to calculate the critical control rod insertion. On the level of cell calculations a good agreement has been obtained between the cross-sections and the spectra as prepared by the SCALE system and as prepared by WIMS. Calculations with detailed geometry converged to very good agreement between the results of PANTHER and the results of KENO with an exact geometrical model. The results are summarised in Table 2, in which the measured values are also listed.

In Phase 2 of the benchmark, only KENO calculations are performed to calculate the scram reactivities of the core and reflector control rods and the isothermal temperature coefficients for which measured data are also available. KENO results efficiently provided the measured values of the scram reactivities as well as the estimation of the isothermal temperature coefficient within the requested temperature interval. Therefore, we can conclude that the benchmark calculations of the start-up physics calculations were successful and in agreement with the results of the reactivity measurements.

NRG/IRI also participated in the start-up measurements of reactivity and reactor noise. During the start-up phase of the HTTR at different core configurations, reactivity and the reactor noise measurements were carried out in parallel with measurements of the HTTR physics group. For these measurements two temporary compensated ionisation chambers were used. During the on-line reactivity experiments, measured DC signals were digitised and the reactivity is calculated from these signals using the inverse kinetics (IK) method.

Long-term behaviour of disposed spent HTR fuel (NRG)

One of the activities in the EU 5th Framework Programme projects "HTR-N" and "HTR-N1" is the investigation of the long-term behaviour of disposed spent HTR fuel. Although the behaviour of HTR fuel elements in a salt environment has been experimentally investigated for a long time, there is still a considerable lack of reliable experimental data. It has been shown that no radionuclides can be released from the fuel kernels as long as the coating layers are intact. As a small fraction of the coated particles might fail dependent on burn-up, it can be expected that the long-term source term of HTR spent fuel elements will be controlled by the retention capabilities of the intact particles. After failure of the coatings due to chemical or mechanical interaction, fuel dissolution and the interaction of the radionuclides with the graphite become the dominant factor. NRG is performing leaching experiments on SiC and fuel kernels of different compositions. Whereas the experimental work of the "HTR-N" project will be performed only with un-irradiated materials, the main focus of the "HTR-N1" project will be the investigation and prediction of the behaviour of irradiated spent HTR fuel in different aquatic phases on the basis of experimental data from un-irradiated material and the experimental validation of the predictions.

Model calculations and software development

Besides in the HFR irradiation and other experiments and measurements, the organisations in the Netherlands are involved in model calculations and the development of software for the simulation of several aspects of HTR systems. These model calculations mainly concern the PBMR and the ACACIA.

The PBMR is seen as the typical pebble bed nuclear power plant for utilities. In 2000, electric utility EPZ and NRG conducted a design review of the heating, ventilation and air conditioning (HVAC) system of the PBMR demonstration module [9]. NRG conducted calculational analyses for PBMR on four areas: simulator validation [10], pipe rupture in the PBMR reactor cavity [11], core physics [12], radiation shielding [13] and system dynamics [36-38].

Forecasts concerning the future consumption and production of energy indicate an expanding world market for the combined generation of heat and power. This market for energy efficient combined heat and power (CHP) with an overall capacity of 10-150 MW is particularly well developed in the Netherlands. In 1998, decentralised units and auto producers generated 27% of the total electricity supply in the Netherlands. Another 26% was generated by large natural gas fired power plants. Given the expected further depletion of the indigenous resources of natural gas (the fuel for CHP), a potential market could emerge for an alternative primary energy source within the next two decades. Nuclear energy could be one of the substitutes, if competitive prices and public acceptance for this new nuclear application can be achieved. The ACACIA unit [14], a small 40 MWth pebble bed HTR, is designed for this sector of the energy market. It could also find its way to regions in the world, which are deprived from reliable natural gas networks.

A selection of these activities is presented below. A more elaborate presentation of the computational analyses performed on PBMR and ACACIA can be found in [15].

PBMR core physics (NRG)

The PANTHERMIX code system [16,17] has been under development at NRG for a number of years. This code system, which combines the 3-D core simulation code PANTHER [18] and the 2-D R-Z HTR thermal hydraulics code THERMIX [19], can be used to simulate the steady-state, burn-up and transient behaviour of a pebble-bed HTR core.

Recently, the reactor physics code PANTHERMIX has been extended in such a way that pebble bed reactors with fuel re-circulation can be analysed with it [12]. For a cylinder symmetric reactor model, power per pebble, power density, solid structure temperature, gas temperature, thermal flux, fast flux and mean burn-up throughout the core in equilibrium fuelling condition have been calculated for the PBMR. A possible burn-in scenario was also investigated, simulating the transition from the initial core loading until equilibrium composition of the core. It was found that, in order to attain a real equilibrium composition, i.e. constant distribution of burn-up throughout the core, the fuel must have circulated at least 10 times through the core, which takes approximately three years [12].

PBMR shielding (NRG)

Dose rate calculations on several selected locations in and around the PBMR primary system have been performed. Radiation sources were direct fission, decay within the fuel, plate-out and activation. As PBMR specified radiation sources and geometries, we built models of reactor, primary system and building in the codes MCNP and FISPACT. As intermediate results flux spectra and nuclide activities within activated components were calculated. As an alternative design, the outer carbon reflector layer was borated with B_4C. As a result, neutron fluxes and even more the gamma fluxes directly outside the reactor vessel decreased significantly, and thus the radiation dose rates. Whether the PBMR design will be adapted this way remains to be seen. Follow-up activities concentrate on detailed reactor design and the fuel handling system.

PBMR CFD analysis of primary pipe rupture (NRG)

For the structural design analysis of the PBMR reactor building the analysis of a rupture of a pipe directly connected to the reactor pressure vessel was needed. This pipe rupture would give rise to pressure build-up in the building containing the reactor pressure vessel. NRG has determined detailed information of the pressure load on the walls of this building resulting from considered potential pipe ruptures. Because of the high pressure and high temperature in the system, a supersonic jet escapes from the leak opening and almost immediately reaches a wall of the building. CFD calculations have been performed in order to determine the largest pressure and thermal load on the walls of the reactor cavity resulting from direct jet impingement. It was found that immediately following the pipe rupture, a high speed jet escapes and almost immediately hits the ceiling of the reactor cavity.

ACACIA safety analysis (NRG)

For the ACACIA plant, both normal operation/control, and incident conditions have been analysed. In the first case, operation of the entire cogeneration plant has been modelled in detail, whereas the reactor model has been limited to a point kinetic one [20,21]. This is a dedicated code, implemented in the simulation tool Aspen Custom Modeler (ACM). For the second case, detailed analysis of reactor behaviour becomes important; the analysis was thus performed time dependent in 3-D, whereas the energy conversion model has been limited to the primary helium cycle [22,23]. For these analyses the pebble bed core simulation code PANTHERMIX and the thermal hydraulic code RELAP5 for the energy conversion system have been coupled. However, for the analysis of loss of cooling incidents (LOCI) and loss of flow incidents (LOFI) the transient simulation of the core by PANTHERMIX is sufficient. Simulation calculations for both un-irradiated and irradiated graphite (difference in heat conductivity) have been performed and it was found that the maximal fuel temperatures will exceed the 1 600°C (or 1 873 K) temperature limit for the irradiated LOFI scenario and reach 1 900 K. The irradiated LOCI scenario reaches exactly 1 873 K. The maximal fuel temperature for the un-irradiated scenarios remains under the 1 600°C limit.

Besides these well-established incident situations the question remains as to whether fuel temperatures remain within acceptable limits during all operational transients. This can be answered by systematically calculating the final states for a large number of load following transients. In the case of a load reduction, the final states of the transient will represent the worst-case scenario in terms of fuel and coolant outlet temperature. High fuel temperatures must then compensate the decrease of the negative xenon reactivity.

Two important conclusions can be drawn from these analyses: firstly, the LOCI and LOFI do not represent a situation where the maximal fuel temperature is reached, and secondly, according to this ACACIA design the reactor must be prevented from operating in the region below mass flow rates of 1.0 kg/s as the 1 600°C limit will then be exceeded. The fact that the maximal fuel temperature is not reached during a LOFI or LOCI, but during low mass flow rate transients, is not specific for this design.

ACACIA fission product transport (NRG)

Release, transport and deposition of fission products in the primary system of ACACIA have been investigated [24-26]. Four radionuclides have been identified in literature as most relevant because of volatility and radiotoxicity: ^{137}Cs, ^{90}Sr, ^{110m}Ag, ^{131}I.

For normal operation, the coated particle failure fraction has been assumed to be 10^{-5} consistent with German experience. With the codes PANAMA and FRESCO, transport of fission products through the fuel element has been analysed. Transport and deposition of the fission products within the primary system has been analysed with the code MELCOR.

The total activity caused by the nuclides depends on the average burn-up time. A ten-year operation period corresponds to an average burn-up of 848 full power days. The activity in each component at the end of the ten-year operation period (average burn-up of 848 full power days) was calculated. The total activity is about 59 GBq. The highest activity is produced by ^{137}Cs, followed by ^{131}I and ^{110m}Ag. The contribution of ^{90}Sr is very low.

The highest activity is found in the precooler: 56 GBq. The main reason is the condensation of the volatile species CsOH and CsI in this component. Other components with high activities are the recuperator (1 400 MBq) and the compressor (7 MBq). These components are mainly contaminated by ^{110m}Ag. The gas ducts in the energy conversion unit are also mainly contaminated by ^{110m}Ag (43 MBq) and by ^{131}I (11 MBq).

Pu incineration in HTR (NRG, IRI)

Besides the activities on long-term behaviour of disposed spent HTR fuel, both NRG and IRI are involved, together with other European partners, in HTR core physics simulation activities in the EU 5th Framework Programme projects "HTR-N" and "HTR-N1". On the basis of earlier studies [27], NRG is investigating the possibility to use pure Pu fuel in a HTR-MODUL type pebble bed reactor with continuous fuel recycling. IRI is investigating in detail the use of burnable poison (see next section).

In order to validate the code systems and nuclear data used in the investigations, benchmarking activities have also been incorporated into "HTR-N" and "HTR-N1" on the basis of data available from the ongoing HTR-10 and HTTR projects. In addition to these activities based on data from reactors using low-enriched uranium, a Pu benchmark is being set up in the "HTR-N1" project.

HTR with burnable poison (IRI, NRG)

During the operation of an HTR, the reactivity loss due to fuel burn-up and fission product poisoning must be compensated by some means of long-term reactivity control. This can be achieved by adding burnable poison to the fuel. In this study by IRI the focus is on heterogeneous poisoning in which burnable particles (TRISO-coated particles containing burnable poison only) are mixed with the fuel particles in the graphite matrix.

The aim is to minimise the reactivity swing as a function of the irradiation time. The SCALE-4.2 system [28] together with JEF-2.2 libraries [29] are being used for burn-up calculations on a macro-cell that consists of a burnable particle surrounded with a layer containing a homogeneous mixture of fuel particles and graphite. For the fuel considered first, the ESKOM PMBR has been chosen with an average core power density of 3 $MW.m^{-3}$. The fuel particles contain 8% enriched UO_2 with a uranium mass of 9 grams per pebble. One pebble contains more than 13 000 fuel particles, each with a radius of 0.5 mm. For the poison in the burnable particle, we have chosen B_4C with either natural boron or 100% enriched ^{10}B.

When we consider a "black" burnable particle, which means that every neutron that hits the surface will be absorbed, the effective absorption cross-section of the particle is related to its geometrical cross-section [34,35]. In spherical geometry, different radii are considered with a constant volume ratio of the fuel and the burnable particles. This means that the number of burnable particles per pebble increase when the radius of each particle decreases. The smaller the burnable particle, the more homogeneous the poison is distributed, and the faster the particles will burn. The reactivity as a function of time was calculated for the reference case without poison and for two radii of the burnable particles. It was found that the reactivity swing is smallest for the burnable particle with a radius of 0.3 mm.

In the case of a spherical burnable particle with a radius of 0.5 mm, both natural boron and enriched boron (100% of ^{10}B) have been considered. The initial reactivity is lower for the case with enriched boron, which indicates that the natural boron particle cannot be considered as a "black" one. Its diameter is about 3 absorption mean free paths.

Using burnable particles that contain a graphite kernel surrounded with a B_4C layer with 100% of ^{10}B reduces further the reactivity loss. Again the reactivity as a function of time was calculated for the reference case without poison and for two burnable particles. (Either with a graphite kernel up to 0.43 mm and B_4C between 0.43 and 0.46 mm, or with a graphite kernel up to 0.273 mm and B_4C between 0.273 and 0.3 mm.) Note that the ratio between the fuel and BP equals 10 300 in this case. It was found that the reactivity swing is smallest for the burnable particle with a radius of 0.3 mm.

For the same radius, the effective absorption cross-section is larger for a cylindrical shape. As a result, a cylindrical particle with a radius of 0.5 mm is already fully burnt after 1 000 equivalent full power days (EFPD), while the spherical particle still contains significant amounts of ^{10}B after 1 600 EFPD. For the same effective absorption cross-section (this means a smaller radius for the cylindrical particle) and the same ratio of fuel volume and poison volume, the reactivity swing is smallest for the cylindrical particle. Of course, a cylindrical particle is very difficult to realise in practice.

As a conclusion, we can state that the heterogeneous poisoning of HTR fuel seems quite promising. We have designed burnable particles containing B_4C that considerably reduce the reactivity swing. Further studies will focus on burnable particles that contain graphite or fuel surrounded with a thin layer of poison, on different poison materials (e.g. Er or Gd), and on different fuel compositions.

NRG recently started a project to investigate the possibility of a pebble-bed HTR (ACACIA) which operates on the same fuel loading during the entire operating cycle of a couple of years. The excess reactivity is compensated by burnable poison.

HTR system dynamics (IRI, NRG)

The dynamic behaviour of a pebble-bed HTR in combination with its energy conversion system has been studied extensively at NRG [22,23,30]. The code system PANTHERMIX-RELAP, developed at NRG, combines 3-D core neutronics, 2-D RZ core thermal hydraulics and ECS thermal hydraulics modelling.

A preliminary model of PBMR system has been prepared [36] for analyses with the SPECTRA code [38]. The model includes the reactor unit, the ECS, as well as the confinement building and various safety systems. The core power calculation is performed using the point kinetics model. Two scenarios have been analysed, namely loss of flow accident (LOFA) and loss of coolant accident (LOCA) [37]. In both cases no active safety systems were assumed to be available. It was found that the maximum fuel temperature remained within the acceptable limits during the analysed period (three days) for both scenarios.

A model of the conceptual design of INCOGEN [31] has been prepared for SPECTRA. Long-term system behaviour has been analysed for two accident scenarios: LOFA and LOCA [38]. Results were compared with the PANTHERMIX results reported in [31]. Good agreement between SPECTRA results and PANTHERMIX results was obtained.

At IRI a project has started to design small HTRs for different purposes and to simulate the dynamic behaviour of these HTRs during incidents. Use will be made of a new program that has been created by a Ph.D student at Delft University [21] in co-operation with NRG. The program is coded in ASPEN Custom Modeler, and can be used to investigate the dynamics and the interaction between neutronics and thermal hydraulics of the reactor and the energy conversion system. The latter system can be modelled to a level of detail that surpasses that of previous models described in literature. The neutronics model will be extended from a point-kinetics model to a two-dimensional model based on diffusion theory.

Auxiliary studies

Since the beginning of the 90s several organisations in the Netherlands have been involved in auxiliary studies concerning HTR design. In the INCOGEN program (Innovative Nuclear COGENeration) several aspects have been investigated of the predecessor of the present ACACIA concept, such as plant layout, design of the energy conversion system, control philosophy, inspection and maintenance, licensing, economics and market potential [31].

Presently, NRG and JRC-IE are still involved in auxiliary studies concerning HTR concepts, e.g. JRC-IE, together with CEA, is contributing to the RPV materials database activities in the "HTR-M" project with their expertise and understanding of the behaviour of materials under creep and non-creep both under nuclear and non-nuclear application. NRG, on the other hand, is studying aspects of the gas-cooled fast reactor [32] and revising the cost assessments for the ACACIA direct nuclear cogeneration plant [33]. One of the conclusions from this study is that, although the cost figure of 8 US$c/kWh is too high to be competitive in areas where low-priced natural gas is readily available, which is currently still the case in the Netherlands, ACACIA could still be an alternative for areas in the world where this supply is not existing or unreliable, or when gas prices are fluctuating [15].

Summary

An overview has been presented of past and present research activities of organisations in the Netherlands relevant to the design of high-temperature gas-cooled reactors. An important role is played by JRC's high flux reactor at Petten, which is used for numerous test irradiations of fuels and other materials relevant to HTR design. Besides that, since the beginning of the 90s there is a growing involvement and expertise of Dutch organisations in other HTR-related investigations, such as auxiliary supporting experiments and model calculations and software development. Most of these activities are embedded in an international framework.

REFERENCES

[1]   J. Ahlf, A. Zurita, eds., "High Flux Materials Testing Reactor, HFR Petten, Characteristics of Facilities and Standard Irradiation Devices", EUR 15151EN, 1993.

[2]   R. Conrad, K. Bakker, "Irradiation of Fuels for High-temperature Gas-cooled Reactors at the HFR Petten", paper presented at the International HTR Fuel Seminar, 1-2 February 2001, Brussels, Belgium.

[3]   B. Schröder, W. Schenk, Z. Alkan, R. Conrad, "Ceramic Coatings for HTR Graphite Structures: Tests and Experiments with SiC-coated Graphitic Specimens", OECD/NEA/NSC, First Information Exchange Meeting on Survey of Basic Studies in the Field of High-temperature Engineering, 27-29 September 1999, Paris.

[4]   J. Ahlf, R. Conrad, M. Cundy, H. Scheurer, "Irradiation Experiments on High-temperature Gas-cooled Reactor Fuels and Graphites at the High Flux Reactor Petten", *J. Nucl. Mater.*, 171 (1991) 31-36[6].

[5]   R. Conrad, A.I. van Heek, "Present Status in the Netherlands of Research Relevant to Irradiation of Fuels and Graphite for High-temperature Gas-cooled Reactors", OECD/NEA/NSC, First Information Exchange Meeting on Survey of Basic Studies in the Field of High-temperature Engineering, 27-29 September 1999, Paris.

[6]   K. Verfondern, H. Nabielek, R. Conrad, K. Bakker, "Test Specification for the Irradiation Experiment HFR-EU1 within HTR-F/WP2", November 2000, to be published.

[7]   E.J.M. Wallerbos, "Reactivity Effects in a Pebble-bed Type Nuclear Reactor", Ph.D Thesis, Delft University of Technology, Delft, Netherlands, ISBN 90-407-1662-5, 1998.

[8]   J.B.M. de Haas, E. Türkcan, "HTTR Criticality, Physical Parameters Calculations and Experimental Results", Proc. International Conference on Emerging Nuclear Energy Systems ICENES 2000, 24-28, ISBN 90-805906-2-2, September 2000, Petten, the Netherlands.

[9] A. Greiner, H.W. de Wit, "Design Review of the HVAC System of the PBMR Demonstration Module, Design Report and Specification", NRG, Arnhem, 20511/00.52391C/C, 6 Sept. 2000 (*confidential*).

[10] F. Roelofs, H. Koning, "PBMR: CHARME Validation Calculations", NRG, Petten, 910134/99.26813/C, 4. August 1999 (*confidential*).

[11] E.M.J. Komen, F. Roelofs, C.J.J. Beemsterboer, "Analysis of Postulated Pipe Ruptures in the PBMR Reactor Cavity", NRG, Petten, 20080/99.23620/C, 27 April 1999 (*confidential*).

[12] H.T. Klippel, J.C. Kuijper, J.B.M. de Haas, J. Oppe, E.C. Verkerk, "PBMR Burn-in Core Analysis, PBMR Benchmark with PANTHERMIX", NRG, Petten, 20429/00.38175/C, 2001.

[13] A.I. van Heek, A. Hogenbirk, R.C.L. van der Stad, "PBMR Shielding Analysis", Contract No. 2PG 000019 – Part I, II and III, NRG, Petten, 20441/00.33335/C, 22 May 2000; 20441/00.34923/C, 22 December 2000; 20441/00.38555/C, 26 January 2001.

[14] A.I. van Heek, B.R.W. Haverkate en J.N.T. Jehee, "Nuclear Cogeneration Based on HTR Technology", ENC'98, Nice, France, 25-28 October 1998.

[15] A.I. van Heek, "The Pebble Bed HTR, the Nuclear Option in the Netherlands", paper presented at the Seminar of HTGR Applications and Development, 19-21 March 2001, Beijing, China, NRG report 20631/01.40976/P.

[16] J.C. Kuijper, *et al.*, "Reactor Physics Calculations on the Dutch Small HTR Concept", IAEA Technical Committee Meeting on High-temperature Gas-cooled Reactor Development, Johannesburg, Republic of South Africa, 13-15 November 1996.

[17] J. Oppe, J.B.M. de Haas, J.C. Kuijper, "PANTHERMIX – a PANTHER-THERMIX Interaction", Report ECN-I—96-022, May 1996.

[18] P.K. Hutt *et al.*, "The UK Core Performance Package", *Nucl. Energy*, 30, No. 5, 291, Oct. 1991.

[19] S. Struth, " THERMIX-DIREKT: ein Rechenprogramm zur instationaeren, zweidimensionale Simulation thermohydraulischer Transienten", ISR-KFA Jülich, August 1994 (*private communication*).

[20] J.F. Kikstra, A.I. van Heek, A.H.M. Verkooijen, "Transient Behaviour and Control of the ACACIA Plant", IAEA Technical Committee Meeting on Gas Turbine Power Conversion Systems for Modular HTGRs, Palo Alto, 14-16 November 2000.

[21] J.F. Kikstra, "Modelling, Design and Control of a Cogenerating Nuclear Gas Turbine Plant", Ph.D Thesis, Delft University of Technology, April 2001, ISBN 90-9014631-9.

[22] E.C. Verkerk, "Dynamics of the Pebble-bed Nuclear Reactor in the Direct Brayton Cycle", Ph.D Thesis, Delft University of Technology, November 2000, ISBN 90-9014203-7.

[23] E.C. Verkerk, A.I. van Heek, "Dynamics of a Small Direct Cycle Pebble Bed HTR", IAEA Technical Committee Meeting on Gas Turbine Power Conversion Systems for Modular HTGRs, Palo Alto, 14-16 November 2000.

[24] P.H. Wakker, "Fission Product Release in the ACACIA Reactor", Report 22951-NUC 98-2644, NRG, Arnhem, the Netherlands, 1998.

[25] N.B. Siccama, "Fission Product Retention in the ACACIA Reactor Primary System", Report ECN-R—98-019, NRG, Petten, the Netherlands, 1998.

[26] A.I. van Heek, N.B. Siccama, P.H.Wakker, "Fission Product Transport in the Primary System of a Pebble Bed High-temperature Reactor with Direct Cycle", IAEA Technical Committee Meeting on Gas Turbine Power Conversion Systems for Modular HTGRs, Palo Alto, 14-16 November 2000.

[27] E.E. Bende, "Plutonium Burning in a Pebble-bed Type High-temperature Nuclear Reactor", Ph.D Thesis, Delft University of Technology, December 1999, ISBN 90-9013168.

[28] SCALE-4.2, "Modular Code System for Performing Standardized Computer Analyses for Licensing Evaluations", Oak Ridge National Laboratory, Oak Ridge, Tennessee, 1994.

[29] JEF-2.2, "The JEF-2.2. Nuclear Data Library", Report JEFF17, Nuclear Energy Agency, Paris, 2000.

[30] E.C. Verkerk, A.I. van Heek, "Transient Behaviour of Small HTR for Cogeneration", OECD/NEA/NSC, First Information Exchange Meeting on Survey of Basic Studies in the Field of High-temperature Engineering, 27-29 September 1999, Paris.

[31] "INCOGEN Pre-feasibility Study, Nuclear Cogeneration", A.I. van Heek, ed., ECN, Sept. 1997.

[32] A.I. van Heek, "Review of the Use of Coated Particle Fuel in Gas-cooled Fast Reactor Concepts", NRG Report 20568/01.41833, June 2001.

[33] J.N.T. Jehee, K. Lievense, J. Hart, P. Lako, A.I. van Heek, "ACACIA Direct Cycle Nuclear Cogeneration Plant – Review of Cost Assessments", Report 20286/01.42382/P, NRG, Petten, the Netherlands, to be published as NRG report in 2001.

[34] H. Van Dam, "Long-term Control of Excess Reactivity by Burnable Particles", *Annals of Nuclear Energy*, 27, 2000, 733-743, 2000.

[35] H. Van Dam, "Long-term Control of Excess Reactivity by Burnable Poison in Reflector Regions", *Annals of Nuclear Energy*, 27, 63-69, 2000.

[36] M.M. Stempniewicz, K. Spijker, "Development and Verification of SPECTRA Model of PBMR", NRG report 20353/01.52091/C, Arnhem, February 2001.

[37] K. Spijker, J.H. Koers, "PMBR LOCA and LOFA Analysis with SPECTRA – SPECTRA as a Tool for PSA", NRG report 20354/01.52517/C, Arnhem, 16 January 2002.

[38] M.M. Stempniewicz, "SPECTRA – Sophisticated Plant Evaluation Code for Thermal-hydraulic Response Assessment", Version 1.00, December 1999; Volume 1 – Description of Models; Volume 2 – User's Guide; Volume 3 – Description of Subroutines; Volume 4 – Verification", NRG/PPT 26094/99.52612/C, Arnhem, October 1999.

Table 1. Survey of irradiation experiments performed at the HFR Petten in support of R&D for German modular HTR and US HTGR

Project name	Contractor	Characteristics of specimens	Irradiation performance data					
			Irradiation period	Irradiation time [fpd]	Temperature [°C]	Burn-up [% FIMA]	Fluence E > 0.1 MeV [10^{25} m^{-2}]	EOI fractional fission gas release [R/B Kr^{85m}]
HFR-K 3	FZJ, NUKEM, HRB	4 spherical fuel elements 60 mm UO$_2$ LEU reference particle	1982-1983	359	1 210 central	10.6	6.0	$2.2 \ 10^{-7}$
HFR-P 4	FZJ, NUKEM, HRB	36 small spheres 20 mm diameter UO$_2$ LEU reference particle	1982-1983	351	1 000 and 1 200 central	14.5	8.2	$8.5 \ 10^{-8}$
HFR-P 5	FZJ, NUKEM	116 coupons UO$_2$ LEU reference particle	1983	142	1 450	7.2	5.5	$1.1 \ 10^{-4}$
HFR-K 4	FZJ	2 spherical fuel elements 60 mm UO$_2$ LEU	1985-1986	667	600 < T < 1 150	13	10.0	Defective coated particles due to drilled holes for thermocouples
HFR-K 4	FZJ	2 graphite spheres 60 mm A3-27	1985-1986	667	800	–	8.5	–
HFR-K 5	FZJ, NUKEM, ABB, HRB	4 spherical fuel elements 60 mm UO$_2$ LEU reference particle	1991-1994	564.28	800/1 000 cycle 3 × 1 200 for 5 h	6.7-9.1	2.85-4.25	$3.0 \ 10^{-7}$
HFR-K 6	FZJ, NUKEM, ABB, HRB	4 spherical fuel elements 60 mm UO$_2$ LEU reference particle	1990-1993	633.55	800/1 000 cycle 3 × 1 200 for 5 h	7.2-9.7	3.2-4.83	$3.0 \ 10^{-7}$
HFR-B 1	General Atomic FZJ	36 fuel rods in 3 capsules UCO LEU & ThO$_2$ Segments of block design	1988-1989	445	900 880 < T < 1 230 820 < T < 1 040	15.4 17.0 14.2	6.1 6.7 5.3	10^{-3} 10^{-3} 10^{-3} and 16 H$_2$O injections
HFR-K 9	FZJ	6 SiC coated graphite spheres 60 mm with and without fuel	1995	93.89	550-680	–	1.0-1.95	–
HFR-K 10	FZJ	5 SiC coated graphite spheres and 3 SiC samples 60 mm with and without fuel	1998	98.01	600-770	–	1.6-1.9	–
HFR-K11	FZJ	6 SiC coated graphite spheres and 6 SiC samples 60 mm with and without fuel	1999	95.88	620-760	–	1.0-1.7	–
HFR-EU1	EU Contract FIKS-CT-2000-00099	4 spherical fuel elements 60 mm UO$_2$ LEU reference particle	2002-2004	(550)	(800-1 100)	(20)	(6.5)	

Table 2. HTTR Benchmark Phase 1

	KENO (IRI/NRG)	BOLD-VENTURE (IRI)	PANTHER (NRG)	Measured (JAERI)
k_{eff} simple core	1.1278 ± 0.0005	1.1592	1.1251	
k_{eff} fully loaded core • Rods withdrawn • Rods inserted	1.1584 ± 0.0005 0.6983 ± 0.0005	1.1974	1.1595 0.7510	0.685 ± 0.010
Critical insertion • Above bottom core	170.5 cm		161.5 cm	178.9 cm

Table 3. HTTR Benchmark Phase 2: Scram reactivities

CR group	Critical position (mm) HTTR-Crit	Position after scram (mm) HTTR-SR	Position after scram (mm) HTTR-SA
C	1 789	1 789	-41
R1	1 789	1 789	-41
R2	1 789	-41	-41
R3	Full out	-41	-41
k_{eff} (average)	1.0093 ± 0.000	0.9178 ± 0.0005	0.6809 ± 0.0005
ρ_{calc}		0.0988 ± 0.0007	0.4778 ± 0.0007
ρ_{meas}		0.120 ± 0.012	0.46 ± 0.04

Figure 1. Standard core configuration of HFR Petten

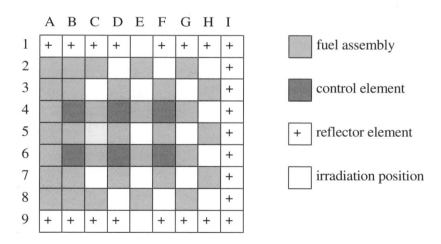

CURRENT STATUS OF HIGH-TEMPERATURE ENGINEERING RESEARCH IN FRANCE

Dominique Hittner
FRAMATOME ANP, France

Franck Carré, Jacques Roualt
CEA, Nuclear Energy Division, France

Abstract

CEA and FRAMATOME ANP researches in the field of high-temperature engineering mainly support high-temperature gas-cooled reactors (HTGCRs or HTRs) technology development. These researches are for the short term directly connected to the support of the FRAMATOME ANP for the industrial development of small or medium sized thermal HTRs with direct cycle and the necessary recovery at a national level of expertise on some associated crucial technological issues. For the longer term, the CEA's main R&D effort is directed toward the study of technologies in support of HTGCRs with hardened neutron spectra, refractory fuel possibly compliant with on-site reprocessing. This appears to be a very promising candidate for future generations of nuclear systems, particularly with regard to the goals of high efficiency, saving of resources, minimisation of long-lived waste production and potential for applications other than electricity production (hydrogen, desalination, etc.). The corresponding programmes concerning high-temperature engineering are reviewed in this paper.

Introduction

The development of nuclear energy in the next decades will depend on the demand of the energy market and the adaptability of the nuclear offer to the demand. For large power generation units, market opportunities will most probably open again in the next decade, in particular in France, with the replacement of the older existing plants. But there is another part of the energy market, the area of small and medium sized power plants and co-generation units, where, in spite of opportunities continuously appearing all over the world, the nuclear industry is remains rather absent. For the first part of the market, FRAMATOME ANP is developing EPR and considers that, for the other part, direct cycle modular HTR is the best candidate for satisfying as soon as possible the market needs with a nuclear offer. Therefore FRAMATOME ANP is involved in the development of this type of reactor with the goal of having an industrial product ready for commercial offers in 2010.

The effort to be made in that perspective is not only an effort of design. There are also important R&D needs in support of the design. The HTR reactor and fuel technology has been developed up to the industrial level in the past, in particular in Europe, but such development has been stopped for quite a while. This technology heritage has to be revived and developed further in order to face more stringent requirements of economic competitiveness and safety. Moreover the direct cycle power conversion system is a new feature of modern HTRs, which is very promising, but which has to be fully developed, even if it partially relies on the existing gas turbine technology. Therefore FRAMATOME ANP and CEA have launched an R&D programme necessary for the industrial development of direct cycle modular HTRs. As there has been recently an increased interest for HTRs in Europe, part of these activities take place in the frame of a European programme involving many research centres and industrial organisations of different countries. With the help of the Joint Research Centre of the European Commission, a HTR Technology Network (HTR-TN) has been created for federating the efforts made in Europe for the development of HTR technology. FRAMATOME ANP is chairing this network. The partners of HTR-TN have submitted several project proposals for the 5th Framework Programme of the European Commission and as most of them have been accepted, they are now starting an R&D programme of about 18 million €, about 50% of it being funded by the European Commission. Co-operation is also starting with non-European partners: JAERI (Japan), INET (China) and General Atomics (USA).

In the longer term, gas coolant (particularly He) and high-temperature reactor technology look sufficiently promising to develop a technological evolutionary series of gas-cooled reactors (GCR) which present attractive features consistent with the goals assigned to new generation nuclear systems that could be deployed by 2030-2050: economy, enhanced safety, minimisation of long-lived radioactive waste production, efficient use of resources, increased resistance to proliferation risk, potentialities for other applications than electricity generation. It is the reason why CEA decided to launch in late 2000 within its programme on "Future Generation Reactors and Fuel Cycles" a strong R&D axis to obtain the necessary knowledge and innovative technological breakthroughs to serve the long term objectives and that may have useful spin off for shorter term options of the GCR technological series:

- For the short term the programme is directly connected to the support of the FRAMATOME ANP programme of industrial development on HTR with direct cycle and the necessary recovery of expertise on some crucial technological issues.

- For the medium term, two evolutions of the HTR technology are considered:

  - A very high-temperature GCR (>1 000°C) for very high thermodynamical efficiency and hydrogen production (thermochemical cycles, high-temperature electrolysis, etc.).

  - A robust GCR with maximised safety features, intrinsic resistance to proliferation for deployment in countries which have little or no experience in nuclear energy production.

- Finally, for the long term, an isogenerating GCR with a fast neutron spectrum, able to transmute any of the actinides so as to comply with objectives of sustainable development.

The main objectives of the strong CEA R&D programme for the GCR series are:

- To develop a fuel capable of fission product retention at sufficiently high temperatures, able to sustain hard neutron spectrum, ultra high burn-up and to be proposed with the options either to be reprocessed or to be very resistant to reprocessing.

- To propose fuel treatment processes (reprocessing and fabrication) compatible with an on-site implementation.

- To develop materials for high temperatures, able to sustain fast neutrons.

- To get expertise in the field of high-temperature He circuits technology.

This entire programme will be as far as possible implemented within the framework of international co-operation (Europe, USA, Japan, Russia) and also in relation with the "GENERATION IV" US DoE initiative.

Fuel development

The first objective is clearly for France to recover the capability of HTR particle fuel fabrication and then to characterise and qualify its behaviour under irradiation with the goal of proving that the necessary level of reliability is achieved in both nominal and high-temperature conditions. This experimental programme is accompanied by a modelling effort in order to understand the particle fuel behaviour under irradiation and to provide a fully qualified calculation tool for the thermomechanical behaviour of the particle and for fission product transport. Key dates are:

2003	Release of a first qualified version of the particle fuel code.
2004	Fabrication of uranium HTR triso particle fuel.
2004-2006	Start of irradiation tests of those particles in the OSIRIS reactor at CEA Saclay.
2006-2010	Characterisation of irradiated particles and qualification of the fabrication process.

The current status in this area is the following:

- The knowledge acquired by the end of the 60s (DRAGON period) and mainly in the 70s (Peach Bottom and Fort Saint Vrain period) essentially in the framework of a contract with General Atomics was in the form of reports and still in the minds of people (retired or not) that were involved at that time. The knowledge concerning fabrication has been recollected (documents, interviews) in the form of a synthesis note. With that background and further analysis, the sol-gel process has been selected for kernel fabrication with still two possible routes (external or internal gelification process) and a CVD process for coatings.

- Currently PyC dense and porous coatings on ZrO_2 yttrium stabilised simulant kernel have already been realised and present work now concentrates on SiC coatings. A sol-gel apparatus will be used at Cadarache to restart UO_2 kernel fabrications. Interim milestones are the recovery of the kernel technology and of the coating technology on simulant kernel by the end of 2002.

- The specification of the device of particle uranium fuel manufacturing is achieved. It is designed for particle batches of 100 to 200 g, which means a capacity of production of several kg per year. This specification will now be used for a preliminary safety assessment, from which, based on the quantities of active effluents generated, the sol-gel preferred route and the design of the laboratory scale fabrication line will be selected.

- Concerning the state of the fuel code development, the following models have already been implemented in a 3-D approach: fission gas and CO pressure built-up, UO_2 swelling, buffer and PyC densification and creep. At present the behaviour under irradiation of both individual particles or particles imbedded in a graphite matrix can be calculated.

As far as GCRs with a harder spectrum are considered, even if the TRISO particle fuel remains a reference, it is clear that different materials and designs will more than probably have to be found. The spectrum hardening requires to limit the C quantity within the core and to use preferably non-absorbing heavier atoms; the effect of fast neutrons on materials resistance will be a crucial issue. Furthermore, considerations of sustainable development will require fuel designs able to incorporate a larger amount of actinide atoms per fuel unit volume. In that area, the key milestones are:

2001-2004	R&D for the selection of a fuel design with appropriate materials.
2005-2015	Characterisation under irradiation and pre-qualification of the fast GCR fuel.

Structural materials for high temperature

The study of high-temperature structural materials, together with the technology of He circuit presented in the next section, clearly show a common body of development necessary for the whole technological series of GCRs. Furthermore, there are also potential benefits from synergies with industry. Of course, requirements for a fast GCR introduce supplementary constraints for inner core structure materials.

Main milestones during the 2002-2015 R&D period for GCRs materials development are:

2005	Materials proposals for an actualised HTR technology (HTR-TN).
2006	Selection of materials for a feasibility study of a fast GCR.

The acquired experience in past R&D programmes on high-temperature structural materials for nuclear and non-nuclear applications makes it possible to gather in the following table the material classes likely to resist to the various operating conditions of the GCRs components:

Reactor component	Fast GCR with direct cycle	HTR
Vessel (250-500°C, irradiation)	Advanced ferritic martensitic steel (AFM), 9-12% of Cr	
Primary circuit (250-850°C, stress)	– Ni based super-alloys – 32Ni–25Cr–20Fe–12.5W–0.05C (German HTR) – Ni-23Cr-18W-0.2C (Japanese HTTR) – AFMA + thermal barrier	
Turbine (850°C, high stress)	Ni based with spectrum hardening or ODS (oxide dispersion strengthened steels)	
Fuel element and inner core structures (1 000-1 600°C, irradiation)	– AFM ODS – Refractory metals and alloys – Ceramics	Graphite PyC, SiC, ZrC

In this table, it can be seen that the levels of R&D necessary to qualify the different materials can be very different, from a simple re-qualification in the particular operating conditions to the launching of a complete R&D programme, when a technological threshold has to be passed. This is particularly the case for the core structural materials of a fast GCR with the conjunction of high temperatures, high neutron doses and specific requirements in terms of fission product confinement. In such a situation several classes of materials have to be selected first, to be followed by screening tests and then a full characterisation of the selected candidates in the required operating conditions.

The R&D work is focused on two different lines, the content of which is briefly recalled:

- The selection and the establishment of the specifications and the determination of the in-pile and out-of-pile behaviour for the structural materials: state of the art concerning current materials for which a lot of work has been done and establishment of their operating limits (temperature, dose, etc.), development of a generic 9-12% of Cr advanced ferritic martensitic steel, ODS steels validation, irradiation experiments in HFR, Phénix, BOR 60. Irradiations can also be defined in JOYO in the framework of the CEA/JNC collaboration, as this reactor offers unique features with capsules able to irradiate samples at high temperature with *in situ* mechanical solicitation, followed by post-irradiation examination.

- The establishment of the rules for mechanical assessments and for the prediction of the structures and the welded joints lifetime.

These two lines of development are complemented by an upstream effort of fundamental research and modelling.

Some effort is also presently devoted to the oxidation of graphite by air, as this is an important issue for HTR safety evaluations. The main parameters of graphite oxidation can be divided into two groups: intrinsic parameters of graphite (porosity, density, impurities content) and oxidation conditions (temperature, oxidising gas composition, flow rate). It is very difficult to extrapolate the behaviour of a graphite grade from experiments performed using a different graphite grade in different oxidation conditions.

This is why the CEA has undertaken an experimental programme to study the behaviour of graphite in air focused on specific graphite grades eligible for HTR use. This programme is divided into three main parts: characterisation of graphite, parametrical study of graphite oxidation behaviour (thermogravimetry experiments including the influence of temperature, the influence of oxidising gas composition) and accident simulation. This latter programme will be conducted in the COMETHE facility at CEA Cadarache (adapted to high-pressure/high-temperature conditions), equipped with on-line gas phase analysis. After each experiment, the mechanical behaviour of graphite will be checked, with special attention paid to the compressive behaviour.

He technology

The definition of the objectives of a gas technology programme has been established starting from the review of the needs identified for a project such as the GT-MHR with thermal spectrum and direct cycle and extrapolated to hardened or fast spectrum. Once again we are there in a kind of logic similar to fuel studies with a first step that is clearly to re-acquire expertise in this field. The key milestones of the He technology programme are:

2001-2005	Experimentation on small dedicated benches.
2003-2010	Realisation and experimentation on a 1 MW multi-purpose loop.
2005-2015	Realisation and experimentation on a 15 MW system loop.

Among the identified requirements for the GCR's He technology are:

- The purification and control of the coolant quality.

- The management of gas inventories.

- Some generic points of technology such as tribology, leak tightness, thermal insulation.

- The heat transfers and fluid flows in the core, the circuits and the heat exchangers (pressure drops, flow distributions, heat transfer coefficients, etc.).

- The dynamics of the circuit and structures.

To achieve this He technology programme, the following facilities and programmes are planned:

- In 2002-2003, small dedicated benches for:

  - Tribology studies (large diversity of contacts and situations in GCRs: fretting, gliding, rotation, control rods mechanisms, stator seals, bearings, structure vibrations, quasi-static displacements to ultra-high frequency vibrations for temperatures from 25 to 850°C and contact pressures from 0.5 to 10 MPa); the oxygen/impurities influence on the binding risk will have to be considered as well as the evaluation of coating performances.

  - Thermal insulation studies (thermal protection of structures, vessel and circuits, efficiency with the need to limit the free convection in their thickness, chemical stability, erosion and thermo-mechanical resistance in particular in case of depressurisation).

  - Leak tightness studies concerning wall diffusion, and leaks through valves, seals and shaft penetration in the case of an alternator located outside the turbine vessel.

- In 2004-2005, dynamic loops for:

  - Purification studies in order to develop and validate the methods and instrumentation to guarantee the He quality, to monitor the level of chemical impurities and radio-chemical impurities, activated corrosion products and fission products and to determine the efficiency and purification rates of the various considered designs.

  - Small power (1 MW) multi-purpose loop designed to operate up to 1 000°C and at pressures of 10 MPa for the qualification of components (or parts of them) such as recuperator, circulator, valves and to perform fuel element and sub-assembly characterisations (heat transfer coefficients, pressure drop, purification and quality control, instrumentation testing, etc.)

- Multi-channel and system multi-purpose loop still to be specified for testing in fully representative conditions large components and GCRs systems (or parts of them) with the particular objective to demonstrate the robustness of the selected technologies in both nominal and accidental situations (depressurisation, etc.).

Components development

The former HTRs in England, Germany or US did not consider direct cycle (steam cycle for the larger reactors THTR and FSV). Thus the experience feedback for the innovative components induced by the direct cycle concept is limited and an important work programme is required for the development of these components. In addition the efficiency, reliability and competitiveness of this kind of reactor are based on the selection of very high performance components in term of efficiency, life time, and thermo-mechanical resistance, and the standard components used in industry cannot be directly transferred to HTRs or gas-cooled reactors applications.

This work programme will be implemented through international co-operation (GT-MHR, HTR-TN).

The main components to be developed and the main critical points to be studied are listed hereafter:

- *The helium turbine.* High efficiency component, creep resistant material for blades and discs, risk of corrosion/erosion, activation by Ag Cs, stator seals, bypass, manufacturing processes for large size elements, maintenance.

- *The recuperator (helium/helium heat exchanger).* Severe loading (high temperature level and pressure difference), high thermal exchange performances, low pressure losses, high compactness, risk of fouling, lifetime, maintenance.

- *The turbo machine support.* By magnetic bearings only (active/permanent), no lubricated mechanical bearings in order to improve safety by suppression of fluid ingress risk in the primary circuit, dynamic control of flexible shafts, resistant catcher bearings under shaft drop, thermal expansion of the shaft, displacement measurement by sensors, diagnosis algorithms.

- *The pre-cooler and inter-cooler (helium/water heat exchangers).* High efficiency component, high pressure difference, low pressure drop on the helium side, compactness, compatibility to co-generation applications, lifetime, maintenance.

For these components the development must be performed in partnership with industrials and some are already identified and involved in the work programme:

- Aubert and Duval for turbine materials.

- Heatric for compact plate type heat exchanger.

- Balcke Duerr for tubular heat exchanger.

- S2M for magnetic and catcher bearings of turbo machine support.

The main objectives of the components work programme (2000-2005) are:

- Recovering of the past experience.

- Review of existing technologies and updating of the state of the art.

- Specification of the component.

- Proposal of a preliminary design.

- Analysis of the design.

- Validation by experimental test on mock-ups.

Concerning the helium turbine an experience feedback exists in Germany on the past facilities of EVO in Oberhausen and HHV in the Jülich centre.

The test facilities of CEA GRETh Grenoble for heat exchangers and Zittau University for magnetic and catcher bearings will be used for experimental tests on mock-ups. For the long term the gas loops planned by the CEA Cadarache will be able to test large scale components.

Conclusion

CEA together with FRAMATOME ANP has set up a comprehensive R&D programme on promising high-temperature technologies for medium term and long term future nuclear systems based on gas-cooled reactors. For the short term, this programme is directly connected to the support of the FRAMATOME ANP for the industrial development of small or medium sized thermal HTRs with direct cycle. It necessitates the recovery at a national level of expertise on some associated crucial technological issues in the field of particle fuel fabrication and He circuits technology. The study of high-temperature structural materials, together with the technology of He circuit and components is a common body of R&D serving as well short and medium term objectives (thermal HTRs) and longer term ones (high-temperature gas-cooled reactors with hardened neutron spectra). Nevertheless, fast neutrons will probably necessitate some breakthroughs in the field of fuel materials and design, and of core structural materials.

OVERVIEW OF HIGH-TEMPERATURE REACTOR ENGINEERING AND RESEARCH

K. Kugeler, P-W. Phlippen, M. Kugeler, H. Hohn
Institute for Safety Research and Reactor Technology, Research Centre Jülich
52425 Jülich, Germany

Abstract

In the future, it is essential that nuclear energy be used under the conditions of sustainability.

Enhanced reactor safety, improved economics, suited waste management and best use of resources are very important topics in this context. The modular high-temperature reactor can contribute to fulfil these requirements because it has excellent safety characteristics and it can be applied not only to produce electricity very efficiently, but also to deliver nuclear process heat for many applications in the energy economy. Broad activities in the field of HTR are currently being undertaken in countries such as Japan and China, which commissioned test reactors, and in South Africa, the USA and Russia, which plan to build such plants. In countries having a long-term tradition of gas-cooled reactors (i.e. Great Britain, France and Germany), interest in this type of reactor persists.

The safety analysis of these nuclear power plants is dedicated to a system which will avoid the problems which are connected to existing reactors. These HTR reactors cannot melt even in severe accidents of loss of coolant and fuel cannot be destroyed by other severe accidents. Therefore, the release of radioactivity to the environment is very limited. The state of technology and new developments are explained in more detail in this paper. The overall concept of inherent safety, including questions concerning self-acting decay heat removal and hypothetical impacts from outside, are discussed and analysed.

Open questions and future needs for research and development in the field of HTR technology are added to better establish an impression of the state of HTR technology worldwide.

Safety requirements for future nuclear reactors

Worldwide there are many projects to develop and design nuclear reactors for future application. They should offer more safety compared to today's systems. In Germany, for example, the Modified Atomic Power Act from 1994 requires a very high safety standard for future nuclear power plants: evacuation and relocation of people must not be necessary, no intolerable radiological consequences are allowed to occur outside the nuclear power plants after accidents, etc. (Figure 1).

Figure 1. Modified German Atomic Act

Amendment of the <u>Atomic Energy Act</u> as a further precaution against risks to the general public (7th Amendment to the Atomic Energy Act of 27.07.1994):

The amended German Atomic Energy Act (1) (Art. 7, para. 2a) for future plants stipulates "that even such events whose occurrence is practically excluded by the precautions to be taken against damage would not necessitate decisive measures for protection against the damaging effects of ionising radiation outside the plant fencing, ...". In the substantiation of the bill for parliamentary discussion, terms from the text of paragraph (2a) are defined in more detail. <u>It is thus postulated that "accidents with core melt" are controlled and "evacuations" are not necessary.</u>

The new type of technology that fulfils these requirements can be designated "catastrophe-free nuclear technology" (Figure 2). The allowed release of fission products in case of severe accidents is very limited. For the case of a reactor with a thermal power of 300 MW, for example (discussed below), this amount should be less than 10^{-5} to 10^{-6} of the fission product inventory inside the fuel elements.

Figure 2. Requirements for future nuclear reactors with new safety concept: retention of fission products and fissile material inside the plant in all accidents, limitation of the release to less than 10^{-5} to 10^{-6} of the inventory to the environment.

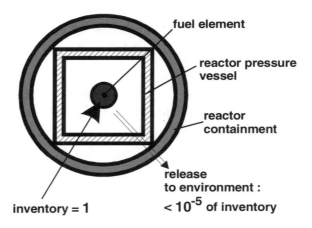

The range of catastrophe-free nuclear technology must be defined. Following Figure 3 accidents from internal and from external sources which can be foreseen and which are covered by the future licensing process have to be considered. Category 1 and 2 contain these accidents. Category 3 contains some accidents which are far beyond today's licensing procedure, which partly require new design measures. One aim of this contribution is to explain the situation so far as the self-acting of decay heat is considered even in extreme situations, i.e. very extreme earthquakes or terrorist attacks.

Figure 3. Accidents considered to be covered by the concept of "catastrophe-free nuclear technology"

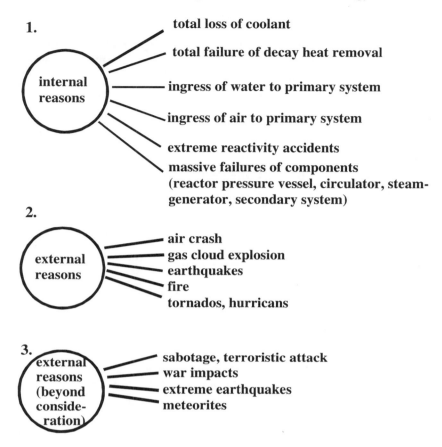

The work to be done in the future is to identify systems in which there is no release of intolerable amounts of radioactivity to the environment through internal or external causes. This safety behaviour of the reactors must be improved by convincing experiments.

Concepts of new pebble bed reactors

All new HTR concepts are designed to integrate inherent safety features as much as possible, the main aspects of this being the self-acting decay heat removal and the limitation of the temperature of fuel elements to allowed values (today 1 600°C for TRISO-coated particles in pebble-bed fuel element).

The concept of self-acting decay heat removal exists – in principle – in all high-temperature reactors, including the old reactor systems. Dependent on power, power density and core dimensions, the maximum fuel temperatures during loss of coolant and loss of active decay heat removal accidents is different for the reactors. Table 1 shows some data resulting from three-dimensional calculations for several layouts of HTR cores.

Especially for large cores (THTR-300, GAC-1160, Fort St. Vrain) high fuel temperatures above 2 000°C would have occurred in decay heat removal accidents, resulting in large fission product releases from the fuel elements. Even these large cores would not have melted down, because graphite stays stable up to 3 600°C.

Table 1. Comparison of data of HTR cores with respect to the accident, loss of coolant and loss of active decay heat removal

Reactor system	Thermal power (MW)	Mean power density (MW/m³)	Type of core	Core diameter (m)	Helium temp.	Max. fuel temp. in operation (°C)	Max. fuel temp. in accident (°C)
AVR	50	2.2	Cylinder	3	950	< 1 100	< 1 300
THTR-300	750	6	Cylinder	5.6	750	< 1 100	< 2 500
HTR-100	250	4.2	Cylinder	3.45	750	< 1 100	< 1 700
HTR-Modul	200	3	Cylinder	3	2 700	< 1 000	< 1 500
HTR-500	1 390	6.6	Cylinder	7.1	750	< 1 100	< 2 800
HHT-1000	2 537	10	Cylinder	8.1	850	< 1 250	< 2 800
HTR-10	10	2	Cylinder	1.8	700	< 900	< 1 100
PBMR	265	2.7	Annular	1/3.2	900	< 1 100	< 1 600
DRAGON	20	14	Cylinder	1.07	700	< 1 200	< 1 400
Peach Bottom	115.5	8.3	Cylinder	2.8	700	< 1 200	< 2 000
Fort St. Vrain	842	6.3	Cylinder	5.9	750	< 1 100	< 2 500
GAC-600	600	6.6	Annular	3/4.8	850	< 1 300	< 1 600
HTTR-30	30	5	Cylinder	~ 3	850/950	1 350	< 1 700

In cylindrical cores up to approximately 3.5 m in diameter the fuel temperatures are restricted below 1 600°C for a power of 200 to 220 MW. If graphite noses are embedded in the core, as in the case of the pebble-bed reactor, the power can be raised to 260 MW. A two-zone cylindrical core with graphite balls in the central column can rise to power levels of 300 MW if the inner diameter of the reactor pressure vessel is restricted to approximately 6.5 m. Raising the vessel diameter to 7.5 m allows annular cores with pebble-shaped fuel elements of 500 MWth power rating, maintaining the requirement of temperature stabilisation.

The annular core with block type fuel elements is designed for a power of 600 MW with a temperature limitation of 1 600°C.

Limiting the accident temperature below 1 600°C, even for higher thermal power ratings up to 1 500 to 2 000 MWth, could be achieved using reactor vessels made of pre-stressed cast iron or cast steel blocks. Using this technology, today's technical limit of 7.5 m could be exceeded.

In Figure 4 the design of a new pebble-bed reactor with an annular core is shown, the ISR-300 (Inherent Safe Reactor 300 MWth). This reactor system is arranged in a pre-stressed primary system made from cast steel. To withstand terrorist attack, nuclear reactors in future must be arranged underground and covered by soil material. This is proposed for the ISR-300.

In this reactor the decay heat is transported from the core, through the core structures and from there to an outer heat sink (water-cooled surface cooler or finally concrete structures) uniquely by radiation, conduction and natural convection of air.

Figure 5 shows the well-known trend over the time of the maximum fuel temperature in the core, if the usual assumption of loss of coolant and loss of active decay heat removal is made.

Figure 4. Concept of a pebble-bed HTR with the principle of self-acting decay heat removal (ISR-300 with annular core, thermal power of 300 MW)

1) Annular core, 2) Pre-stressed reactor pressure vessel, 3) Pre-stressed vessel for steam generator, 4) Pre-stressed connecting vessel, 5) Control and shut off system, 6) Helium circulator, 7) Hot gas duct, 8) Steam generator, 9) Surface cooler, 10) Inner concrete cell, 11) Reactor building, 12) Fuel loading device, 13) Fuel discharge device, 14) Flap, 15) Concrete closure, 16) Storage vessel for spent fuel elements

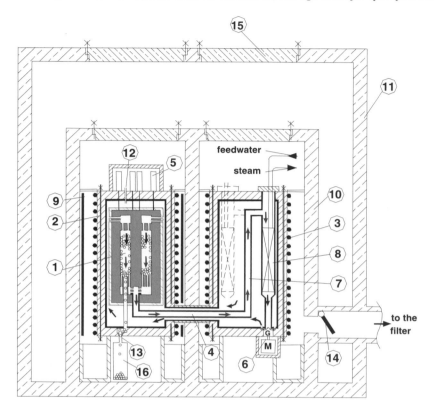

Figure 5. Maximum fuel temperature in ISR-300 core (loss of coolant, loss of active decay heat removal)

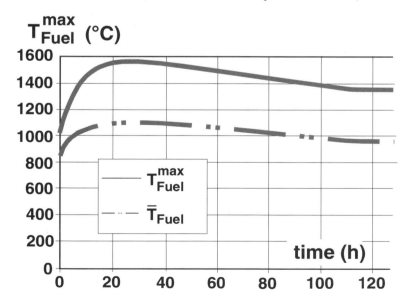

The principle of self-acting decay heat removal and the limitation of fuel temperatures to acceptable values (<1 600°C) is even fulfilled in very extreme situations, as indicated in Figure 6. If the reactor vessel is totally covered with rubble – as a result of a very extreme earthquake or of a terrorist attack – even then the principle of self-acting decay heat works and the maximum fuel temperature is limited. Certainly it takes more time before the temperatures drop again, but the decay heat is finally released to the environment.

Figure 6. Aspects of decay heat removal in very extreme accident situations

a) Rate of occurrence of severe earth quakes (estimation for the site Jülich)
b) Function of damage of buildings depending on strength of earthquakes
c) Situation of a reactor after a very extreme earthquake
d) Maximum temperature of fuel elements, and reactor vessel under different outside conditions

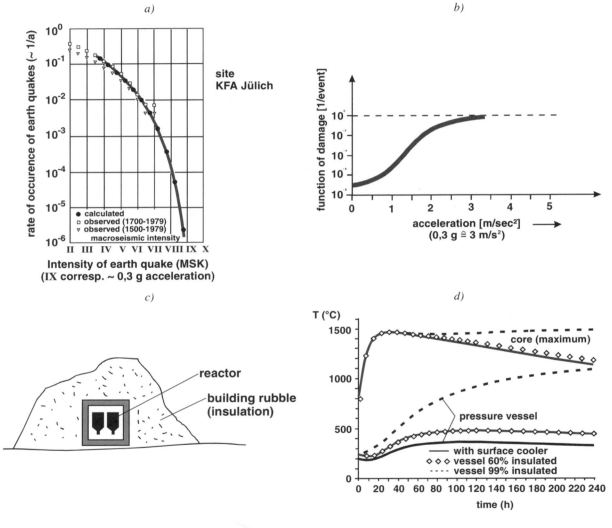

This means that even very extreme accidents do not cause catastrophic releases of fission products. The pre-stressed primary system guarantees that the possible ingress of air is limited to very small values, therefore the burn-up of graphite is very small too.

Experiments to improve the concept of self-acting decay heat removal

Testing the inherent safety features of reactors is one of the most important tasks for further work. Some main aspects of fulfilling the principle self-acting decay heat removal have been tested very well during the development of the pebble-bed HTR. Table 2 provides an overview of experiments which have been carried out and which are planned. Important experiments that have been performed are:

- The measurement of the effective heat conductivity λ_{eff} of the pebble bed is dependent on the temperature inside the reactor core. This parameter directly influences the maximum temperature and the temperature distribution inside the core during a loss of coolant accident. Heat transfer by conduction, thermal radiation and free convection in the core region are summed up in this quantity λ_{eff}, which has been measured in several experiments.

- Experiments to measure the temperature distribution inside a large arrangement of fuel elements, to control the dynamic temperature profiles and to thereby validate computer codes for the calculation of the three-dimensional real temperature distribution in the core, were carried out in the experiment SANA.

- Experiments to measure the heat fluxes and the temperature distribution from the structures of core internals into the reactor vessel and from there to the surrounding concrete cell outside the reactor vessel have been performed in the experiment INWA.

- An experiment to measure the heat fluxes including free convection from the surface of the reactor pressure vessel has been carried out in Japan in the framework of the HTTR project.

- A real reactor experiment has been carried out with the AVR reactor in Jülich. The loss of shut down and of active cooling were simulated by switching off the blower power and measuring the dynamic behaviour of the reactor system.

- In the future a 1/1 large-scale experiment for a "real reactor" can be carried out, in which the decay heat is simulated by electric heating. All accidental situations can be simulated in this model and measurements can indicate the system behaviour as well as the accuracy of codes.

A lot of experiments on the measurement of λ_{eff} (T) have been performed and form the basis to calculate the transport of decay heat inside the core. Figure 7 shows a typical arrangement and results of these measurements.

A large simulation experiment regarding the self-acting afterheat removal from a pebble bed was carried out in the Research Centre Jülich. In the experiment SANA-1, a pebble bed with 9 500 graphite balls of original size was heated by electric resistant heating to maximum temperatures of 1 200°C [see Figure 8(a)]. The temperature transients in the system are measured; they allow to validate computer codes for analysis of accidents and to measure the effective heat conductivity in a pebble bed.

Task and objectives of this experiment up to now include were measurement of time dependent three-dimensional temperature distributions, determination of effective heat conductivity as a function of the temperatures in the core, provision of data sets for the code validation (THERMIX/DIRECT, TINTE) and statements concerning free convection phenomena which are important in case of accidents in the real reactor.

The measured values of λ_{eff} (T) [see Figure 8(b)] are in good agreement with results from smaller test installations and from theoretical considerations on the overall heat transport in pebble bed.

Table 2. Experiments to verify the principle of self-acting decay heat removal from the pebble bed reactor

Experiments	Status	Description	Characteristic of experiments	Characteristic data	Max. temp.	Results
SANA-1	Finished	Pebble bed heated electrically	1 m diameter 1 m height	Pebble-bed with 6 000 balls of 6 cm diameter	1 200	• λ_{eff} (T) up to 1 200°C • Influence of free convection • Transients
SANA-2	Planned	Pebble bed heated electrically	Demonstration of the chain of passive heat transport	Section of core (height 1 m, 20°)	< 1 500	• λ_{eff} (T) • Influence of gaps • Transients
λ_{eff} measurements	Finished	Pebble-bed arrangement	Experiments with some 100 balls	Balls with 6 cm diameter	1 200	• λ_{eff} (T)
INWA	Finished	Section of RPV and outer cell	1/1 section of RPV and cooling system	2 m² area of RPV and cooling system	600	• Heat flux of ~ 4 kW/m² • Heat transfer in gaps
AVR	Finished	Pebble-bed nuclear reactor	Experiments for reactor without and with pressure	50 MW core 3 m diameter	< 1 300	• Concept demonstrated in a real reactor
Large-scale experiment	Planned	Reactor model heated electrically	Demonstration of principle in full scale	3 m diameter 8 m height, 1...2 MW heated electrically	< 1 500	• Demonstration full scale non-nuclear • Other accident demonstration

Figure 7. Measurement of λ_{eff} (T) in a large laboratory experiment

a) Arrangement for measurement of λ_{eff} (T) in a pebble bed
b) Result for the effective heat conductivity in a pebble bed λ_{eff} (T)

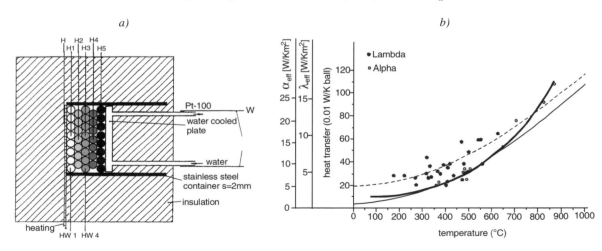

Figure 8. Measurement of λ_{eff} (T) in a large test facility

a) *SANA test facility (diameter: 1.5 m, height: 1 m, 9 500 graphite balls with 6 cm diameter, Tmax = 1 600 °C, power 50 kW).*
b) *Results of measurements in the SANA test facility, for comparison: result of calculations following the theory of Zehner/Bauer/Schluender (dashed and full lines).*

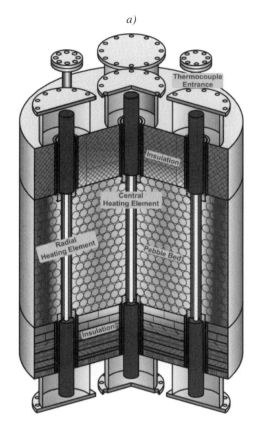

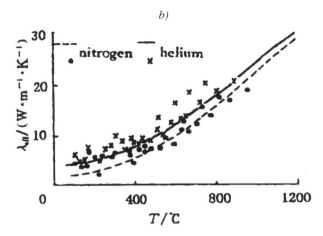

The transport of decay heat from the thermal shield through the wall of the vessel to the concrete cell has been examined in the INWA experiment. The arrangement of this facility is shown in Figure 8. On the inner surface a heat flux of the order of 2 kW/m² is applied, and the temperature distributions are measured. From these data the heat transfer coefficients for different steps in the chain of heat transport are deviated.

Figure 9(a) shows a perspective view of the INWA test facility which corresponds to 20° sector of a proposed pre-stressed cast iron reactor pressure vessel for the 200 MWth Siemens/HTR modular pebble-bed reactor derived and compared with theoretical analysis. The complete INWA facility was embedded within thick thermal insulation. A typical trace at some positions is indicated in Figure 9(b). After start-up, which includes "overheating" to compensate for the thermal capacities of the components, steady state is reached after about 100 h. The simulation of a depressurisation event can then be started at any time. As shown in Figure 9(b) DPE started at about 140 h and lasted approximately 250 h.

The experiment has shown that the decay heat with the heat flux mentioned above can be removed from the surface of the reactor pressure vessel to the surrounding concrete cell. The wall of the cell can be additionally cooled by natural convection of air, if ducts for the air are integrated to the wall of the concrete structure. This is important if the surface cooler around the vessel fails.

Figure 9. INWA test facility for demonstration of self-acting decay heat transport from the reactor pressure vessel to the outer concrete cell

a) Arrangement of facility
b) Temperature profiles for a cast iron wall of a RPV compared to a steel vessel, ① cooling at 38°C, ② no cooling

a)

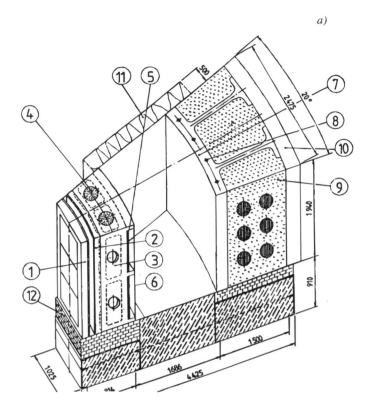

Test set-up (dimensions in mm) with cast iron wall.

① Electric heater and core vessel
② Liner with liner support structure
③ Pressure vessel wall
④ Axial pre-stressing tendon
⑤ Hoop pre-stressing cable
⑥ Cover plate for hoop pressurising cable
⑦ Cast iron block of the reactor cell
⑧ Cooling system
⑨ Reinforced concrete
⑩ Box for sprinkler system
⑪ Insulation
⑫ Refractory support

b)

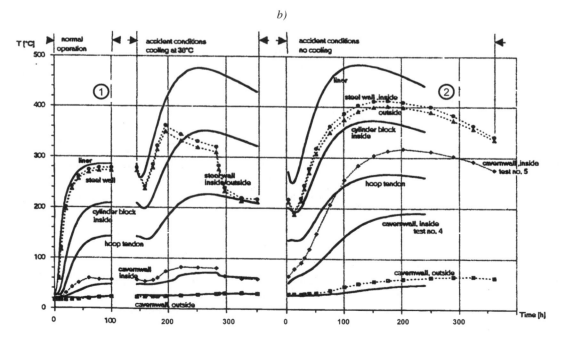

The experimental pebble-bed reactor, AVR, was successfully operated between 1967 and 1988 in the Research Centre Jülich and was used for many safety experiments. One experiment was dedicated to demonstrate the behaviour of this reactor type during a loss of active decay heat removal accident integrally. Therefore, the helium circulators were stopped and the four control rods were blocked in full power position.

As a consequence of the large negative temperature coefficient of the reactivity the nuclear chain reaction stopped immediately. The decay heat was self-actingly transferred from the reactor core through the core internals to the outer structures surrounding the reactor vessel, the temperature in the reflector noses stayed below 700° and the temperature of the fuel elements never rose above 1 200°C, as three-dimensional calculations have shown. It was necessary to calculate these values, because there were no thermocouples inside the pebble bed. The position of the reflector noses, however, is very near the centre of the core and the heat transfer mechanism inside the core is well known. Therefore, the uncertainty of the temperature calculations inside the pebble bed is expected to be small. Figure 10(a) displays some of the calculation results for this case.

Figure 10. Aspects of the loss of decay heat removal in AVR (50 MW)

a) *Temperature distribution inside the AVR reactor in loss of decay heat removal accident calculations for the assumption: loss of control and loss of active decay heat removal, reactor under pressure*
b) *Temperature distribution inside the AVR reactor in loss of decay heat removal accident calculations for the assumption: loss of control and loss of active decay heat removal, reactor without pressure*

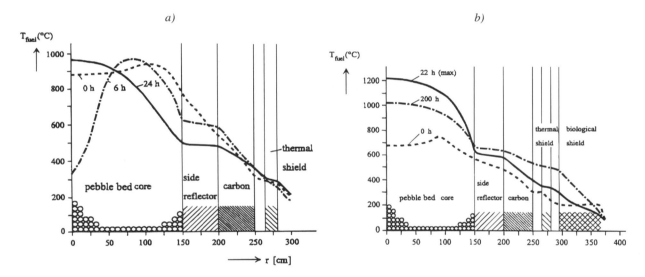

It was stated that the decay heat was partially removed by free convection from the core through the top reflector to the steam generator above the core. The reason was that the reactor was still under pressure, which favours the free convection processes to the "cold" steam generator.

Therefore, an additional experiment was planned and is already licensed: the loss of coolant and failure of all active decay heat removal. The pre-calculation showed that the fuel temperature would stay below 1 400°C even in the hottest part of the reactor in this accident [see Figure 10(b)]. Due to political reasons in Germany and the anti-nuclear politics of some political and local parties, this experiment was not carried out before decommissioning of the AVR.

In a full-scale experiment the self-acting decay heat removal and some other accidents related to an inherently safe HTR can be carried out, if the decay heat is simulated in the pebble bed similar to the technology of the SANA experiment (Figure 11). The planning and realisation of such an experiment could be important from the viewpoint of future public acceptance.

**Figure 11. Proposal for a full-scale experiment for the demonstration
of the self-acting decay heat removal from the pebble-bed reactor core**

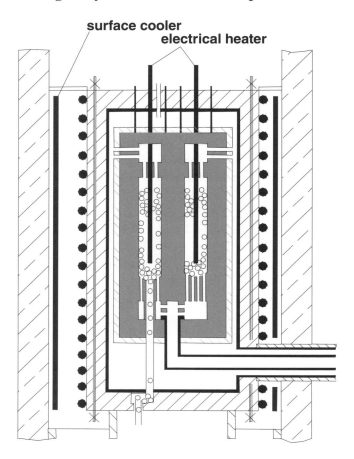

Considerations on the limitations to the principle of self-acting decay heat removal

Of course one has to consider the consequences to be taken into account if, in addition to the loss of coolant and the failure of active decay heat removal accidents, another event occurs. Examples are the ingress of large amounts of air and steam as well as the failure of reactivity control systems. Figure 12 shows the additional effects which are important with respect to HTR design. It is necessary to ensure even in these cases that the fuel temperatures stay below 1 600°C and that the fuel elements stay intact, thus retaining the fission products inside the coated particles.

As an example, the concept of thermal stability must be fulfilled along with the nuclear stability in the case of the pebble-bed core of the ESKOM reactor concept (PBMR). In this reactor a very hypothetical beyond-design accident is considered. It is assumed that there is a loss of coolant along with a loss of active decay heat removal and the rapid loss of the first shut-down system. The fuel temperature stays below 1 600°C in this case too.

Figure 12. Potential impacts on the nuclear power plant which influence the concept of self-acting decay heat removal

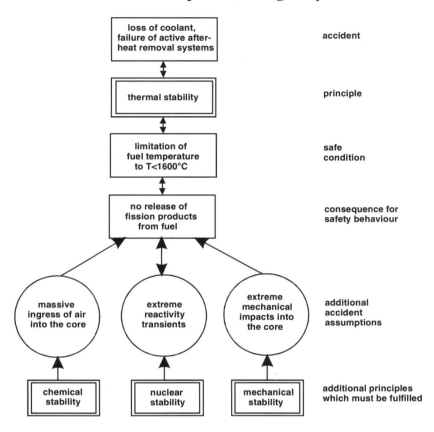

In an ideal way mechanical stability is achieved by burst-protected primary enclosures. This includes gaining additional chemical stability, insofar as air ingress into the primary system is considered. Only very small amounts of air can ingress into the primary system and, therefore, the corrosion of fuel elements and graphite structures is strongly limited to acceptable levels.

Concluding remarks

HTR plants can be realised with high efficiency and inherent safety. The reactor system can be designed to fulfil the requirements of catastrophe-free nuclear technology, that being the restriction of fission product release in all accidents to less than 10^{-5} of the inventory. The core can never melt or overheat to temperatures above 1 600°C. Through the use of an annular core a thermal power rating of 300 MW or even more can be realised.

This safety behaviour can be demonstrated in an integral full-scale experiment. This experiment can be carried out with a real reactor, because the core will not be destroyed by these experiments and thus the consequences of the accident demonstration are insignificant.

In addition to the application for steam generation, the reactor system can be used for coupling with gas turbine cycle, combined cycle and process heat applications. This reactor is especially attractive for cogeneration processes.

The power plant is economically attractive compared to other future options, including world market coal, especially if a series production is realised a modular design of larger power. Furthermore, worldwide fabrication of components will remarkably reduce the cost.

The potential world market for a reactor having such a broad spectrum of applications is large. Nevertheless, international co-operation is necessary to establish this type of reactor technology worldwide.

THE CONTRIBUTION OF UK ORGANISATIONS IN THE DEVELOPMENT OF NEW HIGH-TEMPERATURE REACTORS

A.J. Wickham
Consultant, Cwmchwefru, Llanafanfawr
Builth Wells, Brecknockshire LD2 3PW, UK
(*and* Bath Nuclear Engineering Group, University of Bath
Claverton Down, Bath BA2 7AY, UK)

I.M. Coe
Reactor Systems and Advanced Technology Department
NNC Ltd, Booths Hall
Chelford Road, Knutsford, Cheshire WA16 8QZ, UK

P.J. Bramah
Serco Assurance
Risley, Warrington
Cheshire WA3 6AT, UK

T.J. Abram
BNFL Research and Technology
Springfields, Preston PR4 0XJ, UK

Abstract

The UK has a history of the successful design, construction and operation of graphite-moderated gas-cooled reactors. In particular, a central role was played in the original development of HTR, focused upon the construction of the DRAGON prototype reactor at Winfrith. In this regard there was a particular expertise built up around the performance of the fuel particles, and also in the area of graphite oxidation from potential coolant impurities.

The revival of interest in HTR through the development of GT-MHR and PBMR modular designs, through continuing national programmes such as ACACIA in The Netherlands, HTTR in Japan and HTR-10 in China, and the general potential for co-generation and inherently safe operation, has renewed the interest of the principal UK centres of reactor expertise, which are particularly focused upon the PBMR design. This review covers the present activity of principally NNC Ltd, BNFL and AEA Technology plc; in addition, the activities of some smaller UK companies in support of the programme are also described.

Introduction

At the previous meeting in 1999, Beech and May [1] discussed the experience of the UK in gas-cooled-reactor and high-temperature-engineering technologies, from the particular perspective of NNC Ltd. In the intervening two years, there have been very significant developments in the conception and design of small-scale modular HTRs, both through national programmes in Japan, China and The Netherlands and through international collaborations, notably with the pebble-bed modular design (PBMR) and the GT-MHR concept, taking on board the principle of inherently safe engineering and, particularly in this last case, the capability to utilise weapons-grade plutonium in the fuel. These developments are usefully summarised in an IAEA TECDOC [2].

The UK nuclear industry, both through its established large organisations and also through the involvement of smaller specialist companies, is making a major contribution to these developments, particularly in relation to the PBMR.

This paper has three purposes:

- To briefly review the particular skills and design experience which have arisen through the extensive programme of production and commercial power-producing graphite-moderated reactors, including participation in the DRAGON Project.

- To summarise the present participation of UK organisations in the developing HTR activity.

- To indicate the particular areas where major contributions might be expected in the future.

The paper does not claim to be exhaustive. It has been compiled from largely published information and from the perspective of an independent consultant graphite specialist, previously from the commercial nuclear electricity industry, with the assistance of co-authors from NNC Ltd, Serco Assurance (formerly AEA Technology plc) and BNFL. There are almost certainly some significant omissions, especially in regard to the contributions of smaller organisations, and to these we offer an apology and an invitation to bring the record up to date for any future revision of this review.

The UK gas-cooled reactor experience

The UK programme began at the end of the 1940s with the construction of two small graphite piles at Harwell – the Graphite Low-Energy Experimental Pile (GLEEP) and the British Experimental Pile Zero(BEPO), and these were followed by two large production piles at Windscale. The infamous fire in 1957 was both a set-back and a signal that a successful commercial plant would require to be operated with a higher graphite temperature to limit the accumulation of Wigner energy, and the construction of eight small Magnox reactors followed at Calder Hall and Chapelcross, all of which continue in operation today. The subsequent fully-commercial Magnox programme saw the construction of a further 18 larger reactors, the final ones incorporating the pre-stressed concrete pressure vessel concept which was used for the next generation of advanced gas-cooled reactors (AGRs). Ten of these Magnox reactors continue in operation at this time.

The AGR programme, which saw the graphite temperature raised to approximately 700 K and the maximum coolant temperature at around 930 K, consisted of 14 reactors all of which continue in operation.

This activity within the UK has provided organisations and individuals with skills in two very relevant fields for the new HTR initiatives:

- There is great experience in manufacturing graphite to particular specifications, and in researching and understanding its behaviour under irradiation, particularly with respect to neutron damage, the generation of stresses in complex components through differential dimensional change and the effects thereon of thermal and irradiation creep, and in both thermal and radiolytic oxidation processes in both carbon dioxide and in air.

- Great experience in the requirements for licensing of a gas-cooled graphite-moderated plant, and the production of related safety cases with particular emphasis on life extension in the presence of limited component failure.

In addition, many useful lessons have been learned in other technological areas such as the mechanisms of corrosion of various types of steels, pressure-vessel embrittlement, and the development of sophisticated remote-handling and viewing equipment for the internal inspection of reactors and for the replacement of internal reactor systems. These issues, and many more, are discussed in papers given at two conferences organised by The Institution of Mechanical Engineers (London) on Magnox reactors and AGRs, respectively [3,4].

A separate but very valuable experience for the UK nuclear industry was the participation in the DRAGON Project, culminating in the construction of the prototype HTR at Winfrith Heath. The achievements of all the international partners are summarised in the closing report [5].

In the graphite area, one of the most important lessons was the understand that existing graphites available at the time for the construction of the fuel matrix were much too permeable, and unsuited to control the transport of fission products from failed fuel particles; an extremely high purge flow comparable to the primary circuit flow would have been required, and this led to the development of commercial low-permeability fine-grain graphites and also to the employment of furfuryl alcohol as an impregnant for *in-situ* polymerisation, a process first defined at The Royal Aircraft Establishment, Farnborough. The programme also gave a wide experience of the characterisation of different graphites using the MTR reactor at Petten in The Netherlands, experience further augmented later in support of AGRs using UK MTRs at Harwell as well as BR2 in Belgium and Silöe at Grenoble in France. Such investigations continue in support of the existing reactor programmes even today, with both BNFL Magnox Generation (for the Magnox reactors) and British Energy Generation Ltd (for the AGRs) conducting graphite-irradiation investigations at Idaho Falls (USA) and Halden (Norway) respectively. The DRAGON Project also introduced the first significant experiments on graphite creep to the UK industry.

These examples relate only to the graphite; there were many other examples of the learning process in developing reactor technologies which were to the benefit of all the participating countries. Perhaps one of the most important was an understanding of the so-called "amoeba effect" of kernel transport along a thermal gradient in particulate fuel. One important discipline which is now becoming much more significant is in the area of safe storage of closed plant and the preparations for full decommissioning. This is the subject of another paper to this meeting [6].

We now review some of the current activities within the UK in support of current HTR development.

NNC Ltd

General background

A known and stable regulatory environment was established for light water reactors (LWRs) many years ago. Modular and other current designs of high-temperature reactor are, however, very different from LWR, with a number of inherent safety features. In order to establish a common European safety approach and identify significant issues for this type of reactor, as described in [7], a number of EC sponsored 5th Framework projects have been established. In addition to participation in the HTR co-ordination project (HTR-C) that will be responsible for co-ordinating the activities for the programme as a whole, NNC is a co-ordinator and partner in a number of technical projects, as described below. NNC is also involved in other activities associated with the UK gas-cooled reactor fleet that are of relevance to HTR development.

Licensing

NNC is one of the partners in the EC sponsored project entitled "High-temperature Reactor – Licensing, Safety Approach and Licensing Main Issues" (HTR-L). This project is intended to identify key safety issues, transient and accident scenarios that will help to define a coherent safety approach for modular high-temperature reactors (producing electricity) in a licensing framework. A classification for the design-basis operating conditions and associated acceptance criteria will be derived. NNC is leading the work package that will define rules for the system, structure and component classification and a component qualification level compatible with the economical objectives.

For economic efficiency, the desired simplicity and compactness of design will only be attainable if inherent safety features such as first-barrier efficiency and the core thermal inertia are fully accepted. Many engineered safety features and systems (ESFAS) required for the LWR, such as emergency core-cooling systems, redundant shutdown systems and multiple pressure-barrier containments can, therefore, be eliminated or greatly reduced. This should simplify the accident sequences that will require analysis compared with the complex accident scenarios in LWR. Accident management will also be simpler because of lower system complexity and greater time to react because of increased core thermal inertia.

Although, as indicated above, an HTR will probably be more able to tolerate fault situations than many other reactor types, it is probable that these abilities are differences in emphasis, rather than fundamental differences in behaviour. The safety report underpinning the safety case for an HTR will theoretically need to start by assuming all possible failures (i.e. that which is not unphysical) and by demonstrating during the safety analysis that unquestionable physical phenomena, etc., render the most limiting design basis accidents acceptable. Along with our partners, NNC will be working to propose rules that will make the licensing process for an HTR in Europe acceptable in both safety and economic terms.

Besides its participation in the HTR-L project, NNC is supporting the South African Nuclear Regulator (NNR) in the assessment of the safety case for the pebble-bed modular reactor (PBMR). Aspects of the safety case are being reviewed in order to identify the degree of compliance with NNR requirements and international standards.

General high-temperature engineering

NNC is the co-ordinator for the EC sponsored 5th Framework project entitled "High-temperature Reactor – Materials" (HTR-M) project. This four-year project covers experimental work on graphite oxidation, turbine materials mechanical properties and irradiation effects in RPV steel. Further information on this project is given in [8].

NNC's high-temperature corrosion and metallurgical facilities are located at Risley Laboratories. They play an important role in the extended operation of the UK gas reactor fleet. The knowledge and expertise gained from the operation of these facilities, particularly breakaway oxidation, carburisation and metal dusting, is being used in assessments of the effects of impurities in helium on HTR vessel materials.

Breakaway oxidation is a rapid form of oxidation whereby sub-surface oxidation disrupts the protective oxide film on some steels. Metal dusting is a form of heavy carburisation that can occur in low oxygen, high carbon environments in which carbides form and tunnel around and loosen grains of metal in the steel.

NNC is also one of the partners in the HTR Components (HTR-E) project. This project aims to examine the technologies involved in the key component areas that are expected to influence safety, cost and operability. The involvement will be mainly in the tribology and gas purification areas, where we will provide information from our operating experience of gas reactors.

Fuel and waste management

NNC is a partner and work-package leader in the Reactor Physics and Fuel Cycle Project (HTR-N). The overall objectives of this project are the development of core physics and fuel cycle requirements to meet the needs of the current and future modular concepts of HTR. The management of graphite is a key issue although the single largest task is the analysis of spent fuel arisings. NNC is the work-package leader for developing a strategy for dealing with and minimising HTR wastes, and will also input its experience with the DRAGON reactor and comparative assessments against UK gas-cooled reactors.

UK graphite

In addition to the work on graphite disposal in the HTR-N project described above, NNC is undertaking a number of other tasks associated with graphite behaviour, principally in AGRs, that will be relevant to HTR.

As the graphite core structures in a reactor age, and get increasingly irradiated by fast neutrons, the individual fuel and control rod channels, and the individual graphite blocks within them, become slightly distorted. The safe operation of reactor cores containing graphite requires that the distortion remain within acceptable limits under all normal operating and fault conditions. Sophisticated computer models of the AGR cores have been developed which allow bounding channel distortions to be determined. NNC has designed and built three separate rigs that are being used for the validation of these computer models. The first is a large 3-D rig which can accommodate up to 48 full size fuel bricks as a $4 \times 4 \times 3$ array, which has been used to gain an understanding of the mechanical behaviour of assemblies of bricks (forces, displacements, rotations). The second is a 2-D rig containing a 25 full section brick slice (100 mm thick) arrangement, which has primarily been used to investigate the effect of components with through wall cracks. The third is a one-eighth scale model of a full core (containing up to 5 500 fuel bricks) which is being used to confirm numerical predictions using a substantially larger number of components.

In addition, almost every graphite property is affected by fast neutron irradiation and by radiolytic oxidation when in oxidising coolant compositions. Various computer codes are used to make predictions about the current and future states of the graphite core. Regular monitoring is used to confirm that the cores are behaving as predicted. At each statutory outage, therefore, a number of fuel channels are de-fuelled and a device known as the Channel Bore Measurement Unit is deployed into these channels to make accurate measurements of individual brick bore diameters and brick tilts in orthogonal planes. NNC is comparing the individual brick bore diameter measurements against computer predictions and the tilt measurements are then used to construct the overall shape of the channel. Another device is deployed to remove small cylindrical samples from the bores of fuel bricks by trepanning. These samples are subsequently tested to determine the amount of radiolytic oxidation suffered and important mechanical/physical properties such as strength, Young's modulus and CTE.

A considerable amount of work is being done to assess the effects of brick cracking on core functionality, and to develop improved computer codes relating to core modelling. Once these codes have been validated using data from the NNC test rigs they will be suitable not only for work on the AGR fleet but, with suitable further validation, for the assessment of graphite behaviour at the elevated temperatures to be anticipated in an HTR.

Serco Assurance (formerly AEA Technology Consulting)

Serco Assurance traces its routes through UKAEA to AEA Technology Reactor Services SRD and finally to the recent divestment from AEA Technology Consulting. Obviously, its staff have brought with them the wealth of experience and knowledge gained during their previous work with the precursor organisations. A major part of this experience was gained in the development and assessment of reactor core structures and fuel designs covering all the major UK reactor design projects over the last 30 years, i.e. Magnox, AGR, PWR, SGHWR, PFR and CFR/EFR. The experience spans the full range of skills needed to develop a reactor design from concept to a detailed design. However, Serco Assurance staff have particular specialist knowledge in the detailed assessment of graphite core structures and the effects of irradiation on the material properties and the component distortions. This has been gained through the analysis of Magnox and AGR core components which has been carried out for BNFL and for British Energy Generation Ltd.

In the light of this wide ranging and specialist experience, Eskom invited AEA Technology Consulting in 1999 to develop the PBMR core structure design from the basic conceptual layout through to a detailed design which could be accurately costed and constructed. The scope of the task covered the complete fixed core structure inside the RPV. This includes the core barrel, the core support structure and the graphite reflector i.e. everything in the core except the fuel and the control/shutdown systems.

One of the first tasks was to assess the nuclear graphites currently in production or which would be available for use in the PBMR core and to generate a database of properties which would enable the behaviour of the core structure to be predicted. Very little irradiation data were available on the candidate materials and therefore the database had to be constructed using historical data from graphite with similar microstructures. This was done by suitable manipulation and combination of data from two existing data sets, including the IAEA International Database on Irradiated Nuclear Graphite Properties [9].

The task of the design team was to develop the outline South African concept, which had been based on the German HTR Modular design, to match PBMR operating conditions in terms of fluence levels, temperatures and operational modes. The starting point for much of this process was the core thermal-hydraulic analyses. Analyses of the whole core and of specific regions were performed for both normal operating and loss of forced cooling conditions. Determination and reduction of leakage flows was particularly important. Optimisation of the insulation layout and selection of suitable materials was a particularly complex task as conditions in PBMR in terms of loads and temperatures are significantly more onerous than in earlier HTR designs.

Based upon these thermal data and upon the applied loads, materials for the core barrel and the support structure were selected. Component designs which emphasised ease of manufacture and stiffness were then developed and subjected to stress analysis. For the graphite reflector, the principal tasks were, firstly, to derive a brick layout for which all the components would fit within the envelope of the available raw material sizes and then to assess the component lifetimes. Detailed stressing calculations were performed on critical bricks to determine the component internal stresses and distortions caused by irradiation. The constructed graphite data set, together with the thermal-hydraulic and fluence data, were the basic inputs to these calculations.

In addition to these activities, core instrumentation was selected and its layout was assessed. The core shielding requirements were reviewed and detailed supporting calculations were performed. Graphite and carbon physical properties and compositions were measured.

Finally, a detailed tender exercise was undertaken. Very detailed specifications for all the core components, including the graphite/carbon, the barrel, the assembly, the instrumentation, the shielding, the insulation, etc., were written and tenders for supply were requested from suitable manufacturers. On receipt of responses, the tenders were evaluated and the overall cost of manufacture and assembly was determined.

PBMR used the outputs from Serco Assurance as a major input to the Design Feasibility Study Report, which has been completed recently.

British Nuclear Fuels plc (BNFL)

Background

In recent years, BNFL has grown from being a predominantly UK-based fuel cycle company to a global provider of nuclear services, employing some 23 000 staff across 17 countries. With the acquisition of the Westinghouse Electric Company and the nuclear interests of ABB, BNFL is now involved in all aspects of nuclear power, from the processing of uranium ore, through enrichment, fuel fabrication, transport, reactor supply and maintenance, reactor operation, spent fuel management, and waste treatment and decommissioning. BNFL's historical involvement in HTR systems arises both from its UK experience in the manufacture of fuel for the DRAGON project at its Springfields site (and from investigations into the manufacture of plutonium micro-spheres at the Sellafield site for the UK fast reactor programme, which utilised a similar fabrication technique), and from the unrivalled experience of Westinghouse Reaktor in Germany, whose forerunners designed and constructed both of the German pebble-bed reactors: the AVR at Jülich and the THTR at Schmehausen.

Moreover, through the operation of its Magnox reactor fleet, BNFL has acquired a considerable knowledge of the licensing, operation, inspection and maintenance of gas-cooled graphite-moderated steel pressure vessel reactors. Finally, through the research and experience which underpin the continued operation of the Magnox fleet, together with the design and manufacture of the graphite-sleeved AGR fuel elements, BNFL has acquired considerable knowledge and experience of the behaviour of graphite components during irradiation.

Current HTR activities

BNFL, through its Research and Technology Directorate, is currently one of the four investors in the South African Pebble-bed Modular Reactor (PBMR) project, along with the South African utility Eskom, the Industrial Development Corporation of South Africa (IDC) and the Exelon Corporation of the USA. This consortium has funded the preliminary design of the direct cycle PBMR system and its associated fuel fabrication facility, which culminated in the production of a Detailed Feasibility Study Report in the summer of 2001. BNFL, in association with the other PBMR investors, is currently conducting a thorough review of this DFS report, in order to determine the technical and economic feasibility of the project.

In addition to its role as an investor in the project, BNFL Group companies also supply design and consultancy services. For example, BNFL, together with Nukem of Germany and EMS of South Africa, has collaborated on the preliminary design of the PBMR fuel plant, in close association with both the PBMR design team and the Nuclear Energy Commission of South Africa (NECSA). Similarly, Westinghouse Reaktor has provided consultancy services to the project, and has provided key inputs to the design of the core internals and to the reactor control system. There are also numerous direct inputs in small speciality areas; for example, BNFL expertise is being brought to bear to evaluate any risk associated with thermal oxidation of graphite or graphitic components from impurities in the coolant, whilst the Littlebrook engineering facility, which specialises in the design and construction of remote-handling equipment for viewing, maintenance and replacement work on reactor internals, is evaluating the feasibility of a replacement of the PBMR reflector.

In addition to its role in the PBMR project, BNFL is also involved in the European 5th Framework HTR projects, which are described in detail in Ref. [7] and in an earlier section. In the HTR-Fuel project, BNFL is involved in activities to secure historical knowledge of HTR fuel developments in partner countries, which will include a detailed review of the fuel developments in the DRAGON project, as well as contributions to the development of a European HTR fuel modelling code. In the HTR-Neutronics and Fuel Cycle project, BNFL is also contributing to the studies of Pu burning in HTR systems. BNFL is also a partner in the European HTR Technology Network, which co-ordinates and fosters the research activities carried out in the Framework programme.

Finally, BNFL also participates in the gas-cooled reactor working group of the US DOE's Generation IV programme, and in the IAEA's technical working group on gas-cooled reactors.

British Energy plc

British Energy's role in the United Kingdom is essentially limited to the operation of 14 AGRs, but the parent company as a whole has much wider international interests. Whilst the principal interests within the UK are obviously associated with the management of reactor ageing and life extension where possible, there are numerous areas in which such matters coincide with issues of immediate relevance for HTR and where British Energy personnel could make valuable contributions, should the company elect to participate. As just one example, the company has developed considerable expertise in the prediction of graphite behaviour and eventual component failure, as a consequence of which it is highly proficient in irradiation effects in graphite; discussions are in progress about participation in a proposed international collaborative irradiation programme to address perceived deficiencies in the knowledge of creep behaviour which is planned serve the interests both of the new HTR designers and the existing reactor operators.

Smaller companies

When reviewing the major contributions of the very large nuclear engineering and consulting concerns and reactor operators, it is easy to overlook the important contributions of other, sometimes smaller, companies to HTR development. Here we mention just a few other organisations whose contributions have come to the attention of the authors: to those which have been omitted we can offer only an apology.

Rolls Royce, for example, has enormous expertise in the field of gas turbines, and has offered to review the Brayton Cycle as employed on PBMR and compare it with similar cycles such as their WR21 marine application. Initially, this offer has been declined, although there is some possibility of reviving the option as the PBMR design becomes more mature. Alstom, through their Whetstone plant, are also understood to be extremely interested in tendering for the design and supply of turbo machinery.

Other UK expressions of interest have included a novel recuperator design from Heatric, and insulation materials from DarChem, although a later version of the PBMR design appears not to have identified a requirement for insulation.

It is also important to note that not all expressions of interest in HTR which have reached other UK companies relate just to the PBMR. As an example, a small West-Country Company, Bradtec Decon Technologies Ltd, which has specialised in the application of novel chemical decontamination methods to water-reactor circuits, is about to commence an Electrical Power Research Institute (EPRI, California) contract concerning the potential contamination of leading turbine blades with ^{110}Ag. Forward-facing surfaces are considered to be vulnerable to this contamination from fission products, and the half-life of ^{110}Ag is such that there could be a need for a two-year storage period before maintenance. Whilst this is a generic project which would be applicable to both PBMR and GTHMR, it is the latter which is the prime motivation in this case. The objective is to determine whether there is indeed a problem and, if so, to consider both chemical cleaning methods and also the possibility of using coated blades whereby the coating is removed and replaced.

Resume

As may be seen from this brief review of current activity, UK companies, both large and small, are making major contributions to the development of a new generation of HTRs in a wide variety of fields from engineering design through fuel development to performance and licensing. These inputs are possible both as a result of a wealth of experience on other gas-cooled and graphite-moderated nuclear plants, and also through innovative approaches to particular problem areas. It is expected that this activity is likely to broaden in scope as the two principal HTR projects (PBMR and GT-MHR) mature.

REFERENCES

[1] D.R. Beech and R. May, "Gas Reactor and Associated Nuclear Experience in the UK Relevant to High-temperature Engineering", Proc. 1st Information Exchange Meeting on Basic Studies on High-temperature Engineering, Paris France, 27-29 September 1999, OECD/NEA 2000, pp. 57-67.

[2] International Atomic Energy Agency, "Current Status and Future Development of Modular High-temperature Gas-cooled Reactor Technology", IAEA-TECDOC-1198, February 2001.

[3] Institution of Mechanical Engineers, "Seminar: The Review of Safety at Magnox Nuclear Installations", March 1989, pub. Inst. Mech. Eng., London [11 papers].

[4] Institution of Mechanical Engineers, "The Commissioning and Operation of AGRs", May 1990, pub. Inst. Mech. Eng., London [10 papers].

[5] F.P.O. Ashworth, et al., "A Summary and Evaluation of the Achievements of the DRAGON Project and its Contribution to the Development of the High-temperature Reactor", OECD DRAGON Project Report 1000, November 1978.

[6] B. McEnaney, A.J. Wickham, D. Hicks and M. Dubourg, "Planning for Disposal of Irradiated Graphite: Issues for the New Generation of HTRs", Proc. 2nd Information Exchange Meeting on Basic Studies in the Field of High-temperature Engineering, Paris, France, October 2001, OECD/NEA, these proceedings.

[7] J. Martin-Bermejo and M. Hugon, "Research Activities on High-temperature Gas-cooled Reactors (HTRs) in the 5th EURATOM RTD Framework Programme", Proc. 2nd Information Exchange Meeting on Basic Studies in the Field of High-temperature Engineering, Paris, France, October 2001, OECD/NEA, these proceedings.

[8] D. Buckthorpe, R. Couturier, B. van der Schaaf, B. Riou, H. Rantala, R. Moorman, F. Alonso and B.C. Friedrich, "Investigation of High-temperature Reactor (HTR) Materials", Proc. 2nd Information Exchange Meeting on Basic Studies in the Field of High-temperature Engineering, Paris, France, October 2001, OECD/NEA, these proceedings.

[9] B. McEnaney, P. Hacker, A.J. Wickham and G. Haag, "Status of IAEA International Database on Irradiated Nuclear Graphite Properties with Respect to HTR Engineering Issues", Proc. 2nd Information Exchange Meeting on Basic Studies in the Field of High-temperature Engineering, Paris, France, October 2001, OECD/NEA, these proceedings.

SESSION II

Improvement in
Material Properties by
High-temperature Irradiation

Chairs: T. Shikama, B. Marsden

IMPROVEMENT IN DUCTILITY OF REFRACTORY METALS BY NEUTRON IRRADIATION

Hiroaki Kurishita

The Oarai Branch, Institute for Materials Research, Tohoku University
Oarai, Ibaraki 311-1313, Japan

Abstract

Embrittlement caused by high-energy particle irradiation is the critical concern for structural materials exposed to radiation environments. In particular, refractory metals such as molybdenum and tungsten are known to exhibit serious embrittlement even after low-level irradiation. It is believed that radiation embrittlement arises from radiation hardening and thus inevitably occurs since irradiation always causes significant hardening. Radiation hardening is mainly due to irradiation-produced defects and therefore it has long been believed that the only way to relieve radiation embrittlement is to suppress radiation hardening by introducing a large number of sinks for irradiation-produced defects.

Recent progress in both the development of irradiation rigs [mainly performed in the Japan Materials Testing Reactor (JMTR)] and materials fabrication techniques has enabled to alter this understanding of radiation embrittlement. It was found that fast neutron irradiation caused a remarkable increase in low-temperature ductility of refractory metal by approximately five times as measured with impact three-point bending tests, regardless of significant radiation hardening. This ductility increase due to irradiation is directly opposite to radiation embrittlement and is called radiation-induced ductilisation (RIDU). The only possible explanation for the occurrence of RIDU is that since many materials have microstructural inhomogeneity including grain boundaries and interfaces that may act as crack initiation and/or propagation sites, such weak places are strengthened (i.e. become more resistant to fracture) by irradiation and its beneficial effect is much larger than the detrimental effect of radiation hardening. The strengthening of weak places through irradiation was attributed to radiation-enhanced precipitation and segregation which occur preferentially at weak grain boundaries under high-energy irradiation.

Effects of precipitation and segregation on mechanical properties of materials depend strongly on the species and contents of precipitates and segregated elements. It is therefore important to control the compositions and contents of constituents so that radiation-enhanced precipitation and segregation may have the beneficial effect of strengthening the weak places. For the control, we can make good use of the data obtained so far for un-irradiated samples. As radiation-enhanced precipitation and segregation may occur easily as the irradiation temperature is increased, it is expected that high-temperature irradiation can be a good tool for improvement in mechanical properties of materials.

Introduction

Molybdenum and tungsten, the Group VIa refractory metals, have many advantages such as high melting point, low thermal expansion coefficient, low vapour pressure, high thermal stress factor, low sputtering yield, low tritium inventory and excellent compatibility with liquid metals. However, they are known to exhibit serious embrittling effects, including low-temperature embrittlement, re-crystallisation embrittlement and radiation embrittlement. Among these, radiation embrittlement may be the most crucial problem, which restricts the use of refractory metals as components exposed to radiation environments [1-21].

It is believed that radiation embrittlement inevitably occurs because it arises mainly from radiation hardening and high-energy particle irradiations always cause significant hardening except for high-temperature irradiation, where recovery of radiation hardening may occur. Radiation hardening is mainly due to irradiation-produced defects and therefore it has long been believed that the only way to relieve radiation embrittlement is to introduce a large number of sinks for irradiation-produced defects.

On the other hand, materials have weak places due to heterogeneity in microstructure, such as grain boundaries and phase boundaries which will act as crack initiation and propagation sites and be responsible for embrittlement. Since high-energy particle irradiations can cause a variety of changes in microstructure, it is expected that in some cases irradiations may produce a microstructure having improved resistance to fracture by strengthening the weak places. This may lead to improvement in ductility and toughness of materials as long as the beneficial effect of strengthening the weak places exceeds the detrimental effect of radiation hardening.

In this review, first, the degree of radiation embrittlement in commercially available molybdenum alloys is shown taking TZM as an example. TZM is most widely used among molybdenum alloys and is known to exhibit the highest low-temperature toughness in the un-irradiated condition. Then, it is shown that the radiation embrittlement is much improved by controlling microstructures. Emphasis is placed on the occurrence of radiation-induced improvement in ductility in a microstructure-controlled molybdenum alloy and its mechanism.

Experimental

The most effective microstructure for simultaneous improvement in the low-temperature embrittlement, re-crystallisation embrittlement and radiation embrittlement will be fine grains and finely dispersed particles of transition metal carbides [22], the explanation of which is given in Figure 1. In order to achieve such microstructures, the powder metallurgical method including mechanical alloying (MA) and hot isostatic pressing (HIP) processes were used with the starting materials of commercially available powders of pure molybdenum (an average particle size 5 μm and purity 99.9%) and TiC (0.57 μm, 98%). Since gaseous interstitial impurities of oxygen and nitrogen have a detrimental effect of promoting embrittlement, all powder treatments were conducted in a specially designed glove box with purified argon. Hipped compacts were hot forged and hot and warm rolled to approximately 1 or 2 mm thick sheets. The chemical analysis of the sheets showed that carbon added as TiC are all retained and the impurity contents of oxygen and nitrogen were approximately 1 000 and 100 wt. ppm, respectively.

From the sheets of developed molybdenum alloys and TZM (Mo-0.5Ti-0.1Zr), miniaturised smooth bend bar specimens having dimensions of 1 mm by 1 mm by 20 mm and the longitudinal direction parallel to the rolling direction were prepared. The specimens were irradiated with fast

Figure 1 Explanation of the microstructure required to simultaneously improve the low temperature embrittlement, re-crystallisation embrittlement and radiation embrittlement in refractory metals

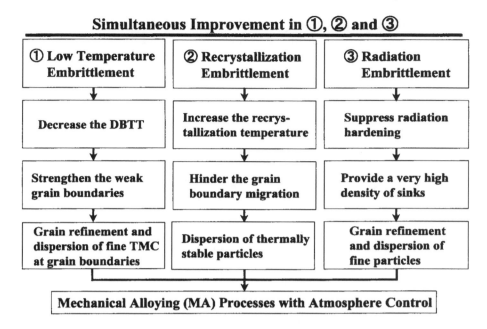

neutrons from 0.01 to 0.1 dpa (displacement per atom) at 573 K and at controlled cyclic temperatures between 473 and 673 K and between 573 and 773 K with one and five cycles in the Japan Materials Testing Reactor (JMTR). The details of the temperature-cycle irradiation technique were described elsewhere [23].

The un-irradiated and irradiated bend bar specimens were impact three-point bending tested to examine the effect of irradiation on impact ductility and strength using a specially designed, electrically controlled hydraulic machine [24] with the span of 12.5 mm and the velocity of approximately 5 m/s. The impact tests are more useful to evaluate the toughness of molybdenum and its alloys than the static three-point bending tests [25]. Signals of load time and displacement time were recorded by transient memories and then read by a personal computer to analyse the data. Fracture surfaces of impact tested specimens were examined by scanning electron microscopy (SEM) to determine the fracture mode. Transmission electron microscopy (TEM) observation and energy dispersive X-ray (EDX) analysis were made with JEM 4000FX operating at 400 kV in the Oarai Facility of Institute for Materials Research (IMR) at Tohoku University.

Results and discussion

Resistance to radiation embrittlement in TZM

The studies conducted so far on the effect of high-energy neutron or proton irradiation on the ductility of molybdenum and its alloys showed that significant radiation embrittlement occurs, however, the degree of radiation embrittlement depends strongly on pre-irradiation heat treatment and irradiation condition [1-7,10-21]. Therefore, at first the effect of heat treatment on radiation embrittlement was examined using TZM specimens with three different microstructures, i.e. the as-rolled, stress-relieved and fully re-crystallised conditions. It was found that although all specimens show an increase in the ductile-to-brittle transition temperature (DBTT) by the very low level of neutron irradiation to 0.01 dpa at 573 K, there is a considerable difference in the magnitude of DBTT shift among the three microstructures. The DBTT shifts for the as-rolled, stress-relieved and fully re-crystallised conditions were 60 K, 140 K and more than 220 K, respectively, indicating that TZM in the as-rolled condition exhibits the highest resistance to radiation embrittlement. Therefore, the effect of irradiation condition on the resistance to radiation embrittlement was conducted for the as-rolled TZM.

Figure 2 shows the test temperature dependence of load-displacement curves for the as-rolled TZM before and after irradiation under four different conditions. In the figure, as the test temperature decreases, the abrupt drop appears on the curve. This drop represents the occurrence of brittle fracture and shows the break of the specimen. When such an abrupt drop was absent, the specimen did not break and showed a full bend, which corresponds to the deflection of approximately 8-9 mm. It is obvious that the present impact test gives a very distinct transition from ductile (full bending) to brittle regions where fracture occurs in a completely brittle manner below the yield stress. It should be noted that the DBTT for un-irradiated TZM is around 250 K, but it significantly increases to around 460 K by a low-level irradiation to 0.1 dpa. This result may give a very pessimistic view for the use of molybdenum and its alloys in radiation environments.

Figure 2. Impact load versus displacement curves for TZM in the as-rolled (a) and irradiated with fast neutrons to 0.01 dpa at 573 K (b), 0.08 dpa at 573/773 K for five cycles (c), 0.1 dpa at 473/673 K for one cycle (d) and 0.1 dpa at 573/773 K for one cycle (e)

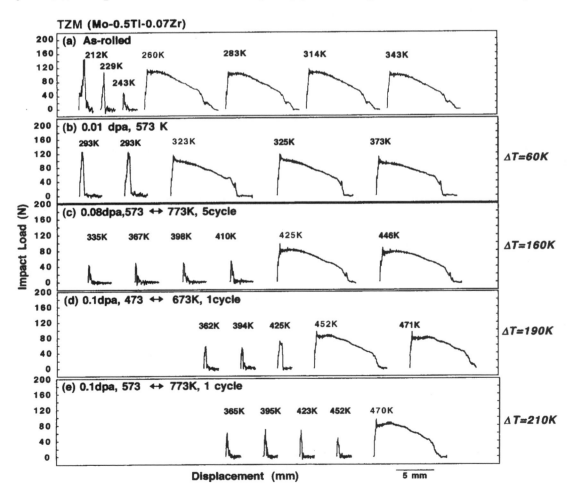

Resistance to radiation embrittlement in developed molybdenum alloys

The effect of irradiation on the low-temperature toughness was examined for the developed molybdenum alloys. Figure 3 shows the test temperature dependence of total absorbed energy for developed alloys and TZM irradiated to 0.08 dpa at controlled cyclic temperatures between 573 and 773 K with five cycles [26]. The total absorbed energy was determined from the integration of impact load displacement records as was shown in Figure 2. Here, MTC-01, MTC-05 and MTC-10A represent Mo-0.1%TiC, Mo-0.5%TiC and Mo-1%TiC alloys which were fabricated with MA treatment, and TC-10 represents Mo-1%TiC alloy fabricated without MA treatment. It is seen that the developed alloys exhibit much higher impact toughness than TZM, i.e. the DBTT is considerably lower and the upper shelf energy is higher than those for TZM. In addition, the impact toughness increases with increasing TiC content and the comparison between the curves of MTC-10A and TC-10 indicates that MA treatment is effective to improve the resistance to irradiation. It should be noted that MTC-10A shows excellent impact toughness after the irradiation. In view of the fact that the test was not a static test but an impact test where the strain rate is approximately five orders higher than the former, the result of MTC-10A seems to be very surprising. Such excellent low-temperature toughness has never been obtained even in the as-rolled and annealed conditions before irradiation.

Figure 3. Test temperature dependence of total absorbed energy, Et, for developed molybdenum alloys and TZM irradiated to 0.08 dpa at 573/773 K for five cycles

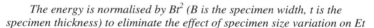

The energy is normalised by Bt^2 (B is the specimen width, t is the specimen thickness) to eliminate the effect of specimen size variation on Et

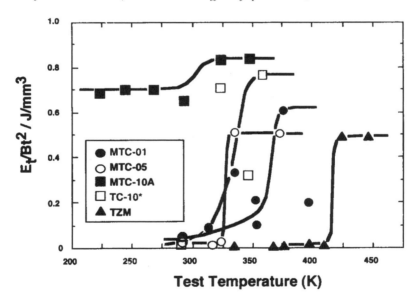

Figure 4 shows the effect of neutron irradiation on the test temperature dependence of total deflection measured by impact three-point bending (left) and Vickers microhardness measured at room temperature (right) for MTC-10A [27]. The low-temperature ductility greatly increases *after irradiation*, regardless of the occurrence of significant radiation hardening. Since no appreciable increase in ductility was observed by similarly thermally treated, but un-irradiated specimens of MTC-10A, such a notable ductilisation is not due to the effect of thermal annealing during irradiation, but to a radiation-induced phenomenon. Therefore the phenomenon is called "radiation-induced ductilisation" (RIDU).

Figure 4. Effect of neutron irradiation on total deflection by impact three-point bending tests and Vickers microhardness at room temperature for MTC-10A

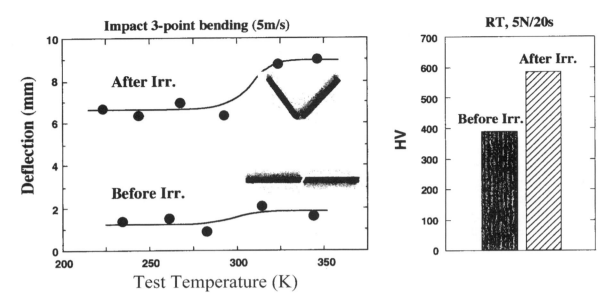

Mechanism for radiation-induced ductilisation (RIDU)

As mentioned previously, the occurrence of RIDU can easily be understood from the viewpoint that the weak places acting as crack initiation and propagation sites are strengthened by irradiation and its beneficial effect is larger than the detrimental effect of radiation hardening. It is well known that in molybdenum high-energy grain boundaries are very weak and are responsible for intergranular embrittlement [28-31]. In addition, SEM observations of fracture surfaces showed that the fracture of MTC-10A before and after the irradiation occurred intergranularly. It is therefore reasonable to say that the occurrence of RIDU is attributable to the strengthening of weak grain boundaries due to the irradiation.

In order to study the mechanism of strengthening of weak grain boundaries due to the irradiation, TEM observations were made. Figure 5 shows a TEM micrograph taken from MTC-10A in the as-irradiated condition [27]. A number of large precipitates are observed. Figure 6 shows the histogram of the size distribution of larger precipitates which were identified by EDX and electron diffraction analysis. It should be noted that the observed precipitates are divided into five kinds of precipitates, i.e. TiC, (Ti,Mo)C and (Mo,Ti)C, which have the cubic structure, and Mo_2C and $(Mo,Ti)_2C$, which have the hexagonal structure. In view of the fact that these carbides are non-stoichiometric and the amount of TiC addition in MTC-10A is only 1 wt.%, these precipitates are presumed to include some amounts of oxygen and nitrogen impurities. It is also seen from Figure 6 that approximately 60% of the observed precipitates exist at grain boundaries.

It was reported that there is a stronger cohesion between molybdenum and (Ti,Mo)C or Mo_2C than that at high-energy weak grain boundaries [31,32]. In addition, radiation-enhanced precipitation may preferentially occur at high-energy grain boundaries. Therefore it may be concluded that in the irradiated MTC-10A the observed grain boundary precipitates of transition metal carbide can strengthen the weak boundaries and thus significantly improve the low-temperature toughness.

Figure 5. TEM micrograph for MTC-10A irradiated with fast neutrons to 0.08 dpa at controlled cyclic temperatures between 573 and 773 K

Precipitates are indicated by arrows

Figure 6. Histogram of the size distribution of precipitates identified to be TiC, (Ti,Mo)C, (Mo,Ti)C, Mo_2C and $(Mo,Ti)_2C$ and the relative ratio of each precipitate existing at grain boundaries and in grain interior

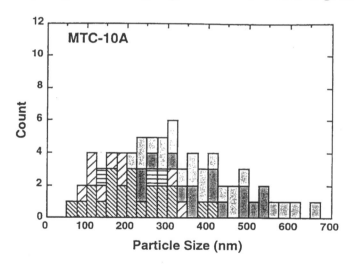

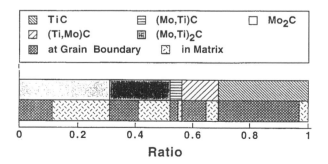

High-temperature irradiation and significance of RIDU

As the irradiation temperature is increased, radiation-enhanced precipitation may occur more easily and radiation hardening becomes less significant because of the occurrence of increased recovery of irradiation-produced defects. This indicates that in the high-temperature irradiation RIDU is expected to occur more frequently than in the present irradiation.

Since the effect of precipitation on mechanical property changes depends strongly on the nature of precipitates, it is important to control compositions so that radiation-enhanced precipitation may have the beneficial effect of strengthening the weak places. For the composition control, we can make good use of the data obtained so far for un-irradiated specimens. We believe that the strengthening of weak places due to irradiation occurs in many materials because materials have microstructural heterogeneity.

Finally it should be stated that the finding of RIDU owes to the recent progress of the developments of irradiation rigs mainly performed in JMTR and materials fabrication techniques and that it has enabled to change our understanding of radiation effects on materials from the pessimistic one to the very challenging one, i.e. irradiation has the beneficial effect of producing new phenomena and/or innovative materials that will not be available without irradiation.

Concluding remarks

TZM, which is in any conditions of the as-rolled, stress relieved and fully re-crystallised one, exhibited significant radiation embrittlement even by the irradiation to 0.1 dpa. On the other hand, the developed molybdenum alloys with fine grains and finely dispersed particles of transition metal carbides showed much improved resistance to radiation embrittlement. In particular, the Mo-1wt.%TiC alloy showed a remarkable increase in ductility by fast neutron irradiation, regardless of the occurrence of significant radiation hardening. This is directly opposite to radiation embrittlment and is called radiation-induced ductilisation (RIDU).

The explanation for the occurrence of RIDU is that the weak places acting as crack initiation and propagation sites are strengthened by irradiation and its beneficial effect is larger than the detrimental effect of radiation hardening. In molybdenum, high-energy grain boundaries are weak places and are strengthened by radiation-enhanced precipitates of transition metal carbides of TiC, $(Ti,Mo)C$, $(Mo,Ti)C$, Mo_2C and $(Mo,Ti)_2C$.

It is expected that RIDU may occur in various materials because any materials have weak places associated with microstructural heterogeneity. As the irradiation temperature is increased, RIDU can occur more frequently because radiation-enhanced precipitation occurs more easily and radiation hardening becomes less significant.

REFERENCES

[1]   J.M. Steichen, *J. Nucl. Mater.*, 60 (1976), 13.

[2]   B.L. Cox and F.W. Wiffen, *J. Nucl. Mater.*, 85 & 86 (1979), 901.

[3]   M. Tanaka, K. Fukuya, K. Fukai and K. Shiraishi, Effects of Radiation on Materials, 10th Conference, D. Kramer, H.R. Brager and J.S. Perrin, eds., ASTM STP 725 (1981), p. 247.

[4]   K. Abe, T. Takeuchi, M. Kikuchi and S. Morozumi, *J. Nucl. Mater.*, 99 (1981), 25.

[5]   R.K. Williams, F.W. Wiffen, J. Bentley and J.O. Steigler, *Metall. Trans.*, 14A (1983), 655.

[6]   K. Abe, M. Kikuchi, K. Take and S. Morozumi, *J. Nucl. Mater.*, 122 & 123, (1984), 671.

[7]   Y. Hiraoka, M. Okada, T. Fujii, M. Tanaka and A. Hishinuma, *J. Nucl. Mater.*, 141-143 (1986), 837.

[8]   P. Krautwasser, H. Derz and E. Key, *High Temperatures-High Pressures*, 22 (1990), 25.

[9]   H. Ullmaier and F. Carsughi, *Nuclear Instr. and Mech. in Phys. Res.*, B, 101 (1995), 406.

[10]   F. Morito and K. Shiraishi, *J. Nucl. Mater.*, 179-181 (1991), 592.

[11]   I.V. Gorynin, V.A. Ignatov, V.V. Rybin, S.A. Fabritsiev, V.A. Kazakov, V.P. Chakin, V.A. Tsykanov, V.R. Barabash and Y.G. Prokofyev, *J. Nucl. Mater.*, 191-194 (1992), 421.

[12]   S.A. Fabritsiev, V.A. Gosudarenkova, V.A. Potapova, V.V. Rybin, L.S. Kosachev, V.P. Chakin, A.S. Pokrovsky and V.R. Barabash, *J. Nucl. Mater.*, 191-194 (1992), 426.

[13]   G.V. Muller, D. Gavillet, M. Victoria and J.L. Martin, *J. Nucl. Mater.*, 212-215 (1994), 1283.

[14]   B.N. Singh, A. Horsewell, P. Toft and J.H. Evans, *J. Nucl. Mater.*, 212-215 (1994), 1292.

[15]   B.N. Singh, J.H. Evans, A. Horsewell, P. Toft and D.J. Edwards, *J. Nucl. Mater.*, 223 (1995), 95.

[16]   A. Hasegaea, K. Abe, M. Satou and C. Namba, *J. Nucl. Mater.*, 225 (1995), 259.

[17]   K. Watanabe, A. Hishinuma, Y. Hiraoka and T. Fujii, *J. Nucl. Mater.*, 258-263 (1998), 466.

[18]   B.N. Singh, J.H. Evans, A. Horsewell, P. Toft and G.V. Muller, *J. Nucl. Mater.*, 258-263 (1998), 865.

[19]   I.V. Gorynin, F. Morito, V.A. Kazakov, Yu.D. Goncharenko and Z.E. Ostrovsky, *J. Nucl. Mater.*, 258-263 (1998), 883.

[20]   A. Hasegaea, K. Ueda, M. Satou and K. Abe, *J. Nucl. Mater.*, 258-263 (1998), 902.

[21]   M. Scibetta, R. Chaouadi and J.L. Puzzolante, *J. Nucl. Mater.*, 283-287 (2000), 455.

[22]   H. Kurishita, Y. Kitsunai, Y. Hiraoka, T. Shibayama and H. Kayano, *Mater. Trans. JIM*, 37 (1996), 90.

[23]   M. Narui, H. Kurishita, H. Kayano, T. Sagawa, N. Yoshida and M. Kiritani, *J. Nucl. Mater.*, 212-215 (1994), 1665.

[24]   H. Kayano, H. Kurishita, M. Narui and M. Yamazaki, *Ann. Chim. Fr.*, 16 (1991), 309.

[25]   Y. Hiraoka, H. Kurishita, M. Narui and H. Kayano, *Mater. Trans. JIM*, 36 (1995), 504.

[26]   Y. Kitsunai, H. Kurishita, M. Narui, H. Kayano and Y. Hiraoka, *J. Nucl. Mater.*, 239 (1996), 253.

[27]   H. Kurishita, Y. Kitsunai, T. Kuwabara, M. Hasegawa, Y. Hiraoka, T. Takida and T. Igarashi, *J. Plasma and Fusion Res.*, 75 (1999), 594 (in Japanese).

[28]   J.B. Brosse, R. Fillit and M. Biscondi, *Scripta Met.*, 15 (1981), 619.

[29]   H. Kurishita, S. Kuba, H. Kubo and H. Yoshinaga, *Mater. Trans. JIM*, 26 (1985), 332.

[30]   H. Kurishita, A. Oishi, H. Kubo and H. Yoshinaga, *Mater. Trans. JIM*, 26 (1985), 341.

[31]   H. Kurishita and H. Yoshinaga, *Materials Forum*, 13 (1989), 161.

[32]   H. Kurishita, M. Asayama, O. Tokunaga and H. Yoshinaga, *Mater. Trans. JIM*, 30 (1989), 1009.

EFFECTS OF SUPERPLASTIC DEFORMATION ON THERMAL AND MECHANICAL PROPERTIES OF 3Y-TZP CERAMICS (REVIEW)

M. Ishihara, T. Shibata, S. Baba, T. Hoshiya
Japan Atomic Energy Research Institute
Oarai Research Institute
3607, Oarai-machi, Higashiibaraki-gun, Ibaraki-ken, 311-1394, Japan

C. Wan, Y. Motohashi
Ibaraki University
4-12-1, Nakanarusawa-cho, Hitachi-shi, Ibaraki-ken, 316-8511, Japan

Abstract

Our recent activities related to thermal and mechanical properties of superplastic ceramics, 3Y-TZP, are presented in this paper. These properties were obtained after the superplastic deformation process, because the effect of superplastic deformation on their properties is one of the important points from an application viewpoint of superplastic deformation such as forming and joining technologies.

As for the thermal properties, specific heat was measured using the DSC method over a wide range of temperatures, from 473 K to 1 273 K, with a nominal superplastic deformation strain of 70%. The crystal structure was characterised by measuring X-ray diffraction patterns, and microstructural observation was carried out using a SEM.

As for the mechanical properties, dynamic hardness and Young's modulus were measured using the dynamic indentation technique with a maximum superplastic deformation of 150%. Microstructural observation was also carried out using a SEM after a thermal etching.

Introduction

It is well known that when a ceramic sample is pulled in tension it usually breaks at fairly low strain. However, some ceramics are capable of pulling out to very large tensile deformations of the order of some hundreds or even more than one thousand of per cent. It is said that this phenomena of so-called superplasticity on ceramics appears when the size of grains decreases at the magnitude of sub-micron order. The discovery of the superplastic phenomenon in ceramic materials expands practical applications of the ceramics, and many researches have been performed in the field of basic understandings of superplasticity of ceramics as well as application technologies of the superplastic deformation, e.g. advanced forming and joining technologies with superplastic deformation, etc.

Authors have also investigated the superplastic zirconia ceramics with the aim of nuclear application in the high-temperature engineering field, and neutron irradiation research on superplastic ceramics at high temperatures has been proposed as an innovative basic research using the high-temperature engineering test reactor (HTTR). A preliminary study has started out using other research reactors as well as accelerators in order to determine the most effective irradiation test conditions using the HTTR.

In this paper, our recent activities on thermal and mechanical properties of 3 mol% yttria stabilised tetragonal zirconia polycrystalline (3Y-TZP) after superplastic deformations are introduced.

From the application viewpoint, changes in thermal and mechanical properties after superplastic deformation is one of the key issues. In the thermal property research 3Y-TZP specimens were prepared after the superplastic deformation process. The nominal deformation strain in this study was about 70%, and changing the strain rate in the deformation process produced a different volume per cent of cavities. The specific heat was measured by a differential scanning calorimetry (DSC) method over a wide range of temperatures, from 473 K to 1 273 K. The crystal structure was characterised by measuring X-ray diffraction patterns, and microstructural observation was carried out using a scanning electron microscope (SEM) after thermal etching.

On the other hand, in the mechanical property research, the 3Y-TZP specimens were also prepared after the superplastic deformation process. In this study, the maximum deformation was about 150%. Changes in the hardness and Young's modulus were measured by dynamic indentation technique. A pyramid-type indentor was used, and microstructural observation was once again carried out using a SEM after a thermal etching.

Specific heat after superplastic deformation [1]

Materials preparation

Chemical compositions of 3Y-TZP specimens used in this study are listed in Table 1. The 3Y-TZP specimens were pulled at the temperature of 1 723 K with different initial strain rates to produce different cavity volumes in the specimen. The nominal superplastic deformation is about 70% in this study. From superplastically elongated specimens, about 4 mm squared samples with 1 mm thickness were fabricated using a diamond cutter. Prepared specimens are summarised in Table 2.

Table 1. Chemical compositions of 3Y-TZP specimens (in wt.%)

Y_2O_3	Al_2O_3	SiO_2	Fe_2O_3	Na_2O	ZrO_2
5.15	0.01	0.02	0.01	0.04	bal.

Table 2. Specimen preparation of 3Y-TZP

	Strain rate (s^{-1})		
	4.2×10^{-4}	1.4×10^{-3}	4.2×10^{-3}
Specimen	No. 1, No. 2	No. 3, No. 4	No. 5, No. 6

Experimental methods

Bulk densities were measured using the Archimedes method at room temperature in order to estimate the deformation-induced cavity percentages. The microstructures of pre- and post-superplastic deformations were observed by the SEM. The phase change in the specimens was observed by X-ray diffraction technique at room temperature. The specific heat was measured by the DSC method over the temperature range 473 K to 1 273 K. The specimens were heated up at a speed of 5 K/min. In this study, the accuracy and reproducibility of the measurements were checked with isotropic graphite (IG-110) powder and platinum powder as standard samples prior to and after the measurements.

Results and discussions

Volume fraction of cavities

The deformation-induced volume fractions of cavities were estimated from the bulk density measurement using the Archimedes method. Obtained results are summarised in Table 3. We can see from this table that the cavity percentage of prepared specimens is 1.0% under low strain rate conditions to 3.4% under high strain rate conditions.

Table 3. Volume fraction of cavities of superplastically deformed specimens (%)

Specimens					
No. 1	No. 2	No. 3	No. 4	No. 5	No. 6
1.1	1.0	1.7	1.6	3.1	3.4

Microstructure observation

Microstructures observed by the SEM are shown in Figure 1. The homogeneous grain distribution is observed in both un-deformed and deformed specimens. It is obvious that the grain size of the un-deformed sample is smaller than those of deformed ones, indicating strain-induced grain growth. However, fine grain size remained, and the average grain size is not more than 0.6 μm. In addition, the grain aspect ratio varied slightly, i.e. the grain was elongated a little along the deformation direction. It is admitted that there are much more cavities with higher strain rate samples. However, large cavities as well as cavity linkage were not observed in the present samples, since the nominal deformation strain is below 70% in this study.

Observation of phase change during superplastic deformation

Microstructural changes due to superplastic deformation have been reported [2-5], and it is said that usually the grain growth as well as nucleation and growth of cavities occur in the 3Y-TZP during

Figure 1. Microstructures of un-deformed and deformed 3Y-TZP specimens

(1) Un-deformed specimen ($\times$ 2 000)   (2) Strain rate: 4.2×10^{-4} s^{-1} ($\times$ 10 000)

(3) Strain rate: 1.4×10^{-3} s^{-1} ($\times$ 10 000)   (4) Strain rate: 4.2×10^{-3} s^{-1} ($\times$ 10 000)

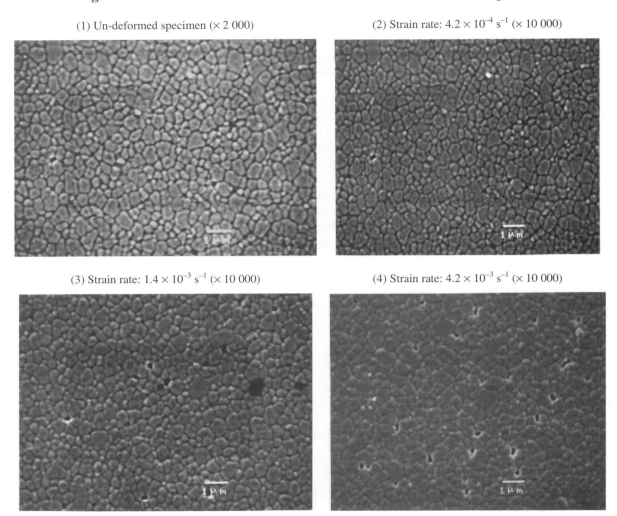

the superplastic elongation process. The X-ray diffraction patterns of the samples deformed with different initial strain rates are shown in Figure 2. There is little difference among them according to the analysis on the peak values in the curves, and an almost tetragonal phase is constituted in each 3Y-TZP sample. It is concluded, therefore, that there is no phase change during the superplastic deformation process in this study, i.e. there is no stress-induced martensitic transformation and so on.

Specific heat

The measured specific heats are plotted in Figure 3 as a function of volume cavity fraction. The inaccuracy in the measurements is within 4%. It can be seen from this figure that the specific heat is not affected by the volume cavity fraction. Since the cavity does not contribute to the mass of the 3Y-TZP specimen, little heat is needed to match the temperature between the cavity and the solid of 3Y-TZP. As a result, the specific heat of cavities can be negligible, which agrees with the fact that the specific heat is independent of volume cavity fraction even if it is a high percentage.

Figure 2. X-ray diffraction patterns of superplastically deformed specimens

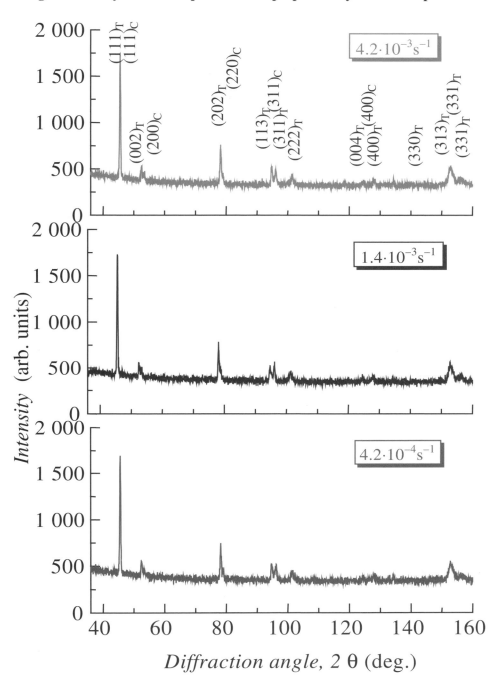

Figure 3. Obtained specific heat after superplastic deformation

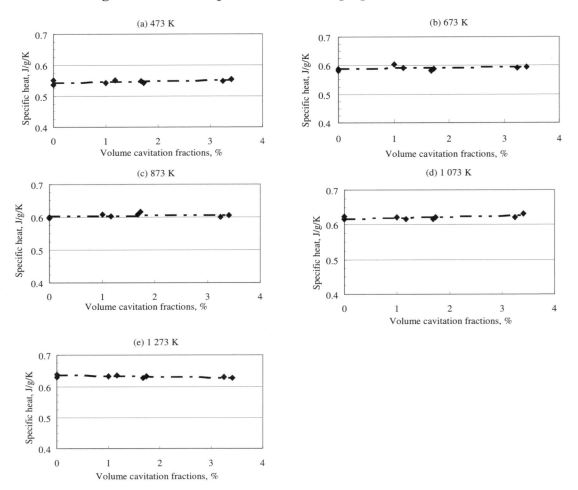

Dynamic hardness and Young's modulus [6,7]

Materials preparation

Typical mechanical and thermal properties of the 3Y-TZP used in this study are listed in Table 4. The 3Y-TZP specimens were pulled at the temperature of 1 693 K with strain rates at 3.3×10^{-4} s^{-1} and 3.3×10^{-3} s^{-1}. To study the superplastic deformation effect on mechanical properties, specimens with different magnitudes of nominal strain were prepared. Table 5 shows the prepared specimens in this study.

Experimental methods

Dynamic hardness testing

A micro-indentation test with the pyramid-type Vickers indentor was carried out. The indentor was pressed on the un-deformed and deformed specimens, and indentation load to depth data were

Table 4. Typical mechanical and thermal properties of 3Y-TZP

Density (g/cm^3)	6.05
Bending strength (MPa) (room temperature)	900~1 500
Young's modulus (MPa) (room temperature)	2.1·10^5
Poisson's ratio	0.31
Fracture toughness (MNm$^{-3/2}$) (room temperature)	8~12
Rockwell hardness (HRA)	92
Thermal expansion coefficient (/°C) (20~400°C) (20~1 000°C)	9.6·10^{-6} 10.4·10^{-6}
Thermal conductivity (W/mK) (room temperature) (600°C)	3.3 2.9
Specific heat (J/g°C) (room temperature)	0.46 (0.11 cal/g°C)

Table 5. Prepared 3Y-TZP specimens for mechanical testing

Specimen	Nominal strain (%)	Strain rate (s^{-1})	Temperature (K)
A	0	0	Room temperature
B	30	3.3×10^{-4}	1 693
C	60	3.3×10^{-4}	1 693
D	100	3.3×10^{-4}	1 693
E	150	3.3×10^{-4}	1 693
F	100	3.3×10^{-3}	1 693

obtained to get the dynamic hardness as well as Young's modulus. The maximum indentation load was selected to be 49 mN, where the maximum indentation depth was about 1 μm. Twenty-eight kinds of indentation data were obtained for each specimen. The dynamic hardness by Vickers indentor, Hv, was calculated by [8]:

$$Hv = c \cdot \frac{P}{h^2}$$ (1)

where c is a constant depending on the shape of the indentor ($c = 3.8548$ for Vickers indentor), and P and h are the indentation load and depth, respectively.

Young's modulus estimation

The indentation load to depth response, as shown in Figure 4, also estimates the Young's modulus as follows [8,9]:

$$\frac{E}{1-v^2} = \frac{1}{1.142\sqrt{S_{max}}} \cdot \frac{dP}{dh}$$ (2)

$$\frac{1-\nu^2}{E} = \frac{1-\nu_s^2}{E_s} + \frac{1-\nu_i^2}{E_i} \qquad (3)$$

where, E and ν are the Young's modulus and Poisson's ratio, and subscripts s and i correspond to the specimen and the indentor, respectively. S_{max} is the projected contact area of the indentor at the maximum load. The dP/dh corresponds to the gradient of indentation load to depth curve at the beginning of unloading process as shown in Figure 4.

Figure 4. Schematic drawing of indentation load to depth curve

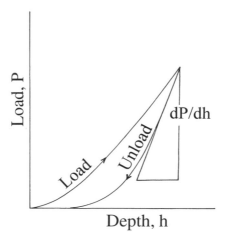

Results and discussion

Microstructure after superplastic deformation

Figure 5 shows the scanning electron micrographs of deformed and un-deformed specimens. From this figure it is found that the grain structure is almost isotropic for the un-deformed specimen, Specimen A, and grain size increases with increasing nominal strain in the superplastic deformation process. The average grain size was analysed from the SEM images as shown in Table 6. Each size was normalised to that of the un-deformed Specimen A. The cavity formation caused by superplastic deformation is also seen in Figure 5.

Dynamic hardness

The dynamic hardness estimated from Eq. (1) is plotted in Figure 6 as a function of normalised average grain size. The hardness decreased rapidly with increasing grain size, i.e. by increasing the superplastic deformation. From a comparison between Specimens B and F, they are almost the same in grain size and hardness in spite of different nominal strain in the superplastic deformation. The hardness may, therefore, be strongly affected by the grain growth caused by the superplastic deformation.

It is probable that the tetragonal phase (t-phase) of the 3Y-TZP changes to monoclinic phase (m-phase) with grain growth caused by superplastic deformation. Table 7 shows the Vickers and Young's modulus of zirconia single crystals with m- and t-phase [10]. The hardness of the m-phase is about a half of that of the t-phase. The decrease in hardness with increasing grain size may be affected by this phase transformation.

Figure 5. Microstructures of un-deformed and deformed 3Y-TZP specimens

(1) Specimen A (deformation: 0%)

(2) Specimen B (deformation: 30%)

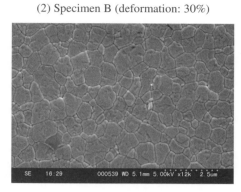

(3) Specimen C (deformation: 60%)

(4) Specimen D (deformation: 100%)

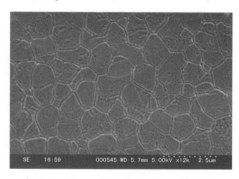

(5) Specimen E (deformation: 150%)

(6) Specimen F (deformation: 100%)

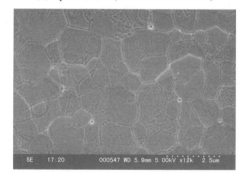

Table 6. Normalised average grain size

Specimen	Normalised average grain size*	Nominal strain (%)
A	1.0	0
B	1.7	30
C	2.6	60
D	3.8	100
E	5.3	150
F	1.7	100

* Each size is normalised by that of Specimen A.

Figure 6. Obtained dynamic hardness as a function of normalised grain size

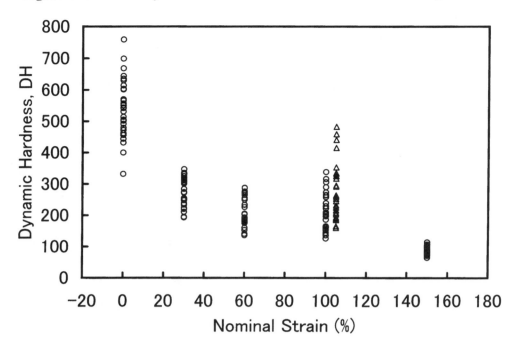

Table 7. Vickers hardness and Young's modulus of zirconia single crystals [10]

Phase	Vickers hardness	Young's modulus (GPa)
Tetragonal phase	13.5	200
Monoclinic phase	7.3	244

Young's modulus

The Young's modulus estimated from Eqs. (2) and (3) is plotted in Figure 7 as a function of normalised average grain size. The Young's modulus decreased slightly with increasing grain size. The grain size dependence of the Young's modulus is no so strong as that of the dynamic hardness, as shown in Figure 6. Since Young's moduli of zirconia single crystals with m- and t-phase are almost the same from Table 7, we believe that the phase change is not a dominant reason for the decrease in Young's modulus with the superplastic deformation. Although Specimens B and F have almost the same grain size, their Young's moduli are quite different. It is probable, therefore, that there are other factors affecting the Young's modulus. One factor of the decrease in the Young's modulus may be the cavity effect. It is well known that the cavity volume fraction increases with increasing superplastic deformation and that the Young's modulus of the porous body depends on the porosity. The decrease in the Young's modulus with the superplastic deformation is, therefore, thought to be affected by the cavity formation. The effect of cavities on the properties should be evaluated more precisely in future.

Summary

Authors have begun the study of superplastic ceramics with the aim of nuclear application in the high-temperature engineering field, and the neutron irradiation research on superplastic ceramics at high temperatures has been planned and proposed as an innovative basic research using the high-temperature

Figure 7. Obtained Young's modulus as a function of normalised grain size

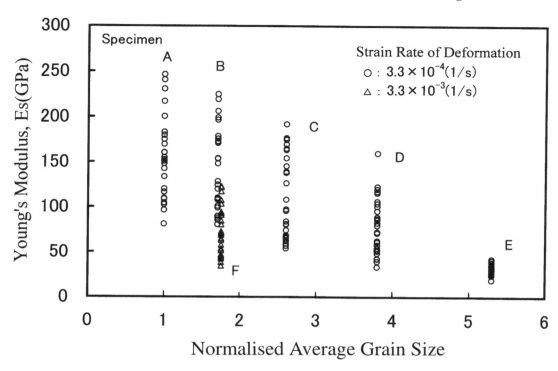

engineering test reactor (HTTR). A preliminary study has started out using other research reactors, the Japan Materials Testing Reactor (JMTR), and accelerators in order to determine the most effective irradiation test conditions for the HTTR.

In this paper, our recent activities related to thermal and mechanical properties of superplastic ceramics are presented. These properties were obtained after the superplastic deformation process, because the effect of superplastic deformation on their properties is one of the important subjects from an application viewpoint of superplastic deformation such as forming and joining technologies.

REFERENCES

[1]   C. Wan, T. Shibata, S. Baba, M. Ishihara, T. Hoshiya and Y. Motohashi, *Japan Journal of Thermophysical Properties*, to be published.

[2]   I-Wei Chen and Liang AN Xue, *J. Am. Ceram. Soc.*, 73 [9] (1990), 2585-2609.

[3]   Y. Motohashi, T. Sekigami, N. Sugeno, *J. Mater. Process. Tech.*, 68 (1997), 229-235.

[4]   T. Sakuma and K. Higashi, *Mater. Trans.*, JIM, 40 [8] (1999), 702-715.

[5]   H. Hosokawa and K. Higashi, *Mater. Sci. Res. Int.*, 6 [3] (2000), 153-160.

[6]   T. Shibata, M. Ishihara, T. Takahashi and Y. Motohashi, presented at the Asian Pacific Conference on Fracture and Strength & International Conference on Advanced Technology in Experimental Mechanics, 20-22 October, Sendai, Japan (2001).

[7]   T. Shibata, M. Ishihara, T. Takahashi, Y. Motohashi and K. Hayashi, JAERI-Research 2001-024 (2001) (in Japanese).

[8]   M. Inamura and T. Suzuki, *Seisann-kenkyu*, 42 [4] (1990), 257-260 (in Japanese).

[9]   J. Alcala, *J. Amer. Ceram. Soc.*, 83 [8] (2000), 1977-1984.

[10]   D.J. Green, R.H.J. Hannink and M.V. Swan, Transformation Toughening of Ceramics, CRC Press (1989).

DEVELOPMENT OF AN INNOVATIVE CARBON-BASED CERAMIC MATERIAL: APPLICATION IN HIGH-TEMPERATURE, NEUTRON AND HYDROGEN ENVIRONMENT

C.H. Wu
EFDA, Max-Planck Institut für Plasmaphysik
Garching,Germany

Abstract

In the framework of the European Fusion Research Programme, a great effort has been made to develop an innovative carbon-based ceramic material to meet all of the operational requirements. After a decade of research and development, an advanced material has finally been developed. This material, a 3-D CFC, contains about 8-10 at.% of silicon and has a porosity of about 3-5%. This advanced ceramic material possesses very high thermal conductivity, dimensional stability under conditions of neutron irradiation, lower chemical erosion (longer lifetime), lower tritium retention and lower reactivity with water and oxygen (safety concern).

This innovative ceramic material seems very promising for application in the high-temperature, neutron and hydrogen environment. A detailed discussion concerning the development, properties and application of the material is presented.

Introduction

Carbon-based materials are considered an attractive choice for the plasma facing components. This is because their low atomic number engenders a certain plasma compatibility. Carbon-based materials pose excellent thermal shock resistance and do not melt. These properties are indispensable for high heat flux components which must be resistant even under off-normal conditions.

The next step fusion devices (fusion reactor) will be long pulse or steady state operation, the wall loading will be $\geq$ 10-20 MW m^{-2} and the expected plasma flux density is $\geq 10^{19}$ cm^{-2} s^{-1}. In addition intense neutron will be produced, the assessed neutron flux will be around $3.5\text{-}9.0 \times 10^{14}$ cm^{-2}s^{-1} for the first wall, whilst the neutron flux for the divertor is around $1\text{-}3 \times 10^{14}$ cm^{-2}s^{-1}, which will lead to change of material properties. From the economic and safety point of view, an adequate long lifetime of plasma facing components and acceptable low tritium retention are required. However, all low-Z materials possess high erosion yields via D$^+$/T$^+$ sputtering and relatively high tritium retention. Consequently, frequent removal of tritium and replacements of components are indispensable.

In addition, high heat loading due to off-normal events (e.g. plasma disruption, slow transients, and ELMs, which can occur during a transition from detached to attached divertor operation) requires high thermal conductivity for the carbon protection material.

To improve the properties of carbon materials, one strives for increasing thermal conductivity (300 W m^{-1}K^{-1} at 20°C, 145 W m^{-1}K^{-1} at 800°C), reduced tritium inventory, reduced chemical erosion through interactions with deuterium/tritium, increased the resistance to water/oxygen at elevated temperatures and possibly an increased neutron stability.

In the framework of the European Fusion Research Programme, a great effort has been made to develop an advanced CFC to meet all of these requirements. In particular, SEP-NS31, which is an Si-doped CFC, was developed and its properties have been investigated.

Systematic investigation of B, Ti and Si doping on carbon properties

It has been observed experimentally that adding small concentrations of impurities (SiC, B, B$_4$C) to the bulk graphite resulted in the significant reduction of hydro-carbon formation for energetic hydrogen ions and for atomic hydrogen.

In the framework of the European Fusion R&D programme, a systematic study on the influence of B, Si and Ti doping on carbon properties has been performed. The following concentrations were investigated: boron 10-23 at.%, silicon 2-40 at.% and titanium 5-10%. Generally, by doping of B, Si and Ti to carbon, the chemical erosion is reduced. However, boron decreases the initial carbon thermal conductivity with increasing B concentration. The reactivity of carbon with H$_2$O/O$_2$ was not reduced, however, with the addition of titanium. On the other hand, the doping of silicon clearly showed that the chemical erosion and H$_2$O/O$_2$ reactivity at elevated temperatures were reduced and the thermal conductivity of carbon did not suffer up to a Si concentration of 12 at.%. Therefore, to improve the properties of carbon, silicon doping is a most promising measure [1,2].

The influence of silicon on the characteristics of CFC

High heat loading due to off-normal events (e.g. plasma disruption, slow transients and ELMs, which can occur during a transition from detached to attached divertor operation) requires high thermal conductivity for the carbon protection materials. Therefore, CFCs with high thermal conductivity (300 W m^{-1}K^{-1} at 20°C, 145 W m^{-1}K^{-1} at 800°C) are favourable.

In the framework of the European Fusion R&D programme, a great effort has been made to develop CFC to meet all operational requirements. In particular, SEP-NS31, which is an Si-doped CFC, was developed and its properties have been investigated. NS31 is a 3-D CFC constituted of P55 ex-pitch fibres (80% vol.) in one direction and ex-PAN fibres (20% vol.) in the perpendicular direction; this CFC undergoes a subsequent needling which gives the fibres orientation in the third direction (z direction). The high thermal conductivity direction is that of the P55 ex-pitch fibres. NS31 is densified by chemical infiltration of pyrocarbon and heat treated at a temperature > 2 500°C. At last, liquid silicon is injected under pressure leading partly to the formation of silicon carbide (SiC). NS31 contains about 8-10 at.% of silicon and its porosity is about 3-5% [3].

Thermal conductivity

Table 1 shows the thermal properties as a function of temperature. It can be seen that thermal conductivity is as high as 327 W K^{-1}m^{-1} at 298 K and 154.0 W K^{-1}m^{-1} at 1 073 K.

Table 1. Thermal properties of Si-doped 3-D CFC NS31

Temperature (K)	Cp (J/kg·K)	Diffusivity (mm^2/s)	Density (g·cm^{-3})	Thermal conductivity (W/m·K)
298	689.7	224.5	2.116	327.6
323	773.4	200.9	2.116	328.7
373	915.7	165.3	2.115	320.1
423	1 031.2	139.9	2.115	305.2
473	1 126.6	121.1	2.114	288.3
523	1 206.6	106.5	2.114	271.6
573	1 274.4	95.0	2.113	255.7
623	1 332.6	85.6	2.112	241.0
673	1 383.1	77.9	2.112	227.5
723	1 427.3	71.4	2.111	215.2
773	1 466.3	65.9	2.111	204.0
823	1 500.9	61.2	2.110	193.8
873	1 531.9	57.1	2.110	184.5
923	1 559.8	53.5	2.109	176.0
973	1 585.0	50.3	2.109	168.2
1 023	1 608.0	47.5	2.108	161.0
1 073	1 628.9	45.0	2.108	154.4

Chemical erosion

The measurements of the chemical erosion yields of the NS31 material were performed at an both ion beam facility (flux density 1-4 × 10^{15} D cm^{-2}s^{-1}) and using a plasma generator (flux density ≤ 2 × 10^{19} cm^{-2}s^{-1}). The results showed that the chemical erosion yield of NS31 was lower by a factor of about 2-3 in comparison with un-doped graphite [4,5] and the yields decrease with increasing flux density.

Oxidation behaviour of CFCs and doped CFCs

Oxidation resistance is an important criterion in the selection of PFCs. PFC oxidation may have severe consequences in the eventuality of loss of vacuum and loss of coolant into vacuum accidents (water/steam ingress into the vacuum vessel). Details of experimental investigations of doped carbon (Ti, Si) and doped CFCs exposed to steam and air are described elsewhere. Comparing steam and air, it was found that oxidation rates in air at 700°C are roughly about the same as the rates in steam at about 1 000°C. In general, un-doped CFCs have a higher oxidation rate. Highly SiC-doped INOX A14 (40% vol. SiC) revealed an oxidation resistance in steam of about 2-3 orders of magnitude better than un-doped materials; reduction of Si content to 8-10 at.% decreases the oxidation resistance in comparison with INOX A14 (40% vol. SiC), which, however, is still better than for un-doped carbons by a factor of about 2-4 [1]. Doping with Ti, however, does not improve oxidation resistance.

Outgassing behaviour of CFCs

A systematic investigation of outgassing behaviour of doped CFCs and un-doped CFCs was carried out in terms of temperature dependence and temperature pre-treatment. The experimental and evaluation procedures have been described in detail elsewhere. For comparison, the characteristics of five materials, SEPcarb N112 (3-D), Dunlop V (2-D), SiC 2.5%, SiC 8% doped CFCs (2-D), 3-D Novotex (No3) and NS31 (Si-doped 3-D) have been investigated. In order to perform an appropriate comparison, all pre-treated samples (1 000°C, 20 hours) were exposed to air for two weeks before the outgassing test. In general the total amount of outgassing of Si-doped CFC is 1-2 orders of magnitude lower than that of other carbon-based materials tested. The results also indicate that the release temperature of H_2O of Si-doped CFC is about 100°C lower than other carbon-based materials [6].

Comparative investigation of tritium retention in C and Si-doped CFC

A comparative investigation of tritium retention in un-doped carbon and Si-doped CFC (NS31) was carried out. Before loading the specimens were conditioned for one hour at 850°C under vacuum to remove air and moisture. Tritium loading was done with a gas mixture (H_2 + 2 ppm T_2), at a pressure of 800 mbar and a temperature of 850°C for 10 hours. The retained tritium is determined with ionisation chamber by heating the specimens with 5°C min-1 to 1 050°C (purging with He + 0.1% H_2) and oxidation at 850°C. To cross check the results, an alternative experiment was performed. Si-doped CFC (NS31) was loaded with a gas mixture (H_2 + 135 ppm T_2), at a pressure of 370 mbar and a temperature of 850°C for six hours. The retained tritium is then determined with ionisation chamber and proportional counter. For comparison, the tritium retention in various carbon materials is given in Table 2. It can be seen that the Si-doped (CFC NS31) has the lowest tritium retention. The results may also show a general trend of decreased tritium retention with an increasing density of carbon materials.

Table 2. Tritium retention in various carbon materials

Materials	Density (g cm^{-3})	T-retention (ppm)*
S1611 (graphite)	1.75-1.78	295
CL5890 (graphite)	1.81-1.83	175
A05 (2-D CFC)	1.81-1.86	163
N112 (3-D CFC)	1.94-2.0	138
NS31 (Si-doped 3-D CFC)	1.94-2.11	71

* Corrected for isotope effects.

Baking temperature

Figure 1 shows the thermal desorption spectrum of molecules CH_4, H_2O, CO and CO_2 as a function of temperature and time. To obtain optimum information, a very slow ramping rate of temperature was chosen (from room temperature to 400°C in six hours). It can be seen that the main desorbed molecules are H_2O and CO_2. The peak temperature of H_2O is around 210°C; in contrast, the peak temperature for un-doped carbon is 380°C. The peak temperature of CO_2 is around 400°C, whilst the peak temperature of un-doped carbon is around 600°C. The silicon has significantly reduced the baking temperature.

Figure 1. The thermal desorption spectrum of molecules
CH_4, H_2O, CO and CO_2 as a function of temperature and time

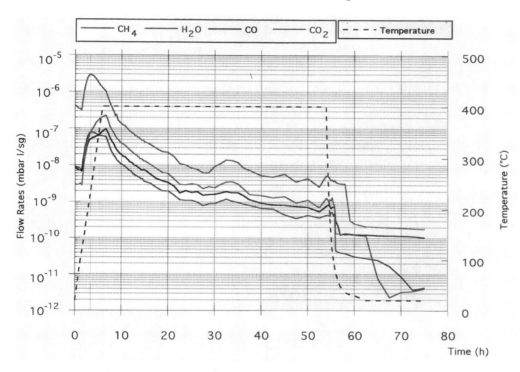

D-diffusivity and retention as a function of Si concentration

Figure 2 shows the diffusivity of deuterium as a function of Si concentration. It can be seen that the diffusivity increases continuously with increasing Si concentration to around 15 at.% and then decreases again. The retained D decreases with increasing Si concentration to around 15 at.% and then increases again (Figure 3).

The neutron effect on Si-contained CFC

8-10 at.% Si-doped CFC (NS11) was irradiated in the high flux reactor (Petten) at temperatures of 335 and 775°C in the range of 0.3/0.35 dpa. The results show that silicon-doped CFC (NS11) has the lowest dimensional change in comparison with un-doped CFCs (Tables 3 and 4). The Si increases the thermal stability under neutron irradiation.

Figure 2. Arrhenius plot for the diffusion coefficient of non-trapped deuterium. Shown are the natural logarithms of the diffusion coefficients vs. 1/kT.

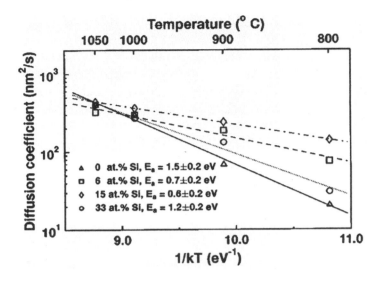

Figure 3. Amount of retained deuterium as a function of temperature in samples with different Si concentrations

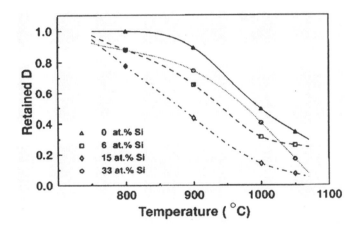

Table 3. Dimensional changes in CFCs and graphite at T_{irr} = 335°C (dL/LO in %)

Damage dpa.g	Dunlop concept 1			Dunlop concept 2			N 112	
	X dir.	Y dir.	Z dir.	X dir.	Y dir.	Z dir.	(//)	(⊥)
0.31	-0.02	0.01	-0.45	0.10	0.25	0.04	-0.08	0.05
Damage dpa.g	N 312 B			NS 11			RGTi (91)	
	X dir.	Y dir.	Z dir.	X dir.	Y dir.	Z dir.	(//)	(⊥)
0.31	-0.55	-0.02	-0.04	-0.07	0.06	0.12	0.30	0.18

Table 4. Dimensional changes in CFCs and graphite at $T_{irr} = 775°C$ (dL/LO in %)

Damage dpa.g	Dunlop concept 1			Dunlop concept 2			N 112	
	X dir.	Y dir.	Z dir.	X dir.	Y dir.	Z dir.	(//)	($\perp$)
0.35	0.04	-0.02	0.12	0.02	0.16	0.27	-0.12	-0.04
Damage dpa.g	N 312 B			NS 11			RGTi (91)	
	X dir.	Y dir.	Z dir.	X dir.	Y dir.	Z dir.	(//)	($\perp$)
0.35	-0.07	-0.02	0.05	-0.07	-0.05	0.00	0.38	0.17

Molecular dynamics modelling of Si effect

Figure 4 shows the carbon sputtering yield as a function Si concentration under 30 eV H bombardment. It is seen that a small fraction of silicon (10-20 at.%) in the initial carbon structure leads to lower sputtering yields. However, at a silicon concentration of 30 at.%, Si does not affect the carbon sputtering yield [7].

Figure 4. Carbon sputtering yield as a function of Si concentration

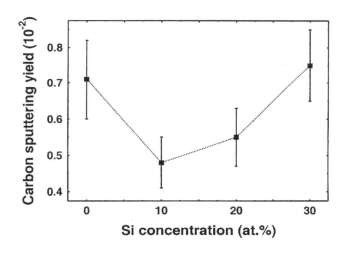

Conclusion

In the present study, a systematic investigation on the properties of an advanced silicon-doped CFC is performed and the following conclusions can be made:

- The thermal conductivity of Si-doped 3-D CFC is as high as 320 W $m^{-1}K^{-1}$ at 300 K.

- It has been demonstrated that doping of Si increases the oxidation resistance in steam/oxygen.

- It seems that doping of silicon decreases the total tritium retention.

- The outgassing rates of gaseous species, H_2, CH_4, H_2O, CO and CO_2 of Si-doped carbon materials are higher than those of un-doped carbon materials.

- The rate of CxHy formation (chemical erosion yield) of carbon materials is decreased by doping of Si.

- The post-neutron irradiation test showed (0.3/0.35 dpa.g, temperature 335 and 775°C) that Si-doping increased the dimensional and thermal stability.

- The results also indicated that the release temperature of H_2O and CO_2 molecules of Si-doped CFC is about 150°C lower than that of un-doped carbon-based materials.

Acknowledgements

The author wishes to thank the following individuals for their contributions: C. Alessandrini, J.P. Bonal, J. Roth, J. Keinonen, W. Bohmeyer, P. Kornejew, R. Moormann and H. Werle.

REFERENCES

[1] C.H. Wu, C. Alessandrini, R. Moormann, M. Rubel, B.M.U. Scherzer, *J. Nucl. Mater.*, 220-222, 860 (1995).

[2] C.H. Wu, *et al.*, in Fusion Technology 1996, Elsevier Science B.V., Amsterdam, the Netherlands, 1997, pp. 327-330.

[3] C.H. Wu. *et al.*, *J. Nucl. Mater.*, 258-263, 833 (1998).

[4] M. Balden, J. Roth, C.H. Wu, *J. Nul. Mater.*, 258-263, 740 (1998).

[5] P. Kornejew, W. Bohmeyer, H-D. Reiner, C.H. Wu, *Physica Scripta*, Vol. T91, 29 (2001).

[6] C.H. Wu, *et al.*, *Fusion Engineering and Design* (2001), in press.

[7] E. Salonen, K. Nordlund, J. Keinonen, C.H. Wu, "Chemical Sputtering of Amorphous Silicon Carbide under Hydrogen Bombardment", Proc. of E-MRS 2001 spring meeting.

A CARBON DIOXIDE PARTIAL CONDENSATION CYCLE FOR HIGH-TEMPERATURE REACTORS

Takeshi Nitawaki, Yasuyoshi Kato
Research Laboratory for Nuclear Reactors
Tokyo Institute of Technology
O-okayama, Meguro-ku, Tokyo 152-8550, Japan
E-mail: Nitawaki@nr.titech.ac.jp, kato@nr.titech.ac.jp

Abstract

A carbon dioxide partial condensation direct cycle concept has been proposed for thermal reactors. This cycle makes it possible to improve cycle efficiency due to low compression work in liquid phase and non-ideal gas behaviour of carbon dioxide, and effective utilisation of recuperative heat. The thermal reactor integrating this concept is expected to be an alternative solution to current high-temperature gas-cooled reactors (HTGRs) with helium gas turbines, allowing comparable cycle efficiency of about 45% at the moderate temperature of 650°C instead of 900°C in PBMR. By using an ultra high-purity Cr-Fe alloy, a reactor outlet temperature of 900°C can be attained, and the cycle efficiency of the direct cycle is about 50% at a pressure of 12.5 MPa.

Introduction

Helium is currently used as a coolant in high-temperature gas-cooled reactors. However, carbon dioxide is used as a coolant for AGRs, as there are certain advantages to be had from its use.

Judging from the data shown in Table 1, carbon dioxide is preferable as a coolant to helium. The higher heat transfer coefficient leads to about a 1.5 times lower temperature difference between coolant and fuel rod cladding surface, and the higher heat transport capacity results in about a 2.5 times more effective core decay heat removal under natural circulation conditions, reference to helium under the same gas velocity. The 3.5 times longer depressurisation time mitigates the depressurisation transient and makes it easier to design a passive decay heat removal system. Carbon dioxide is about 250 times less expensive than helium per unit weight and 24 times per unit volume. In addition, considering the lower leakage rate characteristics of carbon dioxide as compared to helium, coolant leakage problems of gas-cooled reactors in operation are orders of magnitude less sever than with helium.

Table 1. Comparison of main gas coolant properties

Items	CO_2	He	Remarks
Molecular weight (M)	44.0	4.0	
Specific heat (Cp) (kJ/kg·K)	1.11	5.19	
Thermal conductivity (λ) ($\times 10^3$ W/m·K)	46.2	271.0	
Heat transfer coefficient*, fuel cladding to coolant (at 20 MPa, 100 m/s)	1.49	1.0	$= Nu \cdot (\lambda/d)$ $Nu \propto Re^{0.8} \cdot Pr^{0.4}$
Heat transport capacity* (at 20 MPa)	2.72	1.0	$\propto Cp \cdot M$
Pumping power*	0.63	1.0	$\propto 1/Cp^3 \cdot M^2)$
Depressurisation time*	3.6	1.0	**
Cost per unit volume*	0.04	1.0	

*   Relative value.
** From Ref. [10].
D = equivalent hydraulic diameter.

Carbon dioxide might react with structural materials and graphite moderator under high-temperature and radiation fields. The extensive experience in Magnox reactors and advanced gas-cooled reactors (AGRs) for more than twenty years has shown that the corrosion problem of structural materials can be eliminated by appropriate material selection according to service temperature and by reduction of vapour content [1]. For the graphite moderator, it was found that the addition of methane and carbon monoxide in a range of small concentrations (called the coolant "window") inhibited the reaction with carbon dioxide and the satisfactory operation was maintained without excessive corrosion [2]. The graphite corrosion problem does not exist in fast reactors, which have no need for a moderator.

Taking into account the above advantages, we propose a new concept of a carbon-dioxide-cooled direct cycle in the present study. Up to now, the only remotely similar proposal has been an indirect Rankin cycle with a water/steam system which has been applied to AGRs as well as enhanced gas-cooled reactors (EGCRs) in the UK [3,4,5].

Recently, global interest in small and medium-sized reactors has steadily increased, particularly with regard to electricity generation as well as for district heating in cities and islands. The Research

Laboratory for Nuclear Reactors of the Tokyo Institute of Technology started a new programme in December 1999 to develop advanced nuclear reactors of small and medium size. In the framework of this programme, we are designing advanced small fast reactors (FRs) [6,7], high-temperature gas-cooled reactors (HTGRs) and boiling water reactors [8,9] in collaboration with industry.

Carbon dioxide direct cycle

Carbon dioxide is condensable from the gas to the liquid phase through cooling to below the critical temperature of 31°C (304 K). A variety of cycles could be employed for the direct cycles and they are typically categorised into the following three types, depending on the mass fraction condensed to the total carbon dioxide in circulation:

1) Full condensation cycle.

2) Partial condensation cycle.

3) Non-condensation or Brayton cycle.

Thermal efficiencies are evaluated in these three typical cycles.

Full condensation cycle

The full condensation cycle coolant flow circuit and its temperature entropy (T-S) diagram are given in Figure 1. The cycle consists of a reactor, turbine, condenser, circulation pump and recuperator. In this cycle, the total or full quantity of carbon dioxide in circulation is condensed after working at the turbine. The turbine exhaust carbon dioxide is cooled down in the recuperator from point (5) to point (6). The carbon dioxide is further cooled and condensed to point (1) in the condenser at the temperature of 25°C (298 K) and the saturation vapour pressure of 6.4 MPa. The temperature of liquidised carbon dioxide is raised from point (1) to point (2) by pumping work. Then, the liquid carbon dioxide is gasified and heated up to a reactor inlet temperature of point (3) through heat exchange with the turbine exhaust in the recuperator. The gasified carbon dioxide heated in the reactor up to point (4) is introduced in the turbine and the temperature of carbon dioxide is lowered to point (5) after driving the turbine.

Partial condensation cycle

The partial condensation cycle is shown in Figure 2. In this cycle, a part of carbon dioxide is condensed or the compression is performed partly in the liquid phase and partly in the gas phase. Therefore, this cycle can be considered as an intermediate of the full condensation cycle and the non-condensation (Brayton) cycle. While the turbine outlet pressure in the full condensation cycle is determined by the saturation vapour pressure of liquid carbon dioxide at the condensation temperature, e.g. 6.4 MPa at 25°C, in this cycle the turbine outlet pressure is independent of the condensation temperature, allowing a degree of freedom for the turbine outlet pressure or the turbine expansion ratio. The turbine exhaust carbon dioxide is cooled down in the two recuperators and the pre-cooler from point (6) to point (9), then compressed to point (10), and is divided into two flows. A part of the total carbon dioxide in circulation is condensed to point (1), pumped to point (2) and heated up to point (4) in the two recuperators. The remaining carbon dioxide is heated up to point (4) via point (11) in the compressor II and the recuperator I.

Figure 1. Full condensation cycle

(a) Coolant flow circuit   *(b) T-S diagram*

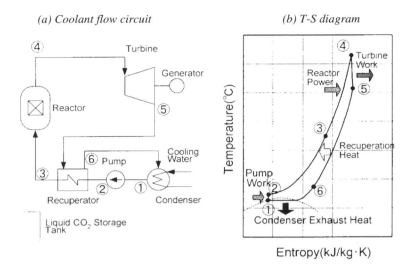

Figure 2. Partial condensation cycle

(a) Coolant flow circuit   *(b) T-S diagram*

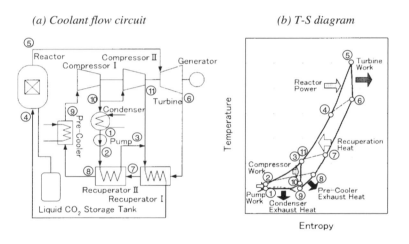

Non-condensation cycle

The non-condensation cycle is shown in Figure 3, which can be considered a Brayton cycle. This cycle consists of a turbine, recuperator, pre-cooler, inter-cooler and two compressors. Its arrangement is the same as that in the typical HTGR helium gas turbine system [11]. The turbine exhaust carbon dioxide gas is cooled down from point (5) to point (7) in the recuperator and the pre-cooler, and heated up from point (7) to point (8) through compression. The heated carbon dioxide is re-cooled down from point (8) to point (1) in the inter-cooler and heated up from point (1) to point (3) through compression and recuperation.

Comparisons of cycle performance

Figure 4 shows the cycle efficiency of the three cycles – the full, partial and non-condensation cycles – with typical values for the current light water reactors (LWRs) and gas-cooled reactors. The assumptions for calculating are summarised in the table to the right of the figure.

Figure 3. Non-condensation cycle

(a) Coolant flow circuit *(b) T-S diagram*

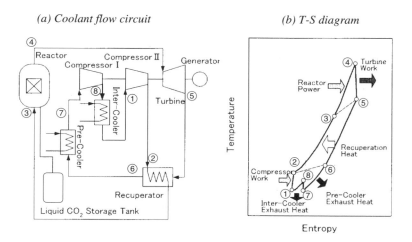

Figure 4. Cycle efficiency of the three cycles

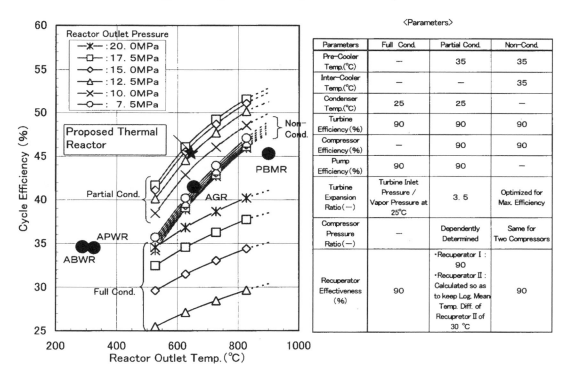

Parameters	Full Cond.	Partial Cond.	Non-Cond.
Pre-Cooler Temp.(℃)	–	35	35
Inter-Cooler Temp.(℃)	–	–	35
Condenser Temp.(℃)	25	25	–
Turbine Efficiency(%)	90	90	90
Compressor Efficiency(%)	–	90	90
Pump Efficiency(%)	90	90	–
Turbine Expansion Ratio(−)	Turbine Inlet Pressure / Vapor Pressure at 25℃	3. 5	Optimized for Max. Efficiency
Compressor Pressure Ratio(−)	–	Dependently Determined	Same for Two Compressors
Recuperator Effectiveness (%)	90	・Recuperator I : 90 ・Recuperator II : Calculated so as to keep Log. Mean Temp. Diff. of Recupretor II of 30 ℃	90

The differences among the cycle performance of the three cycles can be explained by comparing the works in each cycle. Cycle performances are compared in Figure 5, in which they are subdivided into the net turbine work (cycle efficiency), the compressor work, the pump work and the heat rejected from the cycle to cooling water through the heat exchanger surface of the pre-cooler, the inter-cooler and the condenser.

As for the full condensation cycle, its cycle efficiency is lowest among the three cycles in spite of very low pump work done for compression in the liquid phase, mainly due to the large heat rejection loss in the condenser.

137

Figure 5. Heat balance of the three cycles

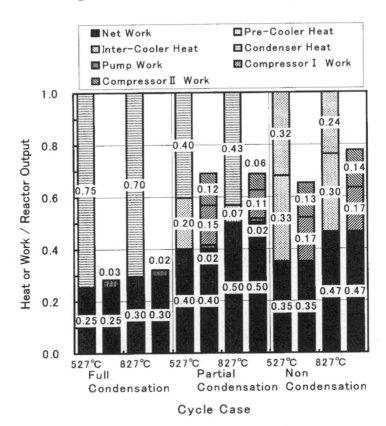

In the non-condensation (Brayton) cycle, cycle efficiency is considerably improved by achieving smaller heat rejection loss through the pre-coolers and intermediate coolers than through the condenser in the full condensation cycle, although the work done in the two compressors is considerably larger than that in the pump of the full condensation cycle.

Cycle efficiency is highest in the partial condensation cycle through minimising the sum of the condenser heat loss, the pump work and the compressor work. The optimum condensation fraction increases with the turbine inlet temperature and the pressure ranging from 35% at 527°C and 7.5 MPa to 64% at 827°C and 17.5 MPa. At a pressure of 12.5 MPa and a core outlet temperature of 650°C, the cycle efficiency is about 45%, which is comparable to that of current HTGRs at the high temperature of 800~900°C, such as PBMR.

The cycle efficiency is higher than that of the HTGR helium gas turbine system [11], resulting from the partial condensation effect described above and the non-ideal gas behaviour of carbon dioxide in the isentropic compression processes. The work W in the isentropic expansion and compression processes of one mol real gas is given by:

$$W = -\int V dP = -\int zRT\,dP/P$$

where V is the gas volume, P is gas pressure, R is gas constant and z is the compressibility factor.

The compressibility factor z is given as a function of the reduced temperature Tr ($Tr = T/Tc$, Tc = critical temperature) and reduced pressure Pr ($Pr = P/Pc$, Pc = critical pressure). Gases with the same z value have the same behaviour according to the law of corresponding states. The factor z gives the fractional deviation of the real gas from the ideal gas ($z = 1$) in the isentropic expansion and compression processes.

For higher pressures and temperatures, z may exceed unity and the real gas produces more work than that of the ideal gas in the expansion process. The z values at high pressure and high temperature are available from the generalised compressibility charts and tables prepared by Hougen and Watson [12]. At the critical temperature and pressure, the z value dips sharply below the ideal line of unity and takes an extremely low value – as low as about 0.2. A low z value indicates that the gas is more compressible than the ideal gas. Since the compression processes in the present carbon dioxide partial cycle and non-condensation cycle are done at the vicinity of Tc and Pc, their z values are estimated to be smaller than that of a helium Brayton cycle with the same temperature and the same pressure conditions (as shown in Table 2).

Table 2. Critical parameters and compressibility factor

Gas	Critical parameters		Typical compression condition (35°C, 5 MPa)			Typical expansion condition (650°C, 12.5 MPa)		
	Tc (K)	*Pc* (MPa)	*Tr*	*Pr*	*z*	*Tr*	*Pr*	*z*
He	5.195	0.275	59.31	18.18	~1	177.70	54.95	~1
CO_2	304.14	7.384	1.01	0.68	~0.7	3.03	1.69	~1

Tc = critical temperature
Pc = critical pressure
Tr = reduced temperature (= T/Tc)
Pr = reduced pressure (= P/Pc)
z = compressibility factor (deviation from ideal gas = 1.0)

Both the z values are nearly equal to unity at the high-temperature and high-pressure condition in the turbine. The z values of carbon dioxide are smaller than those of helium in the compressor. Judging from the z values in compression and expansion, it is understood that the cycle efficiency of the carbon dioxide cycle is higher than that of the helium cycle.

Recently, and ultra-high-purity 70% Cr-Fe alloy has been developed, the tensile strength of which is four times higher at 1 200°C than that of Ni-based alloy [13,14]. Also, the absence of Ni makes it possible to reduce (n,α) and (n,p) reactions in the Cr-Fe alloy to one-fifth compared to 316 stainless steel, which contributes to the swelling of stainless steels. Consequently, if the ultra-high-purity Cr-Fe alloy is used for the core material, the irradiation limit in fast neutron flux might be expected to be two or three times higher at least than that of 316 stainless steels. By using the ultra-high-purity Cr-Fe alloy, the cycle efficiency is estimated from Figure 5 to exceed 50% when the reactor outlet pressure is 12.5 MPa and the reactor outlet temperature is 900°C (cladding hot spot temperature = 1 200°C). The cycle efficiency is 1.5 and 1.1 times higher than those of current LWRs and HTGRs, respectively.

Perspectives

Since the direct cycle simplifies the heat transport system eliminating intermediate and water/steam cooling loops, the plant construction cost or capital cost is expected to be comparable to or less than that of current LWRs. The condensation cycle allows a unique passive decay heat removal system described below.

A passive decay heat removal system is realised by allocating a liquid carbon dioxide storage tank, as shown in Figures 1(a), 2(a) and 3(a). When coolant pressure depression accidents occur, carbon dioxide in the storage tank will be gasified and supplied to the depressurised reactor core and the decay heat is passively removed. The system could work for a short term of several hours after the reactor shut down when the decay heat level is higher than about 1-2% of rated power. After the decay heat level becomes less than about 1-2% of the rated power, the decay heat removal system switches from the short-term system to a reactor auxiliary cooling system, which works passively for the long term though natural circulation of carbon dioxide in the reactor vessel. Natural convection is about 2.5 times more effective than with helium, as described earlier, which might mitigate the depressurisation transients in the carbon dioxide cycles.

Considering various elements such as cycle efficiency, plant safety and plant construction cost, the carbon dioxide-cooled partial condensation direct cycle is expected to be a potential alternative option to LMFRs.

The direct cycle is applicable to thermal reactors. The partial condensation cycle proposed in this study provides cycle efficiency of about 45% at the current AGR core outlet temperature of 650°C. Such efficiency would be obtained at 900°C in PBMR with a helium gas turbine. In other words, our partial condensation concept is possible to provide a comparable cycle efficiency at the moderate temperature of 650°C with that of the PBMR at the high temperature of 900°C [11]. For thermal reactors with the carbon dioxide condensation cycle, it is not possible to add the methane and carbon monoxide as AGR in order to inhibit the reaction of carbon dioxide, because it can affect the condensable performance of the coolant and plant operability. Graphite coating with an inert material such as silicon carbide is expected to be an alternative means. This technique is well established in the non-nuclear field, however, corrosion and/or irradiation tests should be examined in future work. With the graphite coating the ultra-high-purity Cr-Fe used for the core structural materials, the carbon dioxide-cooled reactor might attain a core outlet temperature of 900°C and its cycle efficiency would exceed 50%.

Conclusions

Carbon dioxide can achieve a higher cycle efficiency in a direct gas turbine cycle system due to the real gas effect, especially in the vicinity of the critical points. Analysing the cycle performance of full, partial and non-condensation cycles, the cycle efficiency is highest in the partial condensation cycle. The partial condensation cycle is applicable to both the FR system and the gas-cooled thermal reactors. A gas-cooled thermal reactor employing the partial condensation cycle provides comparable cycle efficiency at the moderate outlet temperature of 650°C with that of PBMR operated at 900°C.

REFERENCES

[1]   G.B. Gibbs, L.A. Popple, "Oxidation of Structural Steels in CO_2 Cooled Reactors", *Nuclear Energy*, Vol. 21, pp. 51-55 (1982).

[2]   G.F. Hewitt, J.G. Collier, "Introduction to Nuclear Power", 2nd edition, Taylor & Francis, pp. 41-43 (1997).

[3]   C.P. Gratton, "The Gas-cooled Fast Reactor in 1981", *Nuclear Energy*, Vol. 20, pp. 287-295 (1981).

[4]   T.A. Lennox, D.M. Banks, J.E. Gilroy and R.E. Sunderland, "Gas-cooled Fast Reactors", Trans. ENC 98, Vol. IV, TAL/005392 (1998).

[5]   T.J. Abram, D.P. Every, B. Farrar, G. Hulme, T.A. Lennox and R.E. Sunderland, "The Enhanced Gas-cooler Reactor (EGCR)", Proc. 8th Int. Conf. on Nuclear Engineering, ICONE-8289 (2000).

[6]   T. Nitawaki, Y. Kato and Y. Yoshizawa, "A Carbon Dioxide-cooled Direct Cycle Fast Reactor", IAEA-SR-218/21 (2001).

[7]   Y. Kato, T. Nitawaki and Y. Yoshizawa, "A Carbon Dioxide Partial Condensation Direct Cycle for Advance Gas-cooled Fast and Thermal Reactors", GLOBAL'2001 (2001).

[8]   J. Yamashita, M. Ootsuka, K. Fujimura and Y. Kato, "An Innovative Conceptual Design of Safe and Simplified Boiling Water Reactor (SSBWR)", Proc. Int. Seminar on Status and Prospects for Small and Medium-sized Reactors, IAEA-SR-218/23 (2001).

[9]   J. Yamashita, M. Ootsuka, M. Hidaka, K. Fujimura and Y. Kato, "An Innovative Conceptual Design of Safe and Simplified Boiling Water Reactor (SSBWR)", GLOBAL'2001, No. 057 (2001).

[10]   E.E. Levis, "Nuclear Power Reactor Safety", John Wiley & Sons (1977).

[11]   International Atomic Energy Agency, "Current Status and Future Development of Modular High-temperature Gas-cooled Reactor Technology", IAEA-TECHDOC-1198, Vienna (2001).

[12]   O.A. Hougen, K.M. Watson and R.A. Ragatz, "Chemical Process Principles, Part II, Thermodynamics", John Wiley & Sons, pp. 579-591 (1960).

[13]   G. Kanou, N. Harima, S. Takaki and K. Abiko, "Mechanical Properties of a High-purity 60 mass% Cr-Fe alloy", *Materials Transactions*, JIM, Vol. 41, No. 1, pp. 197-202 (2000).

[14]   K. Kabiko, *Nikkei Science*, No. 10, pp. 32-41 (2000).

SESSION III

Development of In-core Material Characterisation Methods and Irradiation Facility

Chairs: C. Vitanza, M. Yamawaki

DEVELOPMENT OF THE I-I TYPE IRRADIATION EQUIPMENT FOR THE HTTR

Taiju Shibata, Takayuki Kikuchi
Japan Atomic Energy Research Institute
3607 Oarai-machi, Higashiibaraki-gun, Ibaraki-ken 311-1394 Japan

Satoshi Miyamoto, Kazutomo Ogura
The Japan Atomic Power Company
6-1, 1-chome, Ohtemachi, Chiyoda-ku, Tokyo, 100-0004 Japan

Yoshinobu Ishigaki
Fuji Electric Co., Ltd.
1-1, Tanabeshinden, Kawasaki-ku, Kawasaki-shi, Kanagawa-ken, 210-9530 Japan

Abstract

The High-temperature Engineering Test Reactor (HTTR) is a graphite moderated, helium gas-cooled test reactor with a maximum power of 30 MW. It is the first high-temperature gas-cooled reactor (HTGR) in Japan, and has a unique and superior capability with regard to various irradiation tests due to the advantage of its large and high-temperature irradiation spaces.

The I-I type irradiation equipment, which is the first irradiation rig for the HTTR, is currently under construction after several experiments such as in-core monitoring method. It is served for an in-pile creep test for a stainless steel developed as a structural material of the fast reactor. The creep rate and rupture time of the in-pile creep specimens will be measured at elevated temperatures in the reactor. The equipment can handle a large load of tensile specimens, taking advantage of the HTTR's large, high-temperature irradiation space. Therefore, the specimens with a nominal gauge length of 30 mm and diameter of 6 mm can be adapted for the in-pile test. This specimen size is not small, such as those generally used for in-pile tests, but is the standard size usually used for non-irradiation creep tests. Specimens for out-of-pile creep tests after the irradiation are also installed in the equipment.

This equipment was designed so as to carry out in-pile creep tests in the HTGR. Part of the equipment is installed in the reactor pressure vessel through a standpipe of the HTTR. The installed part is about 8 900 mm long. The upper end of the part is fixed to the standpipe. The lower end is constituted of three tubular parts, two irradiation units and a guide tube, and is installed in the reactor core. The units are held in elevated ambient temperatures. In addition to the ambient temperatures and neutron and gamma heatings, electric heaters surrounding the specimens control the irradiation temperatures of the units to 550 and 600°C, respectively. The weight-loading system located outside the reactor gives a stable and precise tensile load on the specimens by using levers and weights. The maximum load of this system is about 10 kN. The creep behavior of the specimen is detected by a differential transformer developed for high-temperature conditions.

The fabrication of this equipment was started in 1999. Functional tests for each component, such as the weight-loading system, heater and differential transformer, were carried out outside the reactor. The equipment was then fabricated from components. Its performance was also demonstrated outside the reactor after fabrication. The construction is in the final stage. This paper describes the progression of the development of the I-I type equipment.

Introduction

The High-temperature Engineering Test Reactor (HTTR) is a graphite-moderated, helium gas-cooled test reactor with a maximum power of 30 MW [1,2]. It is the first high-temperature gas-cooled reactor (HTGR) in Japan with a maximum coolant outlet temperature of 850°C at rated operation and 950°C at high-temperature test operation. It has a unique and superior capability with regard to various irradiation tests due to the advantage of its large and high-temperature irradiation spaces.

The I-I type irradiation equipment, which is the first irradiation rig for the HTTR, will be used for the in-pile creep test of the stainless steel developed for the structural material of the fast reactor. The specimen size effect on the deformation of metallic material is currently a subject of increasing interest [3,4]. The in-pile creep behaviour of material is generally tested through the use of specimens that are small in size. This is because it can be difficult to use the standard size specimens, usually used for out-of-pile creep testing, due to restrictions such as an irradiation space limit and a big tensile load requirement. The I-I type irradiation equipment is designed to carry out the in-pile creep test for the standard size specimens while taking advantage of the superior irradiation capability of the HTTR. It can handle large loads of specimens in a stable manner. The specimens' creep behaviour, such as the creep rate and rupture time, will be measured at elevated temperatures in the reactor. The fabrication of the equipment was started in 1999 and its performance has already been demonstrated outside the reactor. Presently, the equipment is in the final stages of development. This paper describes the design of the I-I type irradiation equipment and the progress of its development.

Design of I-I type irradiation equipment

In-pile test condition

The targets of the test condition of the in-pile creep test in the HTTR are shown in Table 1. The cylindrical-type creep specimen, which has a nominal gauge length of 30 mm and a diameter of 6 mm, was adopted for the test. This specimen is not a small one generally reserved for in-pile creep tests for other reactors, but the standard one usually reserved for non-irradiation creep tests. The material of the specimen is 316 FR steel (FBR grade type 316ss), which is a structural material developed for the fast reactor [5,6]. The temperatures of 550 and 600°C are the targets of the irradiation temperature. The target fast neutron fluence of 1.2×10^{19} n/cm^2 (E > 0.18 MeV) will be attained in about 70 days of the rated operation of the HTTR. The tensile stresses were determined to allow the investigation of the creep rate and rupture time of the specimens [5], as 343 MPa (at 550°C) and 248 MPa (at 600°C). To apply them to the specimens, tensile loads of 9.7 and 7.0 kN are respectively necessary.

Table 1. Targets of test condition on in-pile creep test in HTTR

Specimen	
Material	316FR steel (FBR grade type 316ss)
Size (gauge)	$\varphi 6 \times 30$ mm
Irradiation temperature	550 and 600°C
Tensile stress/load	343 MPa/9.7 kN at 550°C
	248 MPa/7.0 kN at 600°C
Neutron fluence (E > 0.18 MeV)	1.2×10^{19} n/cm^2

Major specification of the equipment

The I-I type irradiation equipment is designed to meet the requirements of the test condition shown in Table 1. A cutaway drawing of the equipment is shown in Figure 1. It is constituted of in-vessel and out-of-vessel parts. The in-vessel part is installed in the reactor pressure vessel through one of the standpipes of the HTTR. This part is held by the standpipe closure with the standpipe. The pressure boundary of the reactor coolant is held by the closure. The in-vessel part has a length of about 8 900 mm and its lower end is constituted of three tubular parts, which are two irradiation units and a guide tube. Each has a length of about 2 600 mm and a diameter of 113 mm. Three specimens are installed in the irradiation unit at the same level. The units and the guide tube are installed in three holes of the graphite reflector blocks in the column of the reflector region of the core. The level of installation of the specimens corresponds to the fuel blocks in the fuel region. The specimens are mainly heated by heat from the core at the operation. Because of the large heat capacity of the core, it is thought that the irradiation temperature will not change very rapidly during operation. The irradiation temperatures are, therefore, mainly controlled taking advantage of this superior feature of the HTTR. An electric heater with maximum power of 300 W is equipped in each unit surrounding the specimens. It will provide a supplementary temperature control in addition to the heat from the core. The target of temperature deviation is a set value of ±3°C for both the irradiation temperatures of 550 and 600°C.

Figure 1. Cutaway drawing of I-I type irradiation equipment

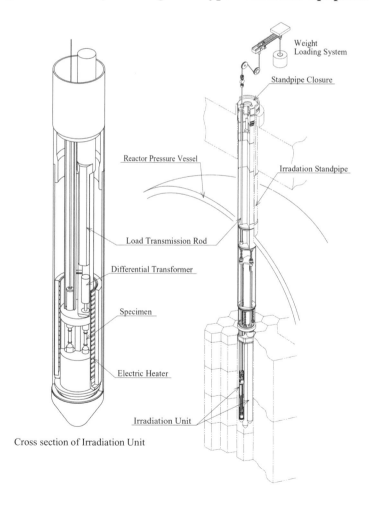

The out-of vessel part of the equipment, shown in Figure 1, is the weight loading system. Figure 2 illustrates this system. The levers and weights generally used for the out-of-pile creep test are used for this system. The weights allow stable loads of 9.7 and 7.0 kN on the specimens. At the in-pile creep test, the selected tensile loads will be applied to the specimens after the attainment of the rated reactor power. Each load is transmitted through the load transmission rod to the specimen in each unit. If the specimen is elongated by the proceeding creep, the level of a pulley will automatically be controlled by a motor so as to keep the balance of the lever. Thus, the applied loads will be continuously and precisely controlled.

Figure 2. Weight loading system of I-I type irradiation equipment

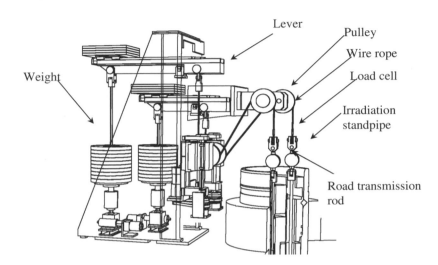

In-core monitoring

Two sets of the in-vessel part, shown in Figure 1, were fabricated. One (Type A) is mainly used for measuring the irradiation condition of the core, though the in-pile creep test will not be carried out by Type A. This measurement will ensure the irradiation condition of the in-pile creep test, next to the Type A irradiation, since it will be the first time the HTTR will measure the condition directly. The installed specimens in Type A are used for the out-of-pile creep test after the irradiation. The other (Type B) is used for the in-pile creep test. The irradiation condition will also be monitored. For this irradiation test, one specimen in the unit is used to study the in-pile creep behaviour and the other two specimens in the unit are not loaded and are used for the out-of-pile creep test after the irradiation. For both tests by Type A and B, the monitoring electric signals are taken out of the reactor vessel through the standpipe closure.

Table 2 shows in-core monitoring instruments for the in-pile creep test. The neutron flux at the irradiation equipment in the core will be monitored by self-powered neutron detectors (SPNDs) installed in it. The size of its sensor part is $\varphi 2 \times 70$ mm and rhodium is used as the emitter material. Some small metallic wires, packed in silica tubes, are also installed in the equipment as neutron fluence monitors. The wire materials were selected for both thermal and fast neutron fluence measurement after the irradiation. The irradiation temperature will be continuously measured by the thermocouples (K-type) in the equipment. The creep elongation of the specimen will be continuously monitored by a differential transformer in the unit. The transformer was developed for this equipment in order to work at elevated temperatures by adopting a fine mineral insulated (MI) type cable to the transformer coils. Its long-term stability at elevated temperatures was demonstrated in advance.

Table 2. In-core monitoring instruments for in-pile creep test

Neutron flux	Self-powered neutron detector (SPND) φ2 × 70 mm (sensor) Emitter: Rhodium
Neutron fluence	Fluence monitor Wire: Al-0.41%Co, Fe, Cu, 86%Cu-12%Mn
Temperature	Thermocouple K-type
Creep elongation	Differential transformer Coil: Fine mineral insulated (MI) cable

Performance demonstration

Functional tests for the components of the I-I type irradiation equipment, such as the weight loading system, heater and differential transformer, were carried out. Afterwards, the equipment was fabricated from the components. Its performance was also demonstrated outside the reactor after fabrication. The main results are described briefly as follows.

Load transmission

The load, produced by the weight loading system, is transmitted to the specimen through the load transmission rod, shown in Figures 1 and 2. The rod is inserted into the reactor vessel through the standpipe closure. The small gap between the rod and the penetration of the closure is sealed by O-rings. The produced load is thought to be slightly reduced and transmitted to the specimen because of the friction at the seal part of the penetration. The effect of the friction on the transmitted load was experimentally evaluated outside the reactor by using the fabricated equipment. Figure 3 shows the relationship between the effective load on the specimen and the produced load using the weight loading system. The produced load was measured by the load cell of the equipment attached to the rod as shown in Figure 2. The load on the specimen was measured by another load cell temporally attached to the specimen. One can see from Figure 3 that the produced load is transmitted to the specimen reduced by about 1.3 kN at this test condition. The effective load on the specimen is linearly proportional to the produced one, ranging from 6 to 11 kN. At the in-pile creep test condition, although the friction loss value may change, the linearity is thought to hold. The value will be measured by a strain gauge attached on the rod near the specimen, and then the effective load will be controlled taking the linearity into consideration.

Temperature control

The electric heater surrounding the specimens in the unit will keep the temperature stable in addition to the heat from the reactor core. The temperature controllability of the heater was tested by using a mock-up of the irradiation unit. The mock-up unit was installed in a furnace which created thermal disturbances. At the disturbance, the power of the electric heater was controlled automatically so as to keep the set value of the specimen temperature. Figure 4 shows the experimental result of the temperature control test. The set value of the specimen temperature is 600°C. The furnace temperature was changed ranging from about 550°C to 570°C as a thermal disturbance. The electric heater power was automatically controlled to keep the set value. As a result, the specimen temperature was maintained around the set value. In this case, the deviation of the specimen temperature was within 3°C. This result confirms that the specimen temperature will be controlled in a stable manner at the in-pile test.

Figure 3. Effective load on specimen by weight loading system

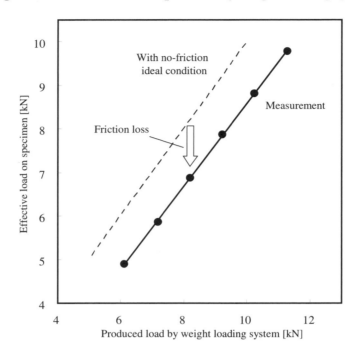

Figure 4. Specimen temperature control by electric heater at disturbance condition. Specimen temperatures were measured at the upper, middle and lower positions of its gauge part.

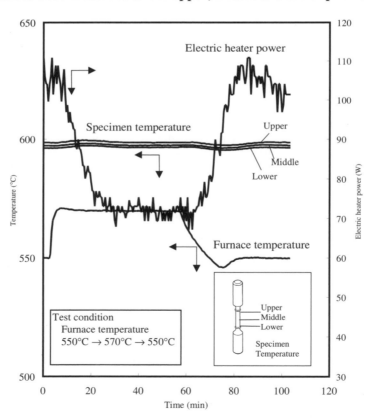

Creep behaviour measurement

The in-pile creep elongation will be detected by the differential transformer with the fine MI cable coil. Figure 5 shows the out-of-pile experimental result of the creep behaviour measurement. The test specimen was elongated at a temperature of 600°C. The elongation of the specimen was detected by both the differential transformer and a strain gauge which was temporally attached to the specimen. We can see from this figure that the both measured values are in good agreement. It is thought, therefore, that the creep behaviour at the in-pile-test will be precisely measured by the developed transformer.

Figure 5. Measurement of creep behaviour by differential transformer

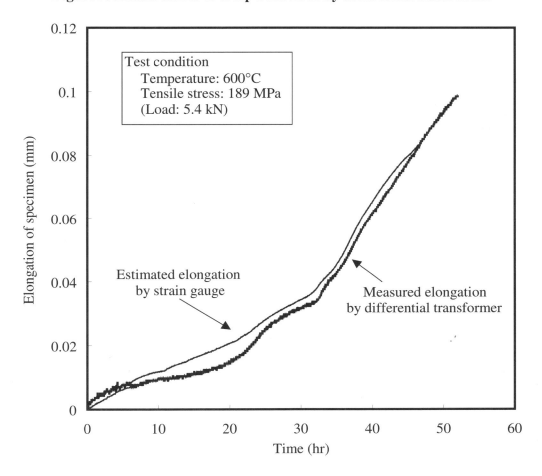

Summary

The I-I type irradiation equipment is the first irradiation rig for the HTTR. It will be used for the in-pile creep test for the standard size specimens of 316 FR steel. Fabrication of the equipment was started in 1999. Its performance, such as the load transmission, temperature control and creep behaviour measurement, has already been demonstrated outside the reactor. It is now temporarily located in the reactor building of the HTTR. After the demonstration of the performance of the HTTR, it will be installed in the reactor for the irradiation test.

Acknowledgements

This study has been carried out as a collaborative effort of the Japan Atomic Energy Research Institute (JAERI) and the Japan Atomic Power Company (JAPC). The authors are deeply indebted to the members of the collaboration, in particular Mr. Minoru Ohkubo and Mr. Tatsuo Iyoku of the HTTR Project of JAERI and Mr. Yuji Kuroda of JAPC.

REFERENCES

[1]   S. Saito, *et al.*, "Design of the High-temperature Engineering Test Reactor", JAERI1332, 1994.

[2]   Japan Atomic Energy Research Institute, "Present Status of HTGR Research & Development", 2000.

[3]   N.A. Fleck, G.M. Muller, M.F. Ashby and J.W. Huchinson, "Strain Gradient Plasticity: Theory and Experiment", *Acta Metall. Mater.*, 2, pp. 475-487, 1994.

[4]   T. Malmberg, I. Tsagrakis, I. Eleftheriadis, E.C. Aifantis, K. Krompholz and G. Solomons, "Gradient Plasticity Approach to Size Effects", Trans. of SMiRT16, Washington DC, USA, F01/1, 2001.

[5]   Y. Wada, E. Yoshida, T. Kobayashi and K. Aoto, "Development of New Materials for LMFBR Components – Evaluation on Mechanical Properties of 316FR Steel", Trans. of Int. Conf. on Fast Reactors and Related Fuel Cycles, Vol. I, pp. 7.2-1-7.2-10, 1991.

[6]   N. Miyaji, Y. Abe, T. Asayama, K. Aoto and S. Ukai, "Effects of Neutron Irradiation on Tensile and Creep Properties of Stainless Steels", *J. Soc. Mat. Sci.*, Japan, 5, pp. 500-505, 1997.

APPLICATION OF OPTICAL DIAGNOSTICS IN
HIGH-TEMPERATURE GAS-COOLED SYSTEMS

T. Shikama, M. Narui
Institute for Materials Research, Tohoku University
Oarai, Ibaraki, 311-1313 Japan

T. Kakuta
Tokai Research Establishment, JAERI
Tokai, 319-1194, Japan

M. Ishitsuka, K. Hayashi, T. Sagawa and T. Hoshiya
Oarai Research Establishment, JAERI
Oarai, 311-1394

Abstract

Recent activity in efforts to develop optical diagnostic techniques for the high-temperature test reactor (HTTR) is reviewed. Radiation resistant optical fibres have been developed, which will withstand radiation damages; these are expected to be deployed in the HTTR core regions within a few years. Temperature, rector power and other operating parameters could be measured by the optical diagnostic systems, which have several advantages over other diagnostic systems.

Introduction

Technologies for high-temperature gas-cooled reactors (HTGR), which have several attractive features, such as more stable and safer nuclear properties than other reactor types, are being extensively developed [1]. Intensified efforts are devoted in Japan to develop technologies related to the stable operation and effective utilisation of a high-temperature test reactor (HTTR) [2]. Among these efforts, the development of optical diagnostic techniques will be highlighted here.

Though the HTGR has several attractive features, it has some disadvantages. Namely, visibility of its core region is very poor in general even at reactor shutdown, as graphite moderator blocks seriously obstruct optical access. A typical example of this shortcoming was experienced when a fire developed in the core region of a so-called Colder Hall type reactor [3]. It was only after desperate and dangerous efforts that the burning position was identified. If remote optical diagnostics were available, it would improve the maintainability of the HTGR, resulting in far better social acceptance of this type of reactor. Ceramic fibre optics are the only tools that could render the solid-made HTGR core accessible.

Other issues are problems related to the highly corrosive environments in an HTGR core and the high-temperature atmosphere. Helium gas is extremely inert even at elevated temperatures. However, once contaminated with a small amount of impurities such as oxygen and carbon, it becomes seriously corrosive [4-6]. At present, among commercially available metallic structural materials, only the so-called alpha-tungsten strengthened "super-alloys" [7], which are expensive and poorly workable, can withstand the mechanical degradation caused by impure helium corrosion in HTGRs [6,7]. Concerning electrical functional ceramic insulators, their electrical insulating ability seriously degrades above 600°C. At best, an electrical resistivity of the order of about 10^6 ohm/m at 1 000°C could be expected for a highly pure alumina (Al_2O_3) [8]. For commercial ceramic insulators, resistance would be of the order of 10^4 ohm/m or less [9]. The poor compatibility of materials with HTGR core environments renders the application of conventional electrical dosimetry, such as thermocouples and self-powered radiation detectors (SPDs), very difficult in HTGR core regions. In the meantime, fibre optics have refractory properties in HTGR cores, as they are made of refractory silicon oxide (silica; SiO_2) and silica showed good compatibility with corrosive helium gases in HTGRs [4].

Until recently, silica core fibre optics have been considered to be vulnerable to nuclear radiation, with the result that the application of fibre optics was judged to be unrealistic in high-power nuclear reactor cores. However, there is evidence demonstrating that recently developed optical fibres can withstand radiation damage in HTGRs [10,11]. The present paper will review recent efforts in the development of radiation-resistant optical fibres and examples of optical diagnostics in radiation environments similar to that of the HTGR.

Development of radiation resistant optical fibres

Two major effects, electronic excitation and atomic displacement, are considered in radiation effects in general. Electronic excitation effects will activate pre-existing structural defects through ionising effects and produce optical absorption centres. In the meantime, atomic displacement effects will introduce new structural defects. In general, radiation effects introduce strong optical absorption bands in the wavelength range shorter than 700 nm, namely visible to near ultra-violet region [10,11]. Also, an increase of optical absorption is observed in the infrared region, in the wavelength range of 800-2 000 nm, by radiation effects. In this case, the strength of radiation-induced optical absorption has a very weak wavelength dependence.

Figure 1 shows the strength of radiation-induced optical absorption in a newly developed radiation-resistant optical fibre, under gamma-ray irradiation, specifically purely electronic excitation radiation. A fluorine doped (F-doped) optical fibre showed far better radiation resistance compared with a pure silica core fibre in the wavelength shorter than 700 nm. Here, the F-doped fibre (FF-4) was developed by Shamoto, *et al.* of Fujikura Co. Ltd of Japan [11]. Radiation-induced optical absorption was 0.5 dB/m at about 600 nm, caused by radiation-induced defects of so-called non-bridging oxygen hole centres (NBOHC), with an electronic excitation dose of 1×10^6 Gy. Better radiation resistance was demonstrated in a F-doped optical fibre (MF-1) developed by Hayami, *et al.* of Mitsubishi Co. Ltd of Japan, as shown in Figure 2 [12]. In this case, the radiation-induced optical absorption at about 600 nm was less than 0.2 dB/m with the same electronic excitation dose.

Figure 1. Growth of optical absorption in newly developed radiation-resistant optical fibre FF under pure electronic excitation irradiation (cobalt-60 gamma-ray irradiation)

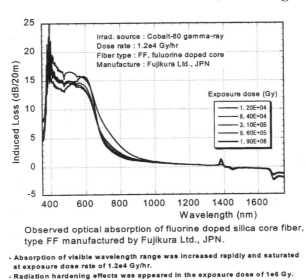

Observed optical absorption of fluorine doped silica core fiber, type FF manufactured by Fujikura Ltd., JPN.

- Absorption of visible wavelength range was increased rapidly and saturated at exposure dose rate of 1.2e4 Gy/hr.
- Radiation hardening effects was appeared in the exposure dose of 1e6 Gy.

Figure 2. Growth of optical absorption in MF-1 developed by Mitsubishi under pure electronic excitation irradiation

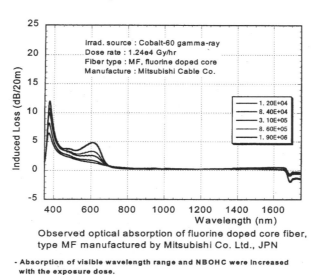

Observed optical absorption of fluorine doped core fiber, type MF manufactured by Mitsubishi Co. Ltd., JPN

- Absorption of visible wavelength range and NBOHC were increased with the exposure dose.
- Radiation hardening effect was not appeared in the large exposure dose.

The electronic excitation dose rate in the core of HTGRs will be not large, due to its moderate nuclear power density during operation and due to low residual radioactivity after reactor shutdown. The dose rate will be about 100 Gy/s at reactor full power and less than 0.1 Gy/s at reactor shutdown in the case of the HTTR. At this point, these newly developed fibres could be used for visible applications for more than 10^7 seconds, namely about 3×10^3 hours at reactor shutdown, without any serious disturbance by radiation-induced optical absorptions.

In the case of reactor full power operation, atomic displacement effects by fast neutrons should be considered. Figure 3 shows an increase of radiation-induced optical loss in FF-4, under irradiation in the Japan Materials Testing Reactor (JMTR). The irradiation temperature was 400 K and fluxes of fast ($E > 1$MeV) and thermal ($E < 0.687$eV) neutrons are 6×10^{16} and 8×10^{16} n/m^{2}s, respectively. An ionising dose rate is calculated to be about 0.5 kGy/s. At reactor start-up, so-called radiation hardening (the increase of optical transmission by radiation effects) was observed as shown in the data at 1 MW. Optical transmission loss was smaller than that before the irradiation in the entire wavelength range of 400-1 600 nm. Then, radiation-induced optical absorption increased in the whole wavelength range. At about 600 nm, the strength of optical absorption due to the NBOHC increased with an increase in the irradiation time. However, the induced optical transmission loss was less than 20 dB/m after 21 days of irradiation. The corresponding fast neutron fluence was about 1×10^{23} n/m^2 and the electronic excitation dose was about 1×10^9 Gy. These values correspond to about 10 years operation of the HTTR. The optical transmission loss of 20 dB/m is acceptable, when a strong light probe is assumed. Concerning the infrared regions, the radiation-induced optical transmission loss was very small, rather it can be said that the reactor irradiation improved optical transmission. However, the optical absorption by OH (oxyhydrate) grew, which would interfere with optical diagnostics in the infrared regions.

Figure 3. Growth of optical absorption in FF-4 under JMTR irradiation

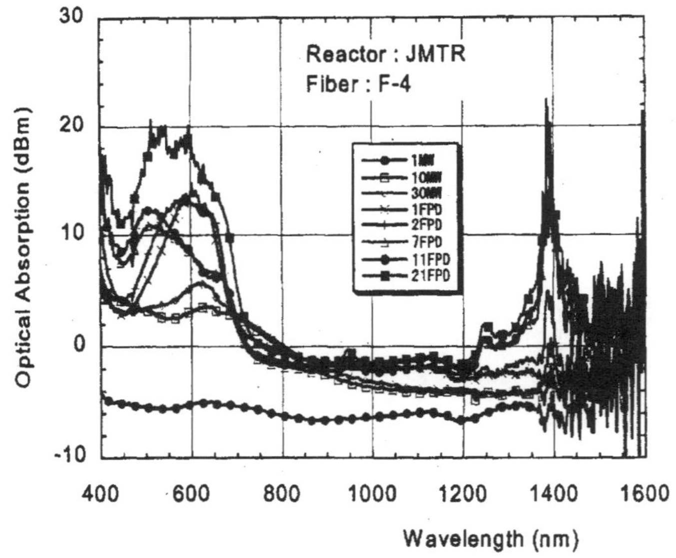

Observed Absorption of Improved Rad-Hard Fibre
Fibre No. F-4

The increase of optical absorption in the infrared range in comparison with the data obtained at 1 MW had a weak wavelength dependence when the absorption by OH was ignored. The introduction of this wavelength-independent absorption could be interpreted as the optical loss being caused by so-called micro-bending of the optical fibres. Specifically, some structural damage at the interface between the core and the clad of optical fibres was introduced over the course of irradiation. F-doping is found to make optical fibres sensitive to micro-bending loss, because the F-doping of the core silica would decrease the difference of reflectivity between the core and the clad. However, the bending loss could be minimised in the HTGR, in which optical fibres would be applied with a large bending curvature.

Visible inspection into HTGR core regions could be undertaken, with fibre bundles of the newly developed radiation-resistant optical fibres such as MF-1 and FF-4, even when the HTGR is operating under full power.

Other possible diagnostic tools include reactor power monitoring and temperature measurement. Radiation effects will generate luminescence in ceramic materials. Figure 4 shows radiation-induced luminescence (radio-luminescence) in fused silica, the material of which an optical fibre is composed. Strong radio-luminescence peaks are observed at 450 nm and at 1 280 nm, as is so-called Cerenkov radiation in the wavelength range of 600-1 800 nm. The intensity of the radio-luminescence peak at 1 270 nm is plotted as a function of the reactor power in Figure 5, which shows linear dependence on the reactor power. The intensity of the Cerenkov radiation also showed good linear dependence on the reactor power. At present, it is assumed that the intensity of the radio-luminescence at 1 270 nm and that of Cerenkov radiation would be proportional to the electronic excitation dose rate, namely, the gamma-ray dose rate in the reactor. In general, the gamma-ray dose rate is proportional to the local nuclear power in the fission reactor. Thus, a local nuclear power could be monitored with fine optical fibres in the HTGRs.

Figure 4. Radiation-induced luminescence (radio-luminescence)
from fused silica core optical fibre irradiated in JMTR

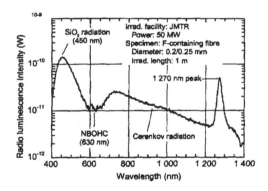

Figure 5. Intensity of Cerenkov radiation observed in a fused
silica core optical fibre as a function of the JMTR nuclear power

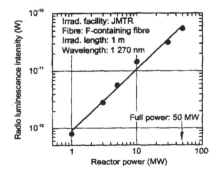

Through radiation-resistant optical fibres, thermo-luminescence and black-body radiation could be observed. Figure 6 shows thermo-luminescence and the black-body radiation from a heated fused silica core of the optical fibres. Figure 7 shows their intensities as a function of temperature with those from alumina (Al_2O_3). As expected, intensities are inversely proportional to the absolute temperatures. Temperatures of alumina were monitored in the JMTR as shown in Figure 8, by a conventional thermocouple of the n-type and by the optical fibre using the relationship shown in Figure 7 [13]. The temperature estimated by optical signals follows that measured by the thermocouples very well over the course of the JMTR operation at about 800°C. The total neutron fluence is more than 1×10^{24} n/m^2, far larger than that expected for the HTGRs over its 10-year operation.

Figure 6. Thermo-luminescence from a fused silica core optical fibre

Figure 7. Intensity of thermo-luminescence as a function of temperature

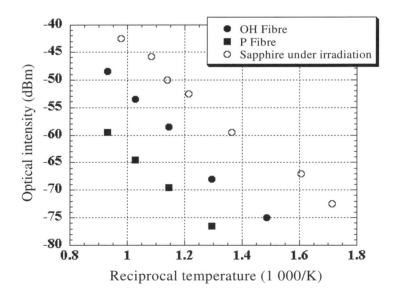

Figure 8. Example of temperature monitoring by fused silica core optical fibre under JMTR high-temperature irradiation

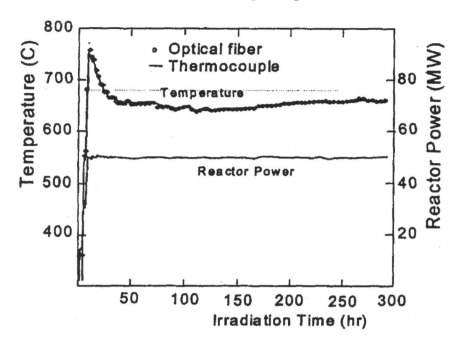

Summary

The present status of the development of optical diagnostics for application in HTGRs in Japan is briefly reviewed. The results strongly demonstrate that some of recently developed radiation-resistant optical fibres could be applied in core regions of the HTGRs for optical diagnostics.

REFERENCES

[1]   Proceedings of the First Information Exchange Meeting on Basic Studies on High-temperature Engineering, 27-29 September 1999, Paris, OECD Proceedings, OECD/NEA, Paris (2000).

[2]   The HTTR Utilisation Research Committee, "Results and Future Plans for the Innovative Basic Research on High-temperature Engineering", JAERI-Review 2001-016, JAERI, Oarai, Japan (2000).

[3]   M. Gowing, "Independence and Deterrence", The Macmillan Press Ltd., London, UK (1974).

[4]   R. Watanabe, *et al.*, "Final Report on Research and Development of Direct Steel Making Utilizing High-temperature Reducive Gas from High-temperature Gas-cooled Reactor", National Research Institute for Metals, Science and Technology Agency, Tokyo, Japan (1981).

[5]   T. Tanabe, Y. Sakai, T. Shikama, M. Fujitsuka, H. Yoshida, R. Watanabe, *Nuclear Technology*, 66, 260 (1984).

[6]   T. Tanabe, T. Shikama, Y. Sakai, M. Fujitsuka, H. Araki, H. Yoshida, R. Watanabe, *Tetsu-to-Hagane*, 69,103 (1983).

[7]   R. Watanabe, Y. Chiba, T. Kokonoe, *Tetsu-to-Hagane*, 61, 2405 (1975).

[8]   G.P. Pells, *Journal of Nuclear Materials*, 155-157, 67 (1988).

[9]   T. Shikama and S.J. Zinkle, *Philosophical Magazine*, B81, 75 (2001).

[10] T. Shikama, T. Kakuta, M. Narui, M. Ishihara, T. Sagawa, T. Arai, in Proceedings of an OECD/NEA Workshop on High-temperature Engineering Research Facilities and Experiments, 12-14 November 1997, Petten, the Netherlands, ECN-R-98-005 (NEA/NSC/DOC(98)4), OECD/NEA, Paris (1998).

[11] T. Shikama, T. Kakuta, N. Shamoto, M. Narui, T. Sagawa, *Fusion Engineering and Design*, 51-52, 179 (2000).

[12] T. Kakuta, T. Shikama, H. Hayami, presented at the 10th International Conference on Fusion Reactor Materials, Baden-Baden, Gemany, October 2001.

[13] F. Jensen, T. Kakuta, T. Shikama, T. Sagawa, M. Narui, M. Nakazawa, *Fusion Engineering and Design*, 41A, 449 (1998).

MEASUREMENT METHOD OF IN-CORE NEUTRON AND GAMMA-RAY DISTRIBUTIONS WITH SCINTILLATOR OPTICAL FIBRE DETECTOR AND SELF-POWERED DETECTOR

Chizuo Mori[1], Akira Uritani[2], Tsunemi Kakuta[3], Masaki Katagiri[3], Tateo Shikama[4], Masahara Nakazawa[5], Tetsuo Iguchi[6], Jun Kawarabayashi[6], Itsuro Kimura[7], Hisao Kobayashi[8], Shuhei Hayashi[8]
[1]Aichi Institute of Technology (Yachigusa 1247, Yagusa, Toyota 470-0392, Japan),
[2]Electrotechnical Laboratory, [3]Japan Atomic Energy Research Institute,
[4]Tohoku University, [5]University of Tokyo, [6]Nagoya University,
[7]Institute for Nuclear Safety Systems, [8]Rikkyo University

Abstract

For the real time measurement of thermal neutron flux and gamma-ray intensity distributions, two methods were developed. One is the method composed of a scintillator optical fibre with a scanning driver and the other is self-powered detector with the driver.

The optical fibre method entails a scintillation powder of ruby, $Al_2O_3(Cr_2O_3)$, being attached to the tip of a fibre and the other end of the fibre connected to a photomultiplier. The current of the photomultiplier anode was measured. When the scintillator was mixed with ^6LiF powder, the neutron flux was measured, and without ^6LiF, gamma-ray intensity was measured. The position distribution was obtained by scanning the position with the fibre through the fibre driver. This method, however, can only be applied at temperatures up to about 100°C. Other scintillators are now under investigation.

For the self-powered detector, the insulator between the central needle electrode, i.e. platinum and the outer cylindrical cathode i.e. Inconel (which can be used at high temperatures) is important. A pure aluminium oxide insulator could be used up to about 400°C, and a quartz insulator was usable up to about 500°C.

Introduction

About 10 years ago, a real time measurement method for neutron flux and gamma-ray intensity distributions was developed [1,2]. In this method, a scintillation powder of ZnS(Ag) was attached to the tip of a fibre and the other end of the fibre was connected to a photomultiplier. Since the thermal neutron flux was less than 10^8 n/cm²s in a critical assembly, the number of pulses from the photomultiplier was counted. When the scintillator is mixed with ^{6}LiF powder, the neutron flux was measured, and without ^{6}LiF, gamma-ray intensity was measured. The position distribution was obtained by scanning the fibre with a fibre driver. With this new method, neutron flux distribution and gamma-ray intensity distribution in a reactor core were obtained over 1 meter in about 10 minutes with a position resolution of about 1mm. This method is very useful for the characterisation of critical assemblies and research reactors.

However, in a usual reactor core with a neutron flux of around $10^{12~13}$ n/cm²s, it was not able to use such detectors, because an uncountably huge number of pulses appeared, and also because the ZnS(Ag) scintillator deteriorated due to radiation damage. We therefore used a ruby, aluminium oxide $Al_2O_3(Cr_2O_3)$ scintillator called "Desmarquest" which is highly resistant to radiation damage, and we measured the electric current intensity from the photomultiplier anode, not the number of pulses.

For the self-powered detector (SPD), we are now trying to find a better insulator, i.e. platinum, between the central needle electrode and the outer Inconel cylindrical cathode, which can be used at high temperatures. A pure aluminium oxide insulator could be used up to about 400°C. A quartz insulator was able to be used up to about 500°C. Other materials are currently under investigation.

Instrument

Measurement system

Figure 1 shows the measurement system. The scintillator fibre detector was inserted into one of the irradiation holes in the reactor core with a length of 550 mm of the TRIGA II of The Institute for Atomic Energy, Rikkyo University or taken out of the hole with a fibre driver which is automatically controlled by a personal computer. The optical signal through the fibre is introduced to a photomultiplier after the transmission of a band pass optical filter. Photomultiplier output current was measured with a nano-ammeter.

Figure 1. Measurement system

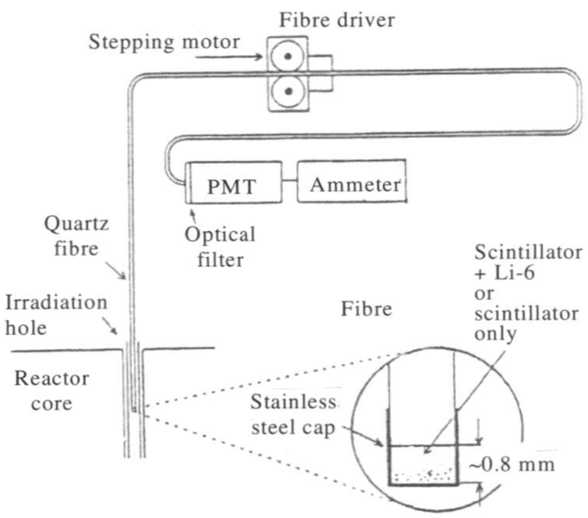

Detectors

Fibre detector

Although a recently developed quartz fibre [3] can be used up to 1 000°C, we tentatively used an ordinary large diameter core quartz fibre, SC•600/750 (Fujikura Co. Ltd.), with a core diameter of 600 µm and an outer diameter of 1.5 mm. The bottom of a stainless steel cap was filled with the powder of scintillator AF995R $Al_2O_3(Cr_2O_3)$ (Desmarquest Co. Ltd.) and then the fibre was inserted into the cap. The scintillator powder was mixed with ^6LiF powder for thermal neutron measurement and with silicon nitride powder for gamma-ray measurement with a ratio of 2/1 for scintillator/chemicals.

To distinguish the scintillation light component with longer wavelength of 690 nm from the Cherencov light component with shorter wavelength, a band pass interference optical filter, AJ46154 (Edmond Scientific Japan Co., Ltd.), was used. With this new method, neutron flux distribution and gamma-ray intensity distribution in a reactor core were obtained over 1 meter in about 10 minutes with a position resolution of about 1 mm. This method is very useful for the characterisation of ordinary research reactors. However, it is applicable only up to about 150°C, because the light emission property of the scintillator deteriorates at high temperatures, as discussed below.

Self-powered detector (SPD)

Figure 2 shows the SPD configuration for gamma-ray measurement. The detector is composed of a central wire electrode with a diameter of 0.7 mm and a length of 70 mm and an outer electrode of Inconel 600 (Ni, Cr, Fe, Mn alloy) with an inner diameter of 2.5 mm and an outer diameter of 4.0 mm. Between the two electrodes, a quartz pipe with a wall thickness of about 0.22 mm is inserted as an electrical insulator. As a central wire electrode, a platinum wire was used as a detector and an Inconel wire was used as another detector for comparison.

Figure 2. Schematic view of self-powered detector

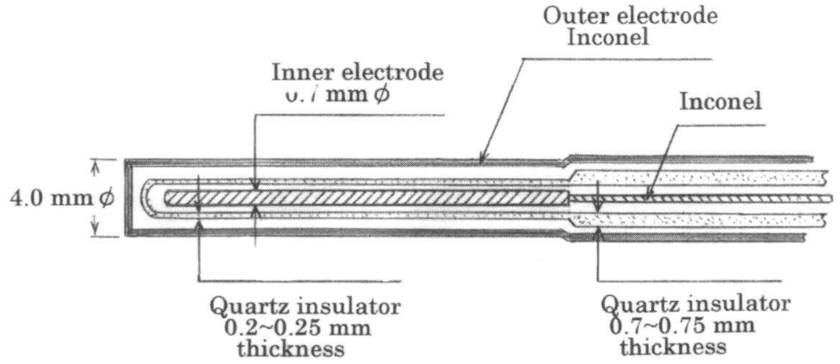

Experimental results

Fibre detector

Figure 3 shows a wavelength spectrum obtained with a ruby scintillator mixed with ^6LiF in the JRR4 reactor, (a) without optical filter showing the light, perhaps Cherencov light, from the fibre and (b) with the band pass filter showing only the light from the scintillator itself.

Figure 3. Wavelength spectra of the light from detector
(a) without optical filter and (b) with optical filter

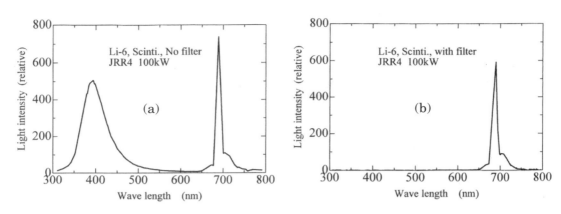

Figure 4 shows the temperature dependence of light intensity. The intensity of 690 nm light from the scintillator drastically decreases with temperature. It disappears above 300°C. This scintillator is difficult to use at temperatures above 100°C. We are now trying to use other scintillators, including a gas scintillator. The intensity of 380 nm light in the Cherencov region does not change much at all.

Figure 4. Temperature dependence of light intensities of scintillation and Cherencov

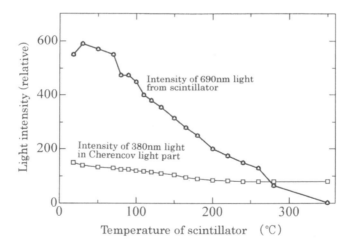

Figure 5 shows the geometrical distributions of neutron flux and gamma rays obtained by moving the fibre detector with the band pass filter by using fibre driver with the scanning speed of 8 min/m in the core of the research reactor TRIGA Mark II of Rikkyo University. Curves (1), (2) and (3) were obtained by the detector with a scintillator mixed with ^{6}LiF, by the detector with only a scintillator and by the detector without scintillator, respectively. The difference between values (1) and (2) was determined to obtain the distribution of thermal neutron flux as shown by the curve (1)-(2). Figure 6 shows the comparison between the neutron flux distribution, the curve (1)-(2), obtained with the fibre detector and that with gold foil activation method. The two are in good agreement. However, the activation method requires four or five hours to obtain the distribution with a position resolution of about 10 mm, whereas the fibre method only requires 10 minutes or so with a position resolution of about 1 mm.

Figure 5. Position distribution of radiation intensities obtained with fibre detectors

(1) Ruby scintillator powder mixed with ^{6}LiF powder, (2) Scintillator powder only,
(1)-(2) The difference, (3) Fibre only without scintillator

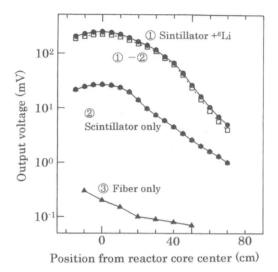

Figure 6. Comparison of position distribution of thermal neutrons obtained with fibre detector and that with Au foil activation method

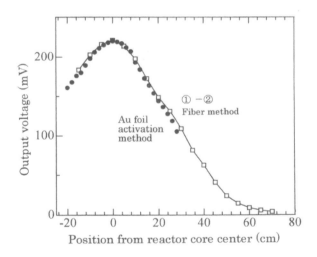

Self-powered detector (SPD)

We tried out two kinds of insulators, a pure alumina (99.5%Al_2O_3) tube insulator with a wall thickness of 0.2 mm and a quartz tube with a wall thickness of 02~0.25 mm. SPDs were irradiated with 5×10^2 Gy/h (for water) of ^{60}Co gamma rays in an electric furnace. With the quartz tube (as shown in Figure 7) the difference between the detector with a central electrode of Pt and that with an Inconel electrode was kept almost constant up to about 500°C. However, with the alumina tube, a rather larger current appeared at temperatures above about 400°C. These results should be confirmed by additional experimentation. Pure quartz, such as optical fibres, appears to be an attractive option.

**Figure 7. Temperature dependence of self-powered detectors
with a Pt central electrode and with an Inconel central electrode**

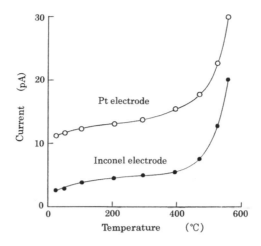

Conclusions

For the measurement of geometrical distribution of neutron flux and gamma-ray intensity in reactor cores, the method of scanning the core with an optical fibre with a scintillator (ruby) attached on the tip was useful. Since there is Cherencov light emission in the fibre, an optical band pass filter was used to pass only the scintillation light. When the scintillation powder was mixed with ^{6}LiF powder, it was possible to measure thermal neutron flux distribution over 1 meter in only about 10 minutes. Without ^{6}LiF, gamma-ray intensity distribution was measured.

However, the light emission of the ruby scintillator decreases at high temperatures, and thus this scintillator can be substantially used only up to about 100°C, rendering this method applicable for research reactors and critical assemblies. We are now trying other scintillators, including gas scintillators.

Self-powered detectors can be used at much higher temperatures. A quartz insulator was used up to the temperature of about 500°C, whereas pure alumina was applicable below about 400°C.

REFERENCES

[1]   C. Mori, T. Osada, K. Yanagida, *et al.*, *J. Nucl. Sci. Technol.*, 31, p. 248 (1994).

[2]   C. Mori, Y. Mito, K. Kageyama, *et al.* Proc. 9th Int. Symp. on Reactor Dosimetry, Prague, 2-6 Sept., 1996,World Scientific, p.159.

[3]   T. Shikama, T. Kakuta, M. Narui, *et al.*, Proc. of the OECD/NEA Workshop held in Petten, the Netherlands, 12-14, November 1997, ECN-R-98-005 (NEA/NSC/DOC(98)4), pp. 125-143.

DEVELOPMENT OF IN-CORE TEST CAPABILITIES
FOR MATERIAL CHARACTERISATION

C. Vitanza
OECD Halden Reactor Project, Norway[1]

Abstract

This paper presents a concise review of some of the test capabilities developed at the OECD Halden Reactor Project for the characterisation of fuel and materials properties under irradiation. Insights on some of the instrumentation utilised in these tests are also presented. The paper addresses some of the irradiation techniques used for gas reactor investigations, notably in the field of AGR fuel and materials.

---

[1] Currently at the OECD Nuclear Energy Agency.

Introduction

The Halden reactor is a test station where a range of investigations is carried out on nuclear materials under a wide variety of conditions. Different types of reactor environments can be simulated covering BWR, PWR, VVER, CANDU and AGR needs.

The key of the Halden irradiation is the ability to monitor on-line the relevant performance aspects. Each experimental rig is, in fact, a compact agglomerate of technology developed to extract data from the test section during the test performance. Most of these instruments are required to work for a long time, from months to many years, under demanding radiation, temperature and pressure conditions.

This paper intends to provide only a brief summary of some of the techniques in use at Halden, addressing in particular some examples of gas reactor applications.

Instrumentation for LWR fuel testing

Investigations of fuel performance under steady state and transient operation conditions have constituted a major part of the experimental work carried out in the Halden reactor since its start-up in 1959. The in-core studies were supported by the development and perfection of instrumentation and experimental rig and loop systems where reactor fuels and materials can be tested under PWR and BWR conditions. Fundamental knowledge and contributions to the understanding of LWR fuel behaviour in different situations could thus be provided in support of a safe and economic nuclear power generation.

Fuel testing at the Halden Reactor Project has for a number of years focused on implications of extended burn-up operation schemes aimed at an improved fuel cycle economy. The experimental programmes are therefore set up to identify long-term property changes with an impact on performance and safety. While PIE ascertains the state existing at the end of irradiation, in-core instrumentation provides a full description of performance history, cross-correlation between performance parameters, on-line monitoring of the status of the test and a direct comparison of different fuels and materials. Trends developing over several years, slow changes occurring on a scale of days or weeks, and transients from seconds to some hours can be monitored. The data generated in the fuels testing programmes originate from in-pile sensors, which allow to assess:

- Fuel centre temperature and thus thermal property changes as function of burn-up.

- Fission gas release as function of power, operational mode and burn-up.

- Fuel swelling as affected by solid and gaseous fission products.

- Pellet-cladding interaction manifested by axial and diametral deformations.

The type of instruments that are put in place for fuel on-line performance monitoring consist mainly of:

- *Refractory metal thermocouples, W-Re type*. These thermocouples can measure fuel temperatures up to 2 000°C-2 200°C. However, the lifetime at very high temperature can be rather limited. Long-term operation (years) can be ensured at temperatures below ~1 400°C. The reason for the limited time endurance in high-temperature fuel environments is believed to be due to chemical attack of fission products. Refractory metal thermocouples normally have a refractory metal sheath and thus can not be directly welded on stainless steel structures. Techniques have been developed at Halden for refractory metal to stainless steel transition.

168

- *Non-refractory metal thermocouples.* For operation up to 1 300°C nicrosil-nisil thermocouples are sometimes used at Halden for fuel temperature measurements when operation allows. For other applications, such as coolant temperature and cladding temperature measurements up to 1 100°C, chromel-alumel thermocouples are normally used.

- *Differential transformers.* These instruments are fundamental for the Halden fuel tests, as they are used for fission gas release, fuel swelling, cladding axial deformation and cladding radial deformation measurements. They are also used for a range of additional applications in the field of materials testing, e.g. creep. A schematic of this type of instrument is shown in Figure 1. In this case the application is to measure cladding axial strain. The Halden operating experience for this type of instrument is ~350°C and it is not believed that it can function at a significantly higher temperature. However, high-temperature phenomena can be detected by this sensor, as long as the sensor itself is kept below ~350°C.

Figure 1. Cladding extensometer (EC)

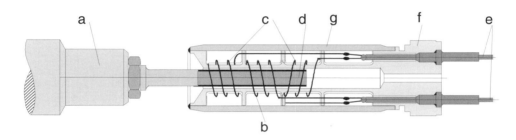

a) Test rod end plug assembly   e) Twin-lead signal cables
b) Primary coil   f) Body
c) Secondary coils   g) Housing
d) Ferritic core

Obviously, there are many other types of sensors, but the above are the most commonly used ones at Halden. The irradiation of instrumented fuel rods is carried out in specialised rigs according to test objectives, e.g. long-term base irradiation, ramps or transient testing. In addition to fuel instrumentation, some rods in experimental rigs have gas lines attached to their end plugs. This allows the exchange of fuel rod fill gas during operation and makes it possible to determine gas transport properties as well as the gap thermal resistance and its influence on fuel temperatures. It is also possible to analyse swept out fission products for assessment of fuel microstructural changes and fission gas release. This is an important experimental technique for the high burn-up programmes currently being executed and defined for the period 2000-2002.

The examples of experimental work and results selected for this paper relate to high burn-up fuel performance with respect to thermal behaviour, fission gas release, PCMI and cladding creep. They can be used for fuel behaviour model development and verification as well as in safety analyses.

Example of fuel temperature and fission gas release data

An example of how Halden measurements are used is given in Figure 2, which shows combined fuel temperature and fission gas release pressure of a MOX fuel rod. The measurement temperature was up to ~1 200°C for most of the operating period, except for the interval 13-20 MWd/kg, where the

Figure 2. Measured and normalised rod pressure histories

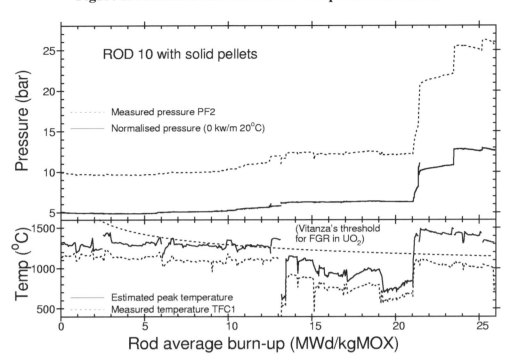

temperature was lower. Since the fuel centre thermocouple requires a 1 to 1.5 mm hole along the fuel pellet axis, the measured temperature is somewhat lower than the actual fuel centre temperature away from the thermocouple. The latter temperature can however be reasonably well estimated based on the measured temperature, as shown in Figure 2.

The rod pressure measurements show that there is a pronounced increase of pressure at a burn-up of ~20 MWd/kg, when the actual fuel temperature exceeds the value of 1 300°C. The operating pressure rises quickly from 10 to 20 bar and, in subsequent steps, up to 27 bar. This gives a clear indication that substantial fission gas is being released from the fuel into the rod free volume. Such coupling between fuel temperature and fission gas release measurements is fundamental for understanding the mechanisms of fuel behaviour and for developing realistic fuel performance models.

Example of cladding creep reversal (and rod overpressure) data

Fission gas release at high burn-up may result in the rod pressure exceeding the coolant pressure. A creep-out of the cladding may then open the fuel-cladding gap and lead to increasing fuel temperatures and further, increased fission gas release. In order to assess the consequences for fuel integrity, the creep characteristics of cladding material must be known.

Cladding creep data at high fluence in the presence of neutron flux were produced in the Halden reactor under representative LWR conditions in a diameter measurement rig. A gas line connected to the cladding tube enabled to change the rod inner pressure. In this way, several stress reversals were produced, and the cladding creep was measured in-pile by a diameter gauge with a relative precision of ±2 μm. Unlike PIE which only provides a single point, the results obtained show in a unique manner the development from primary to secondary creep.

The reaction of pre-irradiated BWR cladding material (fluence 6×10^{21} n/cm^2, E > 1 MeV) to stress reversals is shown in Figure 3. The rod diameter changed in a stepwise manner whenever the applied stress was changed. Figure 3 also shows that immediately following each stress change, although sometimes difficult to detect, primary creep occurred. These data have provided insight in the reaction of cladding material to changing stress conditions and the relation of primary and secondary creep to stress change and stress level. They are used by Halden Project participants for modelling the creep-out behaviour at high burn-up as consequence of rod overpressure. A question posed before conducting the test was whether primary creep would recur with every stress change. This was answered in a direct manner and the result has a bearing on the modelling of cladding failure induced by power changes.

Figure 3. Creep response of cladding tube subjected to compresssive and tensile stress. The recurrence of primary creep can be noted.

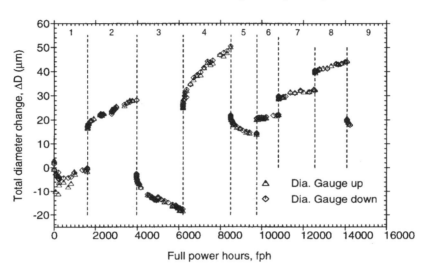

A complementary test, in which a pre-irradiated PWR rod equipped with a fuel centreline thermocouple is subjected to rod overpressure, was executed with the aim to determine the pressure beyond which the fuel temperature will increase due to clad creep-out. The fuel (burn-up 55 MWd/kg) was re-instrumented with a fuel thermocouple and a cladding elongation detector. The rod overpressure was controlled with a high-pressure gas supply system connected to the fuel rod with a gas line.

Re-fabrication and re-instrumentation of test specimen

A range of practical applications requires that materials retrieved form power reactors be re-irradiated in a test reactor. This occurs in particular when fuel properties are to be measured at high burn-up and at conditions that cannot be attained in power reactors. Practical examples are:

- Burn-up extension beyond level allowed in power reactors.

- Power ramps.

- Operations at non-standard water chemistry conditions.

- Special conditions that can only be simulated in a test reactor (e.g. controlled rod overpressure).

- Safety-related tests such as dry-out tests.

Halden has specialised in re-fabrication techniques, which have enabled to perform a variety of tests on representative fuel specimens. A schematic of how this operation is carried out is shown in Figure 4. The re-fabrication (and re-instrumentation) technique has been fundamental for Halden tests in the last decade, when the trend towards high burn-up testing has become important. A similar trend is being experienced in the materials area, where the demand for testing of representative, high-dose materials retrieved from power reactors is increasing.

Figure 4. Re-instrumentation/re-irradiation of fuel rods

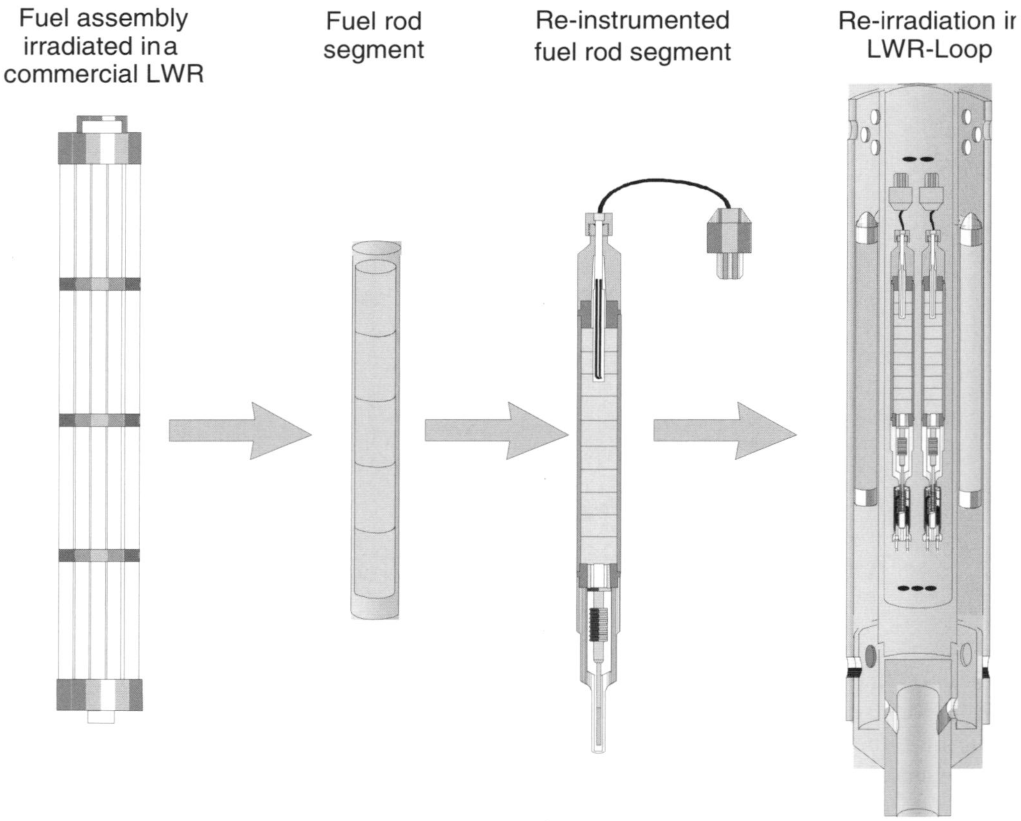

Applications for gas reactors

The Halden instrumented tests have addressed water reactor issues in most cases. However, efforts have been made to broaden the range of applicability to gas reactors. Two such applications are described in the following.

Testing of AGR fuel

During the last decade, tests have been carried out at Halden on AGR fuel. The main objective of such tests was to assess the performance under power uprating conditions, but power oscillation tests were also carried out. Obviously, AGR fuel temperatures are much higher than in water reactor fuel and the question was concerned how an AGR fuel test could be successfully conducted in the Halden reactor. This operates at 240°C coolant temperature, whereas the requirements were that the AGR fuel cladding temperature should be ~700°C at 20 kW/m and 1 000-1 100°C at 50 kW/m.

The required AGR cladding temperature conditions were achieved without utilising expensive gas loop solutions. As shown in Figure 5, the AGR fuel rod is encapsulated into a zircaloy tube such that a controlled gap exists between the AGR fuel rod and the outer tube. The gap site is controlled by means of special spacers placed in a certain manner between the pins of the AGR rod and the filler gas composition is chosen such that the AGR cladding temperature versus power matches the one in actual AGR reactors.

Figure 5. Ramp rig providing data on test temperature and elongation

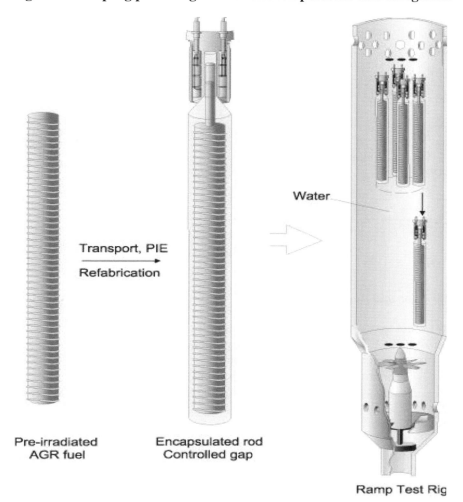

The encapsulation design and procedure were conceptually simple but complex to realise, especially considering that fuel pre-irradiated in a commercial plant had to be tested. This means that the spacer attachment, the encapsulation and the capsule sealing had to be carried out in hot cells. In addition, as shown in Figure 5, differential transformers were attached to the capsule in order to monitor the AGR fuel rod elongation during ramp.

The above tests were preceded by a qualification campaign to qualify the test method. As part of this effort, Figure 6 shows the measured cladding temperatures in two rigs that were used in the early phase of the AGR irradiations. As one can observe, the measured cladding temperatures were exactly those set in the test requirements (e.g. ~700°C and ~1 000-1 100°C at 20 and 50 kW/m).

Figure 6. AGR fuel and clad temperatures in the Halden ramp rig. Clad temperatures were measured, fuel temperatures were inferred from power and clad temperature measurements.

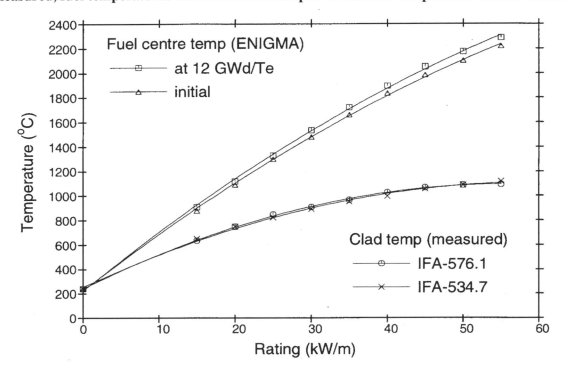

Testing of gas reactor materials

In addition to fuel tests, gas reactor materials are also being irradiated in the Halden reactor. As an example, Figure 7 shows a capsule used for AGR graphite testing. In this case, pre-irradiated samples extracted from power reactors are assembled into capsules where both the gas coolant chemistry (i.e. composition with additives and/or impurities) and the specimens temperature can be varied. Dimensional measurements as well as temperature measurements are carried out on-line, but they are not shown in Figure 7. Rigs with 4 to 6 capsules of the type shown in Figure 7 have been designed and are in service at Halden.

These types of measurements are used to anticipate the behaviour of structural materials and are thus essential for plant life extension assessments. They are also very useful for identifying means to improve the materials long-term performance, for instance by means of coolant chemistry measures. This applies to gas reactor as well as to water reactors, as presented in the next section.

LWR materials testing

In LWR, stainless steel structural materials can crack under the combined effect of stress, chemical environment and radiation. Tests in this field were initiated at Halden a decade ago, aiming primarily at providing basic data on crack propagation and at determining water chemistry conditions that can reduce the stress corrosion cracking rate. In order to achieve these goals, it was essential to put in place reliable means to measure crack propagation with great precision and sensitivity.

**Figure 7. Schematic of the capsules to test HGR graphite materials
in the Halden reactor (specimen temperature measurements by
means of thermocouples not shown in the figure were also performed)**

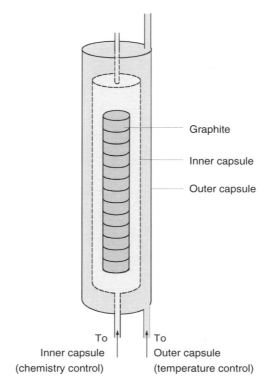

Graphite

Inner capsule

Outer capsule

To
Inner capsule
(chemistry control)

To
Outer capsule
(temperature control)

The most important steps in this development were the following:

- Development of an in-pile potential drop measurement method for crack growth determination. The required sensitivity was of the order of a 1/100 of a millimetre.

- Development of a system for varying the specimens loading on-line. This considerably enhances the test efficiency, in that the range of stress intensity covered by the test can be widened as wished.

- Development of loops able to operate with flexible water chemistry.

- Development of techniques for the utilisation of pre-irradiated specimens such that prototypical material at high dose can be tested at Halden.

An example of test rig for such type of studies is given in Figure 8. Details of the on-line crack propagation monitoring are given in Figure 9.

In addition to the above, other measurements such as in-reactor crack initiation and creep monitoring of stainless steel materials are also carried out at Halden.

Figure 8. Example of test rig for LWR materials test studies (stress corrosion cracking at Halden)

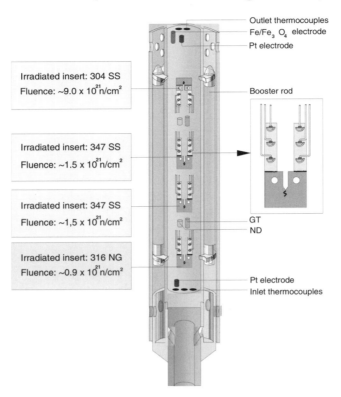

Figure 9. Principle layout of the in-core type CT specimen

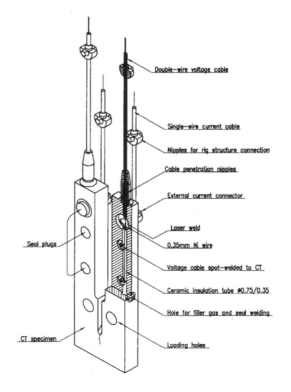

APPLICATION OF WORK FUNCTION MEASURING TECHNIQUE TO MONITORING/ CHARACTERISATION OF MATERIAL SURFACES UNDER IRRADIATION

G-N. Luo, T. Terai, M. Yamawaki
Department of Quantum Engineering and Systems Science
The University of Tokyo

K. Yamaguchi
Department of Materials Science, Tokai Research Establishment
Japan Atomic Energy Research Institute

Abstract

Experimental devices have been developed for examining the work function (WF) change – by means of Kelvin probe (KP) – of metallic and ceramic materials due to ion irradiation in low energy (500 eV) or high energy (MeV) ranges. The charging effect on the performance of the KP was efficiently reduced using appropriate shielding. An error deduction method has been suggested to effectively eliminate the influence due to charging by introducing a reference sample. Polycrystalline Ni samples, each of 99.95% purity, were used in the research. The experiments were performed using He^+ and H^+ ion beams of 1 MeV, 2×10^{16} ions/m^2/s under a pressure of 1×10^{-4} Pa, and a He^+ beam of 500 eV, 2×10^{17} ions/m^2/s under a pressure of about 1×10^{-2} Pa. The results indicated that the irradiation of 500 eV He^+ resulted in a WF decrease, then an increase until saturation, while 1 MeV ions only induced a WF decrease, then saturation. A surface model of a loosely bound adsorbed layer plus a native oxide layer on metals is presented to explain the observed phenomena. The nuclear stopping is responsible for the results in the case of 500 eV He^+ irradiation that is powerful enough to sputter away the whole overlayer from the bulk surface. In the MeV case, the electronic stopping plays a decisive role, which allows the topmost adsorbed layer to be partially removed by He^+ and H^+ of 1 MeV. The application of this technique to monitoring/characterisation of material surfaces in the field of nuclear engineering is to be discussed.

Introduction

Among the engineering issues in realising fusion energy are plasma-facing materials (PFMs) that significantly affect the fraction of the plasma particle flux that is trapped at the surface or in the bulk of materials. This in turn has a strong influence on the density of the edge plasma since recycled particles fuel the plasma [1]. Nickel, capable of trapping more helium than hydrogen, is one of the potential materials to be used to trap the helium ash produced by D-T reactions, the use of which is helpful to maintain the optimum fuel conditions within a fusion reactor [2]. These helium selective pumping materials may be placed near the divertor or the limiter in a tokamak. It will be subjected to heavy radiation from the neutron and various ions existing in tokamaks. Since surface properties significantly influence particle recycling on the surface of the materials, extensive laboratory experiments as well as in-reactor tests are necessary.

In the field of fusion application, the surface of the in-vessel metallic components is by no means clean from beginning to the end, as no strict cleaning procedures are used before the normal operation of the plasma. Even if one wants to do so, it seems to be impossible due to the prohibitive size of the area to be treated in a tokamak. During normal operation, the deposition of the materials eroded elsewhere due to physical or chemical sputtering represents a continuous contamination source. It is of practical importance to study the effects of ion bombardment on the native surface or the surface with possible deposits from within the tokamak. It is these processes that significantly affect the particle recycling on the surface of the materials.

The present authors' group has developed two different types of Kelvin probes (KP) in their lab. One is a "high-temperature" KP (HTKP), constructed with the help of Professor Nowotny, University of New South Wales, Australia. This probe enables contact potential difference (CPD) measurements under a controlled atmosphere around one atm and at elevated temperatures up to 1 000 K. The HTKP has been employed for years in studies on the effect of sweep gases on the surface defects generated on lithium ceramics such as Li_4SiO_4 [3], Li_2TiO_3 [4], Li_2ZrO_3 [5], Li_2O and $LiAlO_2$ [6,7], all potential materials for tritium breeding in a thermonuclear fusion reactor. The other probe has been designed and constructed recently for research under high vacuum conditions, particularly on the influence of ion irradiation on the WF change of metallic materials, with a view to clarifying the interactions between the plasma and the inner surface of the main vessel of the fusion reactor [8,9].

In this paper, construction of the experimental devices is to be described first, including the set-up connecting to one of the beam lines of the van de Graff accelerator of the Research Centre for Nuclear Science and Technology, the University of Tokyo, for conducting studies with an ion beam accelerated to 1 MeV, and the device in the present author's lab that delivers a low energy (<1 keV) ion beam. As the major analyser, a KP is to be used to evaluate the WF change of polycrystalline nickel over the course of the processes. A detrimental charging effect on the normal operation of the KP has been observed. Effective solutions have been taken in the measurements, which ensures the correct acquisition of data. The surface change can then be inferred from all available information. The developed system and method may serve as a powerful tool in *in situ* monitoring of surface changes under ion irradiation.

Construction of devices

Design

The set-up connecting to the van de Graff accelerator is shown in Figure 1. It is a vacuum chamber connected to the beam line of the van de Graff accelerator and evacuated by a pumping system composed of a turbo molecular pump and a rotary pump. Within the chamber, there exists a

Figure 1. Scheme of the device used for MeV ion irradiation

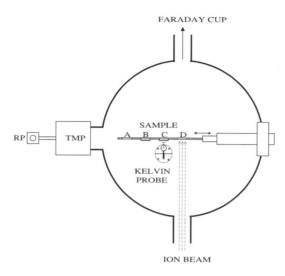

sample holder with four seats for samples. Usually seat D is always kept open to allow the beam from the accelerator to pass though and to reach the subsequent chamber for flux monitoring by a Faraday cup. The holder can be moved by a servomotor in a translation mode to make the samples align to the beam line for irradiation, or the KP for measurement in turn. The mechanism of a KP is based on the formation of a vibrating parallel capacitor between the probe tip and the specimen to be tested [10,11]. Thus the measurement and the irradiation cannot be performed at the same time, otherwise the irradiation would damage the probe and the sample would also be blocked by the probe.

The device used for low energy ion irradiation is a newly established device in the present authors' lab [8], as shown in Figure 2. The new device has been equipped with a KP, and a beam source composed of a dc-arc discharge section, and a beam extraction and focusing section. The ion source is capable of producing both positive ion beam and negative electron beam up to 660 eV. The beam flux is measured with a Faraday cup. The probe can be rotated and moved for fine adjustment of the separation between the probe and the sample. The temperature of the sample surface can be checked occasionally using a rotatable thermocouple. A resistive heater is used for heating or annealing the sample. A quadruple mass spectrometer (QMS) is used for gas analysis. A blank port is prepared for an Auger electron spectrometer/low-energy electron diffraction (AES/LEED) analyser that is intended to be installed in the future. The pumping system consists of a 150 l/s turbo-molecular pump and a 5 l/s rotary pump, and achievable base pressure is nearly 1×10^{-6} Pa after 24 hours bake-out at about 100°C.

Charging effect

The Kelvin probe measures contact potential difference (CPD) between the probe tip and sample, which can be expressed as:

$$CPD = \left(\Phi_s - \Phi_p\right)/e \qquad (1)$$

where *CPD* is in the units of volts, Φ_s and Φ_p, the work functions, of the tip and the sample, respectively, all in the units of eV; and *e* the elementary charge. The probe currently used to measure the CPD was named after Lord Kelvin, who arranged the tip and the sample as two plates of a parallel

Figure 2. Scheme of the device used for low energy

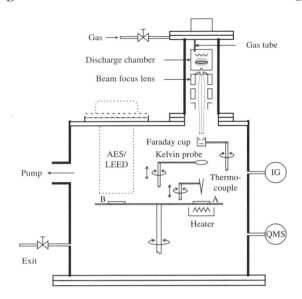

plate capacitor that were in contact via an external circuit [12]. The set-up was greatly improved by Zisman [13] by introducing vibration to one of the electrodes of the capacitor, which made it possible to measure the CPD more exactly, quickly and continuously. Modern Kelvin probes are only variants of the one modified by Zisman, with use different means of vibration and electronics.

Figure 3 illustrates the principle of a Kelvin probe, of which (a) indicates the relationship of the Fermi levels and WFs of the probe tip and the sample in contact; in (b) the surface charge is correlated to the CPD through $Q = CPD \cdot C_{KP}$, where C_{KP} is the Kelvin probe capacity. Charge transfer will not stop until the Fermi levels approach the same value; and (c) shows the situation when the backing voltage V_b compensates the CPD, thus no charges exist on both of electrodes, and $V_b + CPD = 0$. In the following context, $-V_b$ will be used to represent the probe output.

Figure 3. Electron energy level diagrams of two different metals (a) without contact, (b) with external electrical contact, and (c) with inclusion of the backing potential V_b. Here E_F and E_0 denote Fermi and vacuum energy level of an electron, respectively.

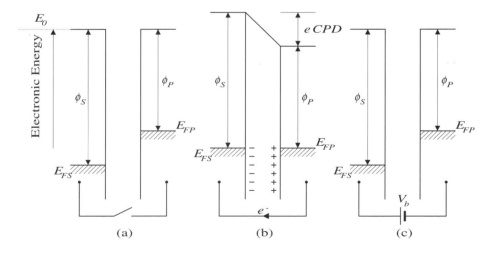

It is well known that the correct measurement of the CPD via the Kelvin probe is vitally dependent on achieving a zero field between the specimen and the probe. Accordingly, one can easily imagine that any external field induced by stray charges in space may damage normal operation of the probe, as was found by the present authors in the course of constructing the devices [14]. Other groups also reported the influence of charges from the heater filament [15] or plasma [16] on the performance of the probe. But their work is rather scattered and unsystematic.

In our work, we observed miscellaneous charging effects [14], e.g. (1) charging on insulating parts due to bakeout process, (2) thermal electrons from the resistive heater, (3) charges from the quadruple mass spectrometer and (4) charges from secondary electrons emitted due to ion or electron irradiation. A model has thus been presented to evaluate the influence of the extra charge-induced electric field on the probe output [14]. Further, direct evidence of such an influence will soon be published [17].

Careful shielding has been used in the experiments, ensuring the correct acquisition of the experimental data via the probe [14]. In the case of the set-up connecting to the van de Graff accelerator, the probe tip and the samples were enclosed with a metallic box, which greatly decreased the incoming secondary electrons and also protected them from the impingement of ions. This measure, along with introduction of a reference sample and adoption of an appropriate method to acquire data and a mathematical post–treatment of the datasets, significantly lowered the detrimental effect to an acceptable level. On the other hand, in the case of new device, the charge sources were carefully enclosed to avoid the escape of the charges into the chamber space. This method also worked quite well. Further measures that may root out any charging effects will soon be tested [18].

Experimental

MeV ion irradiation

The measurement seriously suffered from charging effects at the initial stage of this series of experiments. Reliable results were acquired only after the appropriate shield was put into place to protect the probe and samples, and after application of the mathematic post-treatment on the obtained datasets (see previous section) [14]. The irradiations were carried out intermittently, which means repeated processes of irradiation and measurement. The basic experimental conditions are listed in Table 1. The fluence and time summations for each run are presented in Tables 2 and 3 for He and H ion species, respectively.

Table 1. Basic experimental parameters of MeV ion irradiation

Parameter	Sample	Ion species	Ion energy	Beam flux	Pressure	Temperature
Value	Ni (99.95%)	H and He	1 MeV	2×10^{16} ions/m^2/s	1×10^{-4} Pa	Room T

Table 2. Fluence and irradiation time of MeV H ions (10^{18}/m^2)

Run No.	1	2	3	4	5	6	7	8
Σ (time) (s)	5	25	85	385	2 185	3 985	5 785	7 585
Σ (fluence)	0.1	0.5	1.7	7.7	43.7	79.7	116	152

Table 3. Fluence and irradiation time of MeV He ions ($10^{18}/m^2$)

Run No.	1	2	3	4	5	6
Σ (time) (s)	5	15	45	105	405	2 205
Σ (fluence)	0.1	0.3	0.9	2.1	8.1	44.1

Low energy ion irradiation

Due to the intrinsic configuration of the probe, as mentioned above, the WF measurement could not be made during the irradiation on samples. Once each run of irradiation was completed, the probe had to be rotated over the sample and adjusted to the given distance from the sample. At this point the measurement could be undertaken. The charging effect was also a major obstacle to obtaining reliable data. The shield enclosing the ion beam, along with other measures, greatly decreases the detrimental effect and ensures an acceptable error range in the measurement [18]. The basic conditions and the fluence and irradiation time are given in Tables 4 and 5. The purity of He gas used in the experiments was 99.99995%.

Table 4. Experimental conditions of low energy ion irradiation

Parameter	Sample	Ion species	Ion energy	Beam flux	Base pressure	Working pressure	Temperature
Value	Ni (99.95%)	He	500 eV	2×10^{17} ions/m^2/s	1×10^{-6} Pa	1×10^{-2} Pa (He)	Room T

Table 5. Fluence and irradiation time of low energy He ion ($10^{18}/m^2$)

Run No.	1	2	3	4	5	6
Σ (time) (s)	5	25	85	385	1 285	4 885
Σ (fluence)	1	5	17	77	257	977

For both MeV and low energy cases, the nickel samples from Nilaco Co. were polycrystalline with a purity of 99.95%, and were cleaned using acetone and alcohol prior to transfer into the vacuum chamber. The probe was a commercial product from Besocke Delta Phi GmbH [19]; a schematic drawing is illustrated in Ref. [14]. The probe tip was made of gold mesh with a diameter of 2.5 mm, and the spacing between the probe and the sample was smaller than 1 mm.

Results and discussion

Results

MeV ion irradiation

The work function changes due to the ion irradiations shown in Figure 4, in which "Fluence" signifies the accumulative fluence. It is clearly indicated that the ion irradiations result in a WF decrease coupled with a fluence increase for both of the ion species. The WF change due to H ion irradiation tends to saturate after the fluence reaches about $5 \times 10^{19}/m^2$. Similar behaviour is not clearly observed for the sample under He ion irradiation until a fluence of $5 \times 10^{19}/m^2$. The WF decrease by He ion irradiation is larger than that by H ion irradiation for a given fluence, especially in the low fluence range.

Figure 4. Work function change of Ni due to MeV H and He ion irradiation as a function of irradiation fluence

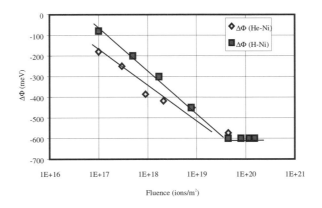

Low energy ion irradiation

The irradiation of 500 eV He$^+$ ions on the nickel sample under the fluence indicated in Table 5 yields results as depicted in Figure 5. The first run decreased the WF by about 220 mV ($\Delta\Phi = 0$ prior to irradiation), and the following runs increased the WF drastically until the fourth run at an accumulated fluence of about 8×10^{19} ions/m^2. The subsequent two runs only induced a slight decrease (saturation?). The process shows an obvious three-stage development of the WF change.

Figure 5. Work function change of Ni due to 500 eV He ion irradiation as a function of irradiation fluence

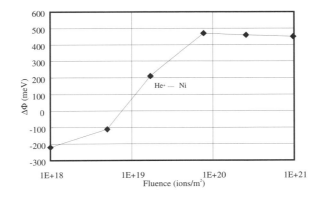

Model of nickel surface

It is well known that there exists a thin oxide layer on the metal surface which naturally forms in air due to the oxidation process which occurs on the surface. The surface of oxides can usually adsorb gas species, as indicated by Henrich and Cox [20]. In the case of interest, NiO on Ni surface, summaries can be found on adsorption of common gas species on the surface. Reactive gas species usually adsorb on the surfaces of NiO as well as Ni in the form of anions or partial transfer of electrons from the surfaces to the adsorbates. A model of a nickel surface (as shown in Figure 6) is thus presented for discussing the experimental results in the following sections. The model advocates a two-layer structure for nickel surface, i.e. oxide layer on the bulk and adsorbed layer on the oxide layer.

Figure 6. Model of nickel surface

Some research groups pointed out that the oxide layer formed on a clean Ni surface is about 1 nm thick [21-23]. No reports are available on the thickness of the adsorbed layer on NiO layer. A thickness of less than 1 monolayer is valid if only chemisorption is taken into consideration.

Since reactive gas species usually adsorb on the surfaces of Ni and NiO in the form of anions or partial transfer of electrons from the surfaces to the adsorbates, the formation of a surface dipole moment towards the bulk can be expected, which leads to a WF increase on the surface according to Ref. [10]:

$$\Delta\phi = \frac{N\theta\mu^*}{\varepsilon_0\left[1+(9/4\pi)\alpha(N\theta)^{3/2}\right]} \qquad (2)$$

where N is the number of substrate atoms per m^2, θ is the fractional coverage of the adsorbate (at $\theta = 1$ the adsorbate density is N), μ^* is the dipole moment of an isolated adsorbate surface moiety in cm (defined as positive if the negative end is toward vacuum). α is the polarisability of the dipole in m^3, and ε_0 is the permittivity of vacuum in SI units; the resulting $\Delta\phi$ is in eV. α is usually treated as a constant. μ^* can be extracted from the slope of the work function change (experimental) as a function of coverage in the limit of zero coverage, and is also a constant.

The desorption/adsorption experiments performed on similar samples [8,17] indicate that $\Delta\phi$ increases with θ monotonically in the present series of experiments, i.e.:

$$d(\Delta\phi)/d\theta > 0 \qquad (3)$$

Discussion

MeV ion irradiation

Calculation using the latest TRIM code (SRIM2000.39 distributed by Ziegler, *et al.* [24]) indicates that in the case of MeV light ions like H^+ and He^+ used in this study, the electronic stopping powers are of the order of 10^2 eV/nm, which is about three orders higher than their nuclear ones. Further, the nuclear sputtering yields are very low, about 10^{-3} level or lower for nickel sample. Therefore, the nuclear sputtering effect is minute, especially at the initial stage where the fluence is low. How did the irradiations change the surface and induce the decrease in WF of hundreds of meV?

Ion irradiation induces temperature rise of the bombarded sample due to energy transfer from the incident ions. A test was carried out to check this effect. The results indicate that an ion beam of the same parameters as indicated in Table 1 led to a temperature increase of less than 10°C within the first

10 seconds. The temperature increase finally saturated at 115°C after 30 minutes irradiation. In view of the results from the test, thermal desorption of the adsorbed species should not be considered the main cause for the decrease in WF in Figure 4, at least in the low fluence range.

Since the electronic stopping powers are much higher than the nuclear ones in the present case, electronic sputtering may play a key role in the processes.

In the regime of electronic stopping, most studies were carried out on insulators such as gas solids, organic and inorganic solids [25-27]. The energy deposited electronically in a solid can led to energetic atomic motion by a number of non-radiative relaxation processes. After the passage of an ion, a roughly cylindrical region of energised atoms and molecules is formed. Below the surface those non-radiative processes that release sufficient energy can produce defects in solids. Non-radiative relaxation can cause the ejection of atoms and molecules at the surface of a solid; this phenomenon is referred to as electronic sputtering.

Electronic sputtering of metals is a completely new field for which few articles in the literature are available [28,29]. A basic requirement is a large enough electronic stopping power, usually of the order of 10 keV/nm, which is always achieved by heavy ions with super-high energy ($10-10^3$ MeV).

Returning to the present experiments, the electronic stopping powers are less than 1 keV/nm based on SRIM calculation. Significant electronic sputtering of the bulk materials should not occur during the irradiations. Nevertheless, the energy deposited electronically at/near the surface may lead to desorption of the adsorbed species on the oxide layer. These adsorbed species may bond to the oxide surface weakly, with low binding energies, which makes the desorption process occur easily. Such a process has not yet been studied in a MeV regime, but similar process in a keV regime have been reported [30], in which a very high sputtering/desorption yield (several molecules/ion) due to Ar ion bombardment of unbaked stainless steel was found.

Thus, it follows reasonably that the WF decreases shown in Figure 4 result from the desorption of the adsorbed gas species from the sample surfaces, which decreased the surface coverages of the adsorbates and in turn led to the decrease in WF, according to Eqs. (2) and (3).

Accordingly, the results that the WF decrease by He ion irradiation is faster than that by H ions is simply because of the difference of about five times in their electronic stopping powers, according to SRIM evaluation. The larger power of He ions certainly led to a faster process.

The saturation tendency is related to the maximum desorption under the present experimental conditions, and the difficulty in adsorption from residual gases since the pressure was kept at 1×10^{-4} Pa or lower during the whole experiment. This is consistent with the result obtained in the oxygen adsorption experiment [8], in which it was found that oxygen adsorption occurred on the heated surface only after the pressure increased to over 1×10^{-3} Pa. Here the maximum desorption does not mean definitely thorough desorption from the surface because the adsorbed phases are different in their binding energies. Which phases will desorb depends on whether the transferred energy from the incident ions is large enough to overcome the corresponding binding energies. For example, O_2 bonds to the defective NiO surface less strongly than does O [20].

The saturation levels may not be the same for two species, depending on the initial surface states and irradiation processes. In our experiments, the details of the saturation level of He ion irradiation are unclear due to insufficient fluence. Moreover, the obvious logarithmic relationship between the WF change and the fluence before saturation needs to be studied further in the future.

Calculation using SRIM2000 revealed a totally different picture of energy transfer from the low energy ions to nickel. The nuclear stopping power is higher than the electronic one, all of the order of 10 eV/nm. Simulation produced a nuclear sputtering yield of about 20%. All this may imply a different mechanism responsible for the irradiation behaviours occurring on the surface.

In a previous section, the decrease in WF was attributed to desorption of the adsorbed species from the surface due to electronic processes. In the present case, the electronic stopping power is too small to function significantly. The nuclear sputtering, however, becomes overwhelming. In view of this, the decrease in WF after the first irradiation as shown in Figure 5 may be attributed to desorption of the adsorbed species due to efficient nuclear desorption processes [30].

Therefore, the following increase in WF can then be attributed to the sputtering of the oxide layer. The opposite process – oxidation – has been studied extensively, which may present a beneficial hint to explaining the present results. Pope, *et al.* [31] studied oxidation process of a clean Ni(100) surface and correlated the surface coverage by nuclear reaction analysis and the WF change by Kelvin probe for O_2 on Ni(100) at 325 K. The clean surface was prepared by sputtering and oxidation/reduction processes, verified by LEED and AES. Oxygen pressure during exposure was about 10^{-5} Pa. The initial $\Delta\phi$ was equal to zero before admitting oxygen. Then a rapid WF increase to its maximum plateau of about +330 meV was observed due to rapid adsorption of oxygen at the initial stage of oxidation up to a coverage of 0.5 ML, followed by a slight decrease to +320 meV with a slow increase of the coverage to 0.57 ML at about 25 L. This probably indicates some penetration of oxygen below the plane of the nickel surface, then an accelerating region of the coverage until saturation at 2.45 ML, due to the nucleation and growth of the oxide layer, accompanied by the drastic WF decrease to $\Delta\phi$ = -470 meV.

Similar behaviour was also reported by Inoue [32] in his work on O_2/Ni(111) systems. A WF increase of $\Delta\phi$ = +0.9 eV was observed first, followed by a decrease to $\Delta\phi$ = -0.7 eV.

They attributed the WF increases to the oxygen adsorption on the clean surface due to the formation of surface dipoles toward the bulk, and the subsequent decreases to the nucleation and growth of the oxide layer.

In our case, the increase from $\Delta\phi$ = -220 meV to $\Delta\phi$ = +460 meV should correspond to the change from oxide layer to adsorbed layer, just opposite to the changes from $\Delta\phi$ = +330 meV to $\Delta\phi$ = -470 meV in Pope's case, and from $\Delta\phi$ = +0.9 eV to $\Delta\phi$ = -0.7 eV in Inoue's case, respectively. The behaviours are similar in these cases, although the absolute values are hard to compare with each other because the samples are different in their crystallographic orientation and initial surface state, and the methods used to measure the WF are also different.

One more point worth indicating is that $\Delta\phi$ = -220 meV might not represent the bottom of the valley due to desorption, due simply to sparse experimental points around the area. The bottom level depends on the maximum desorption that will be investigated in the future, starting from a smaller fluence under smaller fluxes. However $\Delta\phi$ = +460 meV may be the state of nickel surface with adsorbates on after the oxide layer was removed completely. The adsorbates came from the residual gases, taking into account that the base pressure was 1×10^{-6} Pa, together with the residual gases from the working gas and piping system, although the purity of the gas was high. The partial pressure of residual gases was of the order of 10^{-6} Pa, as estimated using quadrupole mass spectrometry. Under this pressure, a clean surface may be covered by the adsorbed species within minutes if each incident species can be adsorbed firmly. Thus in the present study, the surface was by no means clean even after long time sputtering. But the formation of oxide was not observed in the experiment since a drastic drop in WF after the initial adsorption was absent. It seems that the surface changed into the state of the adsorbed surface after the fourth irradiation, and the subsequent irradiations just repeated a desorption-sputtering-adsorption process. This is why the WF tended to saturate after the fourth run. Actually, in Figure 5, a slight decrease in WF is observed, which may be ascribed to a rougher surface induced by irradiation that favours a bit lower WF [33], or to probable penetration of oxygen below the plane of the nickel surface, which leads to a slight decrease in WF [31]. But the slight changes are indeed within the experimental error range (about 30 mV) [18]. It is thus difficult to draw any conclusion about it for the present.

Potential applications in nuclear engineering field

Tritium-breeding lithium ceramics have been studied by the present authors under ion irradiations. Unfortunately, due to charge build-up on the irradiated sample surface, the probe output was greatly affected. But to raise the temperature of the sample so as to increase its electric conductivity and to achieve a charge-free surface may be a possible way to apply the Kelvin probe to the issue, as proved in the use of high-temperature Kelvin probe in the case of oxide ceramics by Suzuki, *et al.* [3-5]. Suzuki found that adsorption/desorption of hydrogen and water vapour, and oxygen vacancies formation on the surface/near surface were considered to be responsible for the corresponding work function changes observed in the experiments.

Studies on poly-Ni may contribute some hints to the fusion society in the issues concerning the particle recycling on the first wall. One indirect example can be presented to demonstrate the possible application in the field. It is relevant to radiation-induced release of hydrogen from C:H films due to MeV-level irradiation of various ion species, studied and summarised by Trakhtenberg, *et al.* [34], who found that the release could be explained well using an ionisation vacancy model (IVM). The model postulates that the electronic stopping is responsible for the C:H bond break, while nuclear stopping is responsible for the generation of vacancy-type radiation-induced defects. The interactions between the resultant mobile hydrogen molecules and the vacancies reasonably interpret evolution of the releases observed.

In view of the surface sensitive nature of the Kelvin probe and its feasibility in studying fusion-related materials as proved by the present author's group, if combined with other surface analysers, it may contribute much more to not only fundamental researches on the issues such as adsorption/desorption, adsorbate phases, retention, and permeation in both tritium breeding and plasma facing materials, but also to the potential future use in fusion reactor as an *in situ* surface analysers.

Conclusions

- Experimental devices have been successfully developed for examining the work function (WF) change by means of Kelvin probe (KP), of metallic materials due to ion irradiation in low energy (some hundreds of eV) or higher energy (MeV) ranges.

- The WF decrease due to 1 MeV light ions He^+ and H^+ irradiation on Ni can be attributed to desorption of the adsorbed species on the surface due to energy transfer via electronic processes (inelastic collision), and the subsequent saturation may correspond to a maximum desorption, according to the simulation by SRIM2000, the two-layer surface model and the model for surface dipole-induced work function change.

- In contrast to MeV light ions, low energy ion irradiation favours elastic collision, ion energy transfers to target atoms mainly via nuclear stopping. The initial WF decrease can then be attributed to efficient nuclear desorption/sputtering of the adsorbed species. The subsequent WF increase should be considered a result of a gradual removal of the oxide layer. The final state of the surface might be adsorbates on the metal bulk, due to the existence of oxygen in the working atmosphere at a partial pressure lower than 10^{-5} Pa. The oxide phase did not form after each irradiation, or a large WF decrease should be observed.

Acknowledgements

The authors gratefully acknowledge the financial support given by Grant-in-Aid for Scientific Research by the Ministry of Education, Science, Sports and Culture, Japan, and the partial support of this study from JAERI as related to the HTTR research project.

REFERENCES

[1]   G. Federici, *et al.*, *Journal of Nuclear Materials*, 283-287, 110 (2000).

[2]   H. Yanagihara, Y. Hirohata and T. Hino, *Journal of Nuclear Materials*, 258-263, 607 (1998).

[3]   A. Suzuki, K. Yamaguchi and M. Yamawaki, *Fusion Engineering and Design*, 39, 699 (1998).

[4]   A. Suzuki, K. Hirosawa, K. Yamaguchi and M. Yamawaki, *Fusion Technology*, 34, 887 (1998).

[5]   A. Suzuki, PhD dissertation, 1999, University of Tokyo, Japan.

[6]   T. Yokota, A. Suzuki, K. Yamaguchi and M. Yamawaki, *Journal of Nuclear Materials*, 283-287, 1366 (2000).

[7]   A. Suzuki, K. Yamaguchi, T. Terai and M. Yamawaki, *Fusion Engineering and Design*, 49-50, 681 (2000).

[8]   G-N. Luo, K. Yamaguchi, T. Terai and M. Yamawaki, *Journal of Nuclear Materials*, 290-293, 116 (2001).

[9]   G-N. Luo, K. Yamaguchi, T. Terai and M. Yamawaki, *Physica Scripta*, T94, 21 (2001).

[10]   J. Hoelzl and F.K. Schulte, in *Solid Surface Physics*, G. Hoehler, ed., Springer, Berlin, 1979, pp. 1-150.

[11]   D.P. Woodruff and T.A. Delchar, *Modern Techniques of Surface Science*, 2nd ed., Cambridge University Press, 1994.

[12]   Lord Kelvin, *Philosophical Magazine*, 46, 82 (1898).

[13]   W.A. Zisman, *Review of Scientific Instruments*, 3, 367 (1932).

[14]   G-N. Luo, K. Yamaguchi, T. Terai and M. Yamawaki, *Review of Scientific Instruments,* 72 (5), 2350 (2001).

[15]   G.R. Brandes and A.P. Mills, Jr., *Physical Review*, B58 (8), 4952 (1998).

[16]   A. Hadjadj, P. Roca i Cabarrocas and B. Equer, *Review of Scientific Instruments,* 66 (11), 5272 (1995).

[17]   G-N. Luo, K. Yamaguchi, T. Terai and M. Yamawaki, submitted to *Surface Science*.

[18]   G-N. Luo, PhD dissertation, 2001, University of Tokyo, Japan.

[19]   K. Besocke and S. Berger, *Review of Scientific Instruments*, 47 (7), 840 (1976).

[20]   V.E. Henrich and P.A. Cox, *The Surface Science of Metal Oxides*, Cambridge University Press, 1994.

[21]   Juan Carlos de Jesus, Pedro Perira, Jose Carrazza and Francisco Zaera, *Surface Science*, 369, 217 (1996).

[22]   H. Oefner and F. Zaera, *Journal of Physical Chemistry*, B101, 9069 (1997).

[23]   M. Lorenz and M. Schulze, *Surface Science*, 454-456, 234 (2000).

[24]   http://www.research.ibm.com/ionbeams/.

[25]   R.E. Johnson, *Review of Modern Physics*, 68 (1), 305 (1996).

[26]   P. Williams and B. Sundqvist, *Physical Review Letters*, 58 (10), 1031 (1987).

[27]   J.A.M. Pereira and E.F. da Silveira, *Physical Review Letters*, 84 (25), 5904 (2000).

[28]   H.D. Mieskers, *et al.*, *Nuclear Instruments and Methods*, B146, 162 (1998).

[29]   A. Gupta, *Vacuum*, 58, 16 (2000).

[30]   E. Taglauer, *Applied Physics*, A51, 238 (1990).

[31]   T.D. Pope, S.J. Bushby, K. Griffiths and P.R. Norton, *Surface Science*, 258, 101 (1991).

[32]   M. Inoue, *Japanese Journal of Applied Physics*, 26 (2), 300 (1987).

[33]   R. Smoluchowski, *Physical Review*, 60, 661 (1941).

[34]   I.Sh. Trakhtenberg, A.P. Rubshtein and A.D. Levin, *Diamond and Related Materials*, 8, 2164 (1999).

DEVELOPMENT OF HIGH-TEMPERATURE IRRADIATION TECHNIQUES UTILISING THE JAPAN MATERIALS TESTING REACTOR

Minoru Narui, Tatsuo Shikama, Masanori Yamasaki and Hideki Matsui
Institute for Materials Research, Tohoku University
Oarai, Ibaraki, 311-1313 Japan

Abstract

Acquisition of systematic irradiation data is essential for understanding the fundamental processes of irradiation effects and for the establishment of a reliable database of irradiation effects in nuclear systems. This will take several years and the great expense of several different irradiation rigs in a fission reactor irradiation. There, it will take quite a long time to carry out the needed iterations between irradiation tests and evaluation and materials developments. An irradiation rig was developed to carry out irradiation under multiple conditions of temperatures and irradiation fluences. Irradiation tests of nuclear materials were successfully carried out in the Japan Materials Testing Reactor.

Introduction

There are increasing demands for high-temperature irradiations in a fission reactor from Japanese university researchers [1]. Under such conditions, a precise temperature control is especially important for fundamental studies of irradiation effects in materials for high-temperature applications.

Below 500°C, a technique of temperature control by electrical heating systems has been established and irradiation temperatures are controlled, being independent of a reactor nuclear power in the Japan Materials Testing Reactor (JMTR) [2]. An irradiation rig composed of several subcapsules that can be retrieved from a reactor core during irradiation makes it possible to study irradiation effects in materials as functions of a temperature and neutron fluence.

Several irradiations have been carried out at temperatures above 800°C for study of irradiation effects in materials for high-temperature gas-cooled reactor (HTGR) applications, such as silicon carbide (SiC) composites. Preliminary studies have also been carried out prior to their planned executions at the high-temperature test reactor (HTTR). In the HTTR, temperature control is carried out by the so-called helium gas pressure control method (GPTC). Thermal conductance between irradiated materials and a capsule wall could be controlled thermal conductance of helium gas filling a capsule.

It is demonstrated that the GPTC method could control irradiation temperatures independent of the JMTR reactor power above 10 MW, being 20% of its full power. Thus, effects of low-temperature irradiation effects could be minimised by the GPTC method. This technique has a large time constant to regulate temperature and it requires a sophisticated program for a computer-assisted automatic system. Recently, the reactor operating group in the JMTR in JAERI Oarai Establishment developed a successful computer program for the GPTC method, making this process applicable under wider ranges of temperature. Also, temperature control by electrical heating, the EHTC method, was also successfully applied to irradiation above 800°C.

This paper will describe the recent status of irradiation tests in JMTR for Japanese university research activities in HTTR related fields. The paper will also describe technical details of temperature control in a fission reactor at elevated temperatures.

Development of irradiation rigs

Figure 1 shows the general structure of the present irradiation system. Ten small subcapsules were accommodated in a temperature-controlled irradiation rig [3], which has two independently regulated temperature zones. The irradiation rig, a protecting tube, a junction box and a lifting device form a pseudo-shroud system, and the small subcapsules can be lifted up from and inserted down into the temperature-controlled rig in the JMTR core during reactor operation. Irradiation temperatures were controlled by electric heaters encased in the rig, as shown in Figure 1. Two different temperature zones were set up above and below the mid-plane of the reactor core, which can be independently controlled irrespective of the reactor power. Figure 2 displays a cross-sectional view of the rig accommodating the subcapsules. Five transfer tubes were installed in the rig, being thermally bonded by aluminium block. The electric heaters were coiled on the outer surface of the aluminium block, which ensures temperature homogeneity. A reflecting tube and a gap between the reflecting tube and a wall of the rig (described as the outer tube in Figure 2) will set up an appropriate heat-removal rate for temperature controls.

Figure 1. Schema of temperature and fluence controlling irradiation rig

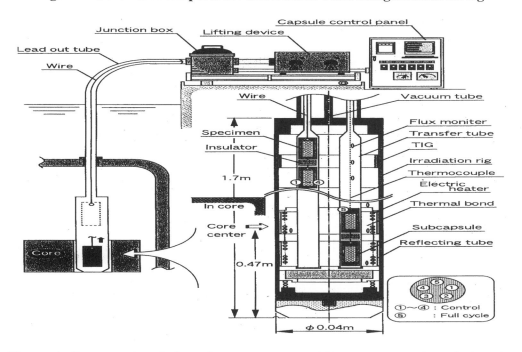

Figure 2. Cross-sectional view of temperature and fluence control irradiation rig

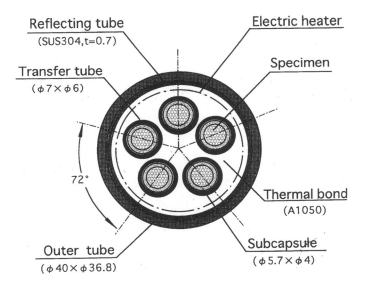

A schematic view of a train of two subcapsules is shown in Figure 3. Two subcapsules were connected through an alumina-made thermal insulator; they can be inserted into or removed from the irradiation rig during reactor operation. The temperature of each subcapsule was monitored by a thermocouple inserted into an aluminium-made top cap of the subcapsule shown in Figure 4. Each of the two subcapsules can be irradiated at different temperatures otherwise under nearly the same irradiation conditions.

Figure 3. A train of two subcapsules for two different irradiation temperatures

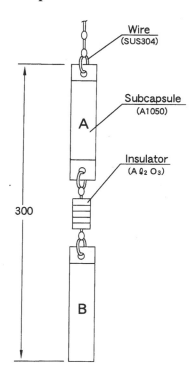

**Figure 4. Structure of a subcapsule in a transfer tube.
Temperature is monitored by a thermocouple on the top cap.**

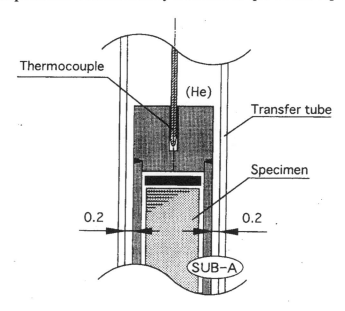

The rig described above was developed for irradiations below 300-400°C. Figure 5 shows details of an irradiation rig for the middle-temperature range, 500-600°C. Electrical heaters were installed inside the inner wall to maintain a good thermal isolation from the reactor coolant. Figure 6 displays the detailed structure of a rig for high-temperature irradiation in the range of 800-1 000°C.

Figure 5. Structure of rig for middle-temperature irradiation

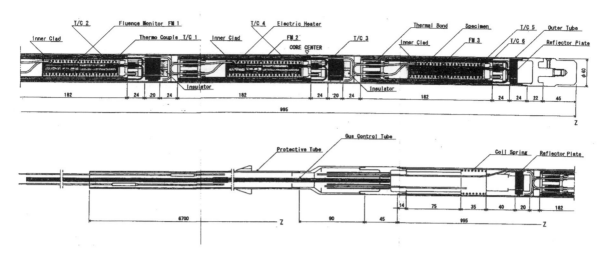

Figure 6. Structure of rig for high-temperature irradiation

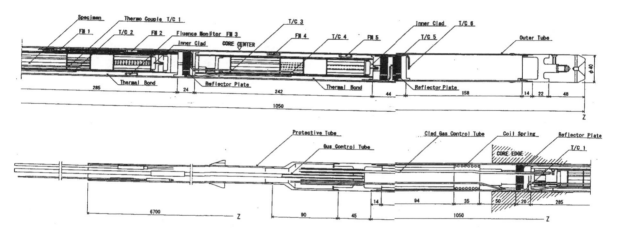

History of irradiation tests

The present irradiation rig has the potential of temperature control by changing helium gas pressure in a gap between a reflecting tube and the outer tube shown in Figure 2, independently of the reactor power (gas pressure temperature control, GPTC). Figure 7 shows one example of temperature control using the GPTC method, without using electric heaters at reactor shutdown. The gamma heating rate changed from about 7 W/g to less than 1 W/g, but the temperature of a subcapsule could be kept constant at about 680 K. However, the reactor power changed fast by abrupt insertion of stopping control rod at the end of reactor shutdown and the temperature could not be controlled well by the GPTC. In general, the GPTC has advantages over electric heater temperature control (EHTC). The rig structure could be simplified and more specimens can be irradiated. The most important issue to point out is that the possibility of failure always exists with an electric heater. Although electrical heaters can occasionally survive more than one year of irradiation in the JMTR (averaging five cycles a year), up to more than 1 021 n/m^2 fast (E > 1 MeV) neutron fluence, their reliable lifetime will only be a few cycles JMTR irradiation.

Figure 7. Temperature control by GPTC method at JMTR shutdown

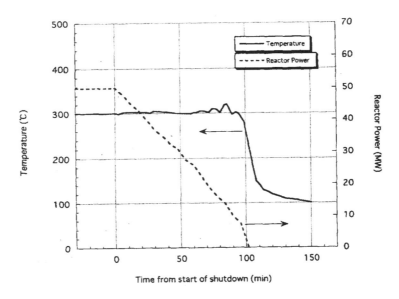

Development of reactor materials sometimes demands irradiation exceeding 10^{25} n/m^2 of fast neutron fluence. The GPTC will not have a life limitation insofar as a geometry of the gas gap does not change drastically due to swelling. However, the GPTC cannot control an abrupt change of reactor power in the present system because it takes several minutes to attain equilibrium gas pressure in the gas gap. The basal temperature control was carried out by the GPTC and the EHTC was used for compensating abrupt but small changes of temperature especially at reactor start-up and shutdown. Figure 8 shows a temperature history of irradiation using the present irradiation rig in 1996.

Figure 8. Example of history for temperature and fluence control irradiation

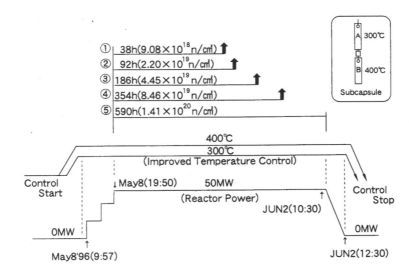

The subcapsules were inserted into the irradiation rig after the reactor power and the temperatures of the rig were stabilised as shown in Figure 7. Then, subcapsules were sequentially removed from the rig during reactor operation. Irradiation with five different levels of neutron fluence, from 9×10^{22} n/m^2 to 1.4×10^{24} n/m^2 and at two different temperatures of 573 and 673 K, were realised in one exertion of irradiation. Previous preliminary results, however, suggested that specimens were exposed to low flux fast neutron irradiation even when they were not inserted in the reactor core. Figure 9 shows induced radioactivity of iron dosimetry foil as a function of a distance along the height of reactor core. The induced radioactivity is roughly proportional to the fluence of fast neutron. It can be seen that substantial neutron flux exists even in an out-of-core region. A length of the rig was enlarged as long as possible and the subcapsules were placed at about 300 mm above the edge of the reactor core to avoid exposure to the low flux neutrons, where the fast neutron flux is about 10^{-4} of that at the core centre.

Figure 9. Distribution of neutron flux at top end of rig

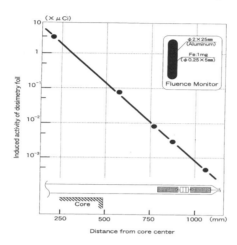

Figures 10 and 11 show examples of temperature histories realised in the JMTR using these irradiation rigs. Temperatures could be controlled as wished in the temperature range of 600-1 000°C, by the GPTC and EHTC methods.

Figure 10. Example of temperature control at middle-temperature range

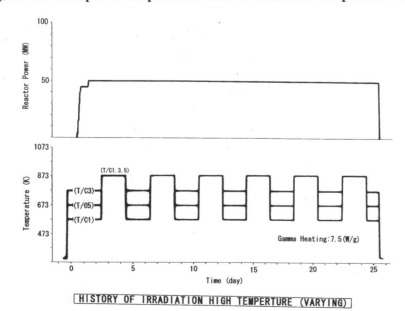

197

Figure 11. Example of temperature control at high-temperature range

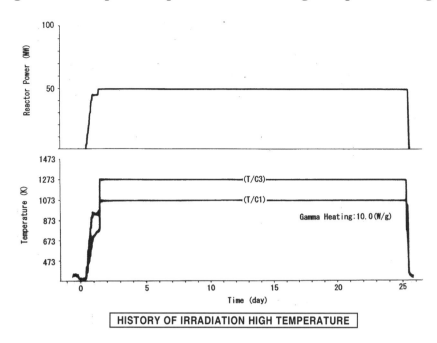

HISTORY OF IRRADIATION HIGH TEMPERATURE

More than a few thousands transmission electron microscope (TEM) specimens of different materials were irradiated. Microstructural modifications due to a fission reactor irradiation were examined in comparison with those under other irradiation sources such as high-energy electrons in high voltage electron microscopes (HVEMs) [3]. Extensive data were accumulated and fundamental processes of irradiation-induced microstructural evolution were analysed as a function of a variety of irradiation conditions.

Conclusion

A rig was developed for controlled irradiation in the JMTR fission reactor. Irradiation of five different fast neutron fluences at two different temperatures could be carried out successfully using the developed rig. The irradiation history confirmed satisfactory control of temperature in the range of 300-1 000°C, as well as of neutron fluence.

Acknowledgements

The work has been extensively supported by the JAERI-JMTR concerned irradiation sections. The authors wish to express their sincere gratitude to the members, especially to Mr. T. Sagawa and Mr. H. Amezawa, for their contributions.

REFERENCES

[1]   M. Kiritani, *Journal of Nuclear Materials*, 191-194, 125 (1992).

[2]   M. Narui, T. Sagawa, Y. Endo, T. Uramoto, T. Shikama, H. Kayano, M. Kiritani, *Journal of Nuclear Materials*, 212-215, 1645 (1994).

[3]   M. Narui, T. Sagawa, T. Shikama, *Journal of Nuclear Materials*, 258-263, 372 (1998).

SESSION IV

Basic Studies on Behaviour of Irradiated Graphite/Carbon and Ceramic Materials Including their Composites under both Operation and Storage Conditions

Chairs: B. McEnaney, A. Wickham

INVESTIGATION OF HIGH-TEMPERATURE REACTOR (HTR) MATERIALS

D. Buckthorpe
NNC Ltd., UK

R. Couturier
Commissariat à l'Énergie Atomique (CEA), France

B. van der Schaaf
Nuclear Research and Consultancy Group (NRG), Netherlands

B. Riou
FRAMATOME ANP, France

H. Rantala
European Commission, Joint Research Centre (JRC), Institute for Advanced Materials, the Netherlands

R. Moormann
Forschungszentrum Jülich GmbH (FZJ.ISR), Germany

F. Alonso
Empresarios Agrupados Internacional S A (EASA.MD), Spain

B-C. Friedrich
FRAMATOME ANP (GmbH), Germany

Abstract

Issues associated with material selection and behaviour are being examined for the main reactor circuit components of the modular high-temperature reactor. The work is being performed as part of a collaborative European development supported by the European Union 5th Framework Programme. The work considers materials for the reactor vessel and certain high-temperature regions including the turbine and the graphite core. This paper reviews the main elements of the materials work programme and reviews some initial findings with regard to material choices and needs.

Introduction

A common European approach to the renewal of HTR technology has been established through the direction of a European HTR Technology Network (HTR-TN) to co-ordinate and encourage work-shared structures and serve as a channel for international collaboration in this technology [1]. The European Commission is supporting a number of research activities on modular HTR technology in its 5th EURATOM Framework Programme [2]. These involve partnerships between principal industrial and research organisations from countries of the European Union working to bring European expertise and experience in the HTR together and to support industry in the design of reactors. Two projects within the Framework Programme – HTR-M and HTR-M1 – consider the selection and development of materials for the key components of the HTR. They investigate the materials for the main components, extend over four years and involve eight partners. This paper provides an overall description of the HTR materials projects, their objectives and some results from initial actions that focus on the needs for feasibility with respect to material choice and structural integrity.

Background to HTR development in Europe

Commercial experience with gas-cooled reactors began in 1956 in the UK, which extended to cover 26 Magnox Stations and 14 advanced gas-cooled reactors using CO_2 gas as a coolant. HTR plants began in Europe in the 1950s. They used helium as the coolant to permit an increase in operating temperature. Coated ceramic fuel particles and a dispersed graphite matrix and moderator were used and in the years 1960 to 1990 many HTR prototypes were built and tested in Europe, i.e. DRAGON (UK, 1964-1975), AVR (Germany, 1966-1988) and THTR 300 (Germany, 1983-1989). These provided a valuable foundation on which to base future HTR development.

There was a continued interest in larger steam cycle plants in Germany, Russia and the US through the 1970s and commercialisation seemed promising in the early 1980s with the medium-sized concept developed in Germany for industrial process heat applications. The commercial development of this HTR-MODUL plant was however terminated in 1990.

Renewed interest in HTRs in Europe came following the development of the HTR coupled to a gas turbine power conversion system which led to today's developments of the industrial prototypes GT-MHR (in Russia) and PBMR (in South Africa) which are supported by international teams with a strong European involvement.

Overall objectives and description of the HTR materials projects

The main objectives of the HTR materials projects are to provide information for the key components and support the development of high-temperature reactor technology in Europe. Two ongoing projects (HTR-M & HTR-M1, managed together) cover three areas: the reactor vessel, high-temperature materials (internal structures and turbine) and graphite structures including oxidation. A description of the main activities and tests to be performed is provided in Figure 1.

Reactor vessel

Design information for the HTR reactor pressure vessel (RPV) is not well established under HTR conditions and further improvement is needed in the areas of design and structural integrity analysis and materials properties data. Information is also needed concerning manufacturing of products and parts, behaviour of welded joints and the influence of environment (neutron-irradiation fluence, operational temperature, and helium environment). The integrity of the welded joints is seen as a particularly important feasibility issue. The objective of this work is therefore to investigate and confirm the choice of candidate material options for different HTR concepts, in relation to both normal and accident conditions, and to compile their design properties.

The work (performed in HTR-M) covers a review of vessel materials and their properties, focusing on existing thermal gas-cooled reactors and previous high-temperature reactors and the setting up of a materials database on design properties. This will lead to a recommendation for appropriate vessel materials, depending on operating conditions, and identification of data omissions and the selection of a material for testing under irradiated and non-irradiated conditions. The work package will consider materials for a "cold" (current PBMR) and "warm" (GT-MHR) vessel option. Specific tests on welded joints are to be performed covering tensile, creep and/or compact tension fracture specimens to determine representative properties including the effects of irradiation. A further review and synthesis is planned on completion of the irradiation test work and examination.

High-temperature materials

The HTR primary circuit components operate at temperature above 600°C and up to 850 to 900°C in order to achieve high energy levels and high cycle efficiency. For such components a number of metallic and ceramic materials are required in the region of the core and reactor internals (i.e. metallic, ceramic and composite materials) and for the core support and other structures. For these materials and components, some data exists but the information and experience are not well established for the HTR environment. Also, metallic materials are required for the turbine components for which the expected working conditions are outside of today's industrial experience and a significant effort of material development is required in this area.

The first step involves putting together a database of properties that takes the experience on existing high-temperature and gas-cooled reactors into account. This is covered in HTR-M. The database will include a reference to material composition, microstructure, manufacture (products and parts, etc.) and extend to materials required to withstand temperature levels experienced in emergency modes (e.g. carbon-based materials reinforced by carbon fibres, Ni-based alloys, etc.). The control rod cladding is investigated for the internal reactor structures and the disc and blades of the turbine. For the turbine, the two most promising grades will be investigated taking different options into account, manufacturing aspects and characterisation of improved alloys made by various manufacturing routes. Short term mechanical tests needed to quantify the materials in the HTR environment will be performed covering tensile, fatigue, short term creep and creep/fatigue tests planned (in air, vacuum, helium) at temperatures up to 1 000°C. The results of the test programme will be stored in the database.

Within HTR-M1 the work on the blade materials has been extended to cover intermediate-term creep effects to confirm the suitability of the chosen alloys for longer-term high-temperature exposure. Testing is planned at temperatures of 850°C with testing times up to 10 000 hours. The expected number of tests will cover specific grades, fabrication routes and heat treatments. Tests will also be carried out on aged material and damage analysis and lifetime modelling will be performed (based on results from notched specimens) and the results incorporated into the developing database.

Graphite

The need for reliable graphite data is a crucial issue for existing and new HTR projects and is an important consideration for future decommissioning activities. Graphite plays an important role as a moderator and structural component and has important safety implications because of structural and other property changes that occur when it is irradiated. Its selection also has important consequences for future decommissioning activities. Oxidation resistance of graphites at high temperature also requires special attention due to its relevance for safety analyses of air and water ingress accidents, licensing procedure and normal operation.

The HTR-M project provides for a limited review of existing graphite information. This is to identify gaps in data and testing requirements. Many of the graphites used in previous core designs are no longer manufactured commercially, and currently available grades of graphite are limited to a few specific sources of raw materials. Very few of the grades may turn out to be suitable or desirable for a European-based modular HTR. The work package also involves test work to investigate the influence of oxidation arising from severe air ingress with core burning and an investigation of protective coatings and innovative C-based materials. Overall this work package is expected to provide a platform of material data for future application and selection of graphites for the HTR.

The HTR-M1 project will focus on initiating an irradiation programme on chosen graphites to determine the variations in their physical and mechanical properties up to low/medium irradiation doses (making use of the results from HTR-M as a basis for estimation and extrapolation to higher doses). Discussions will be held with graphite manufacturers to investigate the possibility of producing small quantities of alternative graphites (in the laboratory) which would have a better combination of desirable properties. Whilst this project falls short of the ultimate selection criteria for the graphite, which will mainly be based on long-term irradiation behaviour, it will establish the groundwork for continuation of the irradiation programme on the most desirable graphite(s). The project will also make use of valuable diminishing experience from experts in design, experiments and testing of graphite within Europe serving to reinstall graphite irradiation and qualification methods within Europe while the available experience and facilities exists.

Material and structural integrity issues

HTR reactor pressure vessel (RPV)

The work on the RPV will establish available information on a few selected vessel steels and a database for design and feasibility investigations. Two types of materials are currently used for RPVs, depending on whether the temperature of the pressure boundary under normal operation or design conditions exceeds the limit for LWR pressure vessels of 370°C.

Most designs of future plants make use of the either SA 508 steel for insulated vessels or modified 9Cr-1Mo steel for "inlet temperature vessels". These materials have similar strength levels at temperatures up to 370°C. The use of modified 9Cr-1Mo steel allows higher temperatures with only a gradual reduction in design strength at temperatures up to 450°C. Above 450°C allowable stresses for all materials fall off rapidly. Modified 9Cr-1Mo steel has an even bigger advantage over the other materials at these higher temperatures. At 500°C for example, RCC-MR indicates that its design strength is twice that of 2¼ Cr-1Mo steel (Figure 2).

Existing LWR pressure vessels in SA 508 and its European derivatives encompass the conditions in proposed HTR vessels with respect to operating temperature, wall thickness, design features (thick flanges, etc.) and irradiation levels. The large database and the extensive fabrication experience from

LWR programmes provide a sound basis for HTR vessel design. The most obvious gap in data relates to possible transients involving temperatures above the usual LWR limit of about 370°C.

A substantial database and fabrication experience exists at least for CO_2-cooled reactors in the UK, with respect to C-Mn steels similar to the SA 516-70 grade specified in Chinese designs. There is European experience of complementary high-temperature materials (2¼ Cr-1Mo grades). There should be relatively few problems in assembling an adequate database extending to temperatures involved during transients.

The above considerations leave modified 9Cr-1Mo steel with the most uncertainties as to the available database and fabrication experience with respect to HTR vessel characteristics. This material also potentially has the wider applicability and since for the Framework Programme it is only possible to test one material within the high flux reactor this represents the most likely candidate for the experiments. The extent of testing of control specimens is expected to cover as received (welded and post-weld heat-treated) and irradiated material (irradiation time about 100-200 hours). Thermal ageing effects in 40 years at 450°C may be significant but useful experiments may not be possible within the time scale of HTR-M.

The main concerns with regard to structural integrity are expected to be at the welds. They carry an additional requirement to satisfy the safety analysis methodology and safety studies (defect location and defect size) and have to be assessed from the point of view of non-destructive examination and potential for failure or leakage.

The main damage mechanisms to be addressed are fracture, fatigue and creep fatigue. These have to be evaluated from an initiation aspect and for crack growth of defects under fluctuating thermal and mechanical loads. The potential effects of environment (temperature, irradiation, ageing) all have to be taken into account. The test programme and its development addresses toughness and creep properties under as-fabricated and simulated end-of-life conditions. This will provide an understanding of the welding processes, weld metals and post-weld heat treatment required for thick welds and weld factors to be applied on the base material properties.

Other sensitive zones that are to be checked are areas that feature thicker sections (potential for reduced properties and strength), hot spots and regions important from a functionality point of view. This includes the flange area and belt line which have to satisfy primary and secondary stress limit requirements and progressive deformation mechanisms.

HTR high-temperature materials

For internal structures austenitic and ferritic steels can be used for some of the non-graphite components (core support plates, grids) where there is an established experience with the steel in gas and other elevated temperature reactors. In general operability and applicability up to temperatures of 550°C has been firmly established for some materials within projects such as the advanced gas-cooled Reactor in the UK, the European Fast Reactor Project and high-temperature reactor projects such as AVR. Some C/C composites have also been investigated for use on HTTR for the control rod. The use of such materials is considered to provide increased thermal resistance offering potential for improved reactivity control during shutdown and for allowing the normal operating temperatures of future reactors to be increased. Experience of materials for operation at much higher temperatures at the moment rely heavily on nickel-based and Fe-Cr-Ni alloys. For the HTR which has the additional effect of a helium environment the PNP and Japanese He/He exchangers provide some important experience (Figure 3).

For the turbine components, especially the turbine discs and blades, criteria for material selection are to be based on a safe operation period, of say 60 000 hours at 3 000 rpm (50 Hz), with upper temperature limits in the range 850 to 950°C. Candidate alloys for the disc must have a sound industrial base and be capable of production of large defect-free ingots with good forging properties and proven thermal stability. Candidate alloys include A286 (Cr Ni Fe), IN 706, IN718 and UDIMET 720. The introduction of HIP material currently being investigated for the fusion reactor also offers the possibility of producing near-finished-sized components suitable for turbine discs. For the turbine, the materials for cooled and non-cooled blades have to be investigated (although the former clearly offers cost advantages). Metallurgical factors such as cobalt content of the material also have to be assessed with regard to activation potential which may impact on maintenance costs.

The main structural integrity issues concern creep and the influence of the environment. Two basic requirements dictate the design of gas turbine blades and discs: cycle temperature and the maintenance of critical dimensions (clearances) throughout the service life. The selection of turbine blade designs are often limited by the high-temperature creep properties of the material (i.e. non-cooled blade).

For turbine discs there is a requirement to limit the permanent growth and distortion to within typical design life targets. This is usually achieved by maintaining a greater part of the disc cross-section within elastic limits (i.e. satisfying progressive deformation or shakedown criteria). Critical regions for the disc from the point of view of potential creep and fatigue failure are the hottest parts. These are at the disc neck because of unrelieved high localised stresses giving rise to undetected growth; and at the rim, due to the additional geometric effects of the blade root slots and possibly cooling holes. Careful material selection or control of localised stress and temperature conditions can avoid creep-fatigue interaction problems in these areas.

For the turbine blades failure modes such as creep, high cycle fatigue and low cycle fatigue have to be taken into account. The gas radial temperature variation often peaks typically in the middle third of the blade with steep temperature gradients across the blade section which vary according to the transient and steady state conditions. All regions of the blade profile therefore undergo complex stress-strain cycling with reversed plasticity in some areas. For material selection a close understanding of some important parameters such as creep rupture, creep ductility and creep rate are needed. Typically for nickel-based super-alloys, time-dependent effects induced by creep and oxidation on cyclic crack growth rates are important for design, and hence an understanding of creep crack growth behaviour and the interaction between creep and fatigue with tensile and compressive dwells is needed. The awareness of the role of grain boundaries in high-temperature fracture has also led to significant developments in single crystal super-alloys that can give beneficial anisotropic material properties.

Corrosion can cause a significant shortening of the material creep life and acceleration of creep crack growth rates. Application of suitable coatings can arrest creep reduction tendencies, however their use requires an understanding of the potential for interfacial cracking at the coating layer to avoid the development of more significant cracking from the interface.

Graphite

a) Properties

The graphite core is a key component that affects safety and operability of the reactor. As well as acting as a neutron moderator and shield, it is a structure that generally provides channels for the passage of fuel and control devices, and coolant flow. The problem is that when irradiated by fast neutrons, graphite suffers damage to the crystallite structure. The resulting damage causes a change in all the physical and mechanical properties of the graphite e.g. strength, coefficient of thermal expansion (CTE) and thermal conductivity.

Ideally, the graphite chosen for HTR designs should be reasonably isotropic, exhibit small dimensional change behaviour, and have a low CTE, a high thermal conductivity, a high irradiation creep constant, a low Young's modulus and a high strength. The problem is that many of these properties are not compatible in practice for normal commercial graphites, e.g. graphites with high strength also have a high Young's modulus. Other factors affecting the choice of graphite would be impurity levels, cost and machinability.

Over the past 60 years, R&D activities have been undertaken in a number of countries to investigate graphite properties. The IAEA International Irradiated Graphite Database Technical Steering Committee is currently compiling a database on graphites past and present and has recently held a seminar [3] to review the current situation.

Many of the graphites used in previous core designs are no longer available in practical terms for a number of reasons. The decline in the ability to manufacture nuclear grade graphite in large quantities in the UK is an example of the above. Two large plants, one owned by Anglo Great Lakes (AGL, now SGL) and the other by British Acheson Electrodes Ltd (BAEL, later known as Union Carbide and presently called UCAR) manufactured all the gilsocarbon graphite used in the UK's AGRs. Neither plant now exists. Today's HTGR projects – HTTR (Japan) and HTR-10 (China) – use a Japanese graphite (IG-110). This graphite, with its high strength, should be taken into account for exchangeable core components of the HTR where low fast neutron fluences and hence low total doses apply. The behaviour for very high doses is not yet fully known.

For a future design it might not be possible to generate a complete irradiated material property database for a particular graphite in advance of the design and construction phases of a reactor. In such cases, existing experience in graphite development together with irradiation test programmes can be used to provide an initial assessment of the suitability of currently available graphites. All available data should be put into a systematic database to find the best candidate materials. These would then have to be irradiated at the appropriate temperatures expected under HTGR conditions and to the peak doses envisaged and screening tests performed, to compare the results with data from the database, and to update the existing models for reliable interpretation and comparison.

Generally, the most important material property change from the point of view of core lifetime integrity is dimensional change. Graphite initially shrinks but then undergoes shrinkage reversal, i.e. growth. This is referred to as "turnaround". The exact form of the dimensional change curve varies with irradiation temperatures. Normally, the higher the temperature the lower the peak shrinkage and the lower the dose at which turnaround occurs. Different graphites can have significantly different dimensional change curves at the same temperatures.

The validation of the through life performance of the graphite core structure is critically dependent on the knowledge of graphite properties and the way they vary under fast neutron irradiation. It is essential therefore that its behaviour be fully understood and investigated up to at least the peak design dose and over the appropriate temperature ranges.

The final choice of graphite for the core should therefore be based on a number of factors, although the most important will be the effects of fast neutron irradiation on its properties up to the peak doses envisaged. Given the best graphite available, it is the task of the core designer to produce a design that will operate safely over the design life of the reactor. The most important considerations are component integrity and changes in core geometry, both of which are affected by the dimensional change. For example, graphite shrinkage could lead to disengagement of individual components and between the core and interfacing structures and to a loss of control of the core geometry; graphite growth could lead to the take-up of design clearances and large forces between structures; and differential shrinkage/growth in a component could lead to high stresses and failure by cracking.

b) Oxidation

The graphite oxidation work involves two principal tasks relevant to safety analysis and licensing of HTRs for normal operation:

- The improvement of experimental database for advanced graphite oxidation models.

- Experimental investigation of innovative C-based materials for application in HTRs.

Graphite burning under severe air ingress accidents is assessed by computer models based (up to now) on isothermally measured kinetic equations which consider in-pore diffusion and chemical reaction mechanisms. The data requirement for such models are burn-off dependent chemical (regime 1) reactivities and in-pore diffusion coefficients. The experimental work to be undertaken deals with the case of severe air ingress, with the diffusion influences of the oxygenation process determined using the thermo-gravimetric facility THERA housed at FZJ. Measurements are made for the German A3 fuel matrix graphite (type A3-27) and for a typical structural graphite (V483T fine grain graphite) in oxygen at temperatures between 823 and 1 023°K and oxygen partial pressures between 2 and 20 kPa.

Although material development for HTRs almost ceased in Germany in 1985, development of C-based materials for fusion reactors continued. These included innovative concepts such as carbon fibre composites and mixed materials (Si, Ti content) which show a high heat conduction and high strength. A second series of experiments are therefore planned to look at the performance of such materials under accident typical temperatures (about 1 273°K) in steam and in air. The experimental work involves the use of induction-heated tube-shaped samples in an induction furnace facility. Oxidised gas flows through the inner bore hole of the tube with flow rates sufficiently high to suppress the influence of boundary layer mass transfer on kinetics. Oxidation effects are measured by mass spectrometric analysis of the product gases and by weight loss of the sample. Experiments are also planned for measurements of CFC materials irradiated and non-irradiated (up to 1 dpa) that will look at strength, dimensional change, thermal diffusivity and heat capacity.

HTR materials database

The need for reliable material data and properties is a key issue in the development of any innovative reactor technology and is especially important for the understanding of structural integrity issues and material behaviour. It is particularly important in those areas where safety considerations are uppermost. For the HTR key component materials a database of properties is required that covers information on manufacturing provisions, products and parts, test data information and design properties. Reliable design property information is one of the main outputs and needs. Work on developing a database is underway with considerations being given to the development of a web-based facility. Design and technological issues plus omissions and shortage of test data will be the basis on which the database will be built up and expanded and on which future R&D needs will be assessed.

Summary and conclusions

The need for reliable material data and properties is a key issue in the development of HTR technology and is especially important for those areas where safety considerations are uppermost. The HTR-M & M1 projects aim to provide a firm materials platform on which to develop the future HTR technological basis within Europe and provide a first step towards building an understanding of the behaviour and manufacturing requirements for some of the new materials that will be needed for such developments. The objective is to improve knowledge of materials for future HTRs for the reactor pressure vessel, the control rod and turbine (blades and discs) and the reactor graphite. The projects extend over a period of four years and are being performed with the support of the European Union 5th Framework Programme. The project is also being co-ordinated through the direction of a European HTR Technology Network (HTR-TN). This paper outlines the work programmes and reviews important issues concerning material needs for feasibility and assurance of structural integrity.

Acknowledgements

The projects mentioned in this paper are being co-sponsored under the 5th Framework Programme of the European Atomic Energy Community ("EURATOM"), contracts FIKI-CT-2000-0032 and FIKI-CT-2001-20135. The information provided herein is the sole responsibility of the authors and does not reflect the Community's opinion. The Community is not responsible for any use that might be made of the data appearing in this publication.

REFERENCES

[1]   W. Von Lensa, D. Hittner, J. Guidez, F. Sevini, A. Chevalier, M.T. Dominguez, G. Brinkmann, T. Abram, J. Martin-Bermejo, "The Role of International Collaboration within the European High-temperature Reactor Technology Network (HTR-TN)", Seminar on HTGR Application and Development, Beijing, China, 19-21 March 2001.

[2]   J. Martin-Bermejo, M. Hugon, G. Van Goethem, "Research Activities on High-temperature Gas-cooled Reactors (HTRs) in the 5th EURATOM RTD Framework Programme", Smirt 16, Washington, August 2001.

[3]   Graphite Specialists Meeting, Oak Ridge National Laboratory, Oak Ridge, Tennessee, USA, 5-6 September 2000.

Figure 1. Summary of HTR-M & HTR-M1 work programmes

Materials for the High-temperature Reactor

Vessel Materials

Review of existing RPV materials used on gas-cooled & other reactors
Establish a database for RPV steels at low and elevated temperatures
Tests on steel RPV welded joints: tensile/creep, cross weld, creep/fatigue, fracture as required.
Tensile/creep fracture tests on specimens irradiated in HFR at NRG
Synthesis of results

High-temperature Materials

Internal Structures
Identification of materials & compiling of existing data.
Selection of most promising grades for further R&D effort.
Development and testing of available alloys for HTR requirements
Test work on control rod cladding materials. Mechanical/creep tests at temperatures up to 1 100°C in facility at CEA

Turbine
Identification of the materials & compiling of existing data
Selection of most promising grades for further R&D
Development and testing of available alloys to meet HTR requirements
Test work on turbine disc and blade materials. Short-term tensile/creep tests (air, vacuum) from 850 to 1300°C, Fatigue tests at 1000°C at CEA. Short-term creep & creep/fatigue tests in helium at JRC on servo hydraulic test rig.
Medium-term creep tests for blade material (10 000 h) at 850°C
Lifetime modelling based on notched specimen tests

Graphite

Properties
State of the art and first version of database of properties
Assess requirement for new graphites and needs of future HTRs
Discussions with manufacturers on producing small quantities of alternative graphites
Irradiation programme on chosen graphites to determine variations in physical and mechanical properties up to low/medium irradiation doses

Oxidation
Tests to obtain kinetic data on fuel matrix & structural graphite in oxygen (823 and 1 023 K – partial pressures 2 & 20 kPa – using thermo-gravimetric facility THERA at FZJ.
Testing of oxygen resistance of advanced C-based materials (CFCs, doped materials, SiC), steam & air using oxidation facility INDEX of FZJ

Figure 2. Allowable stresses for RPV materials

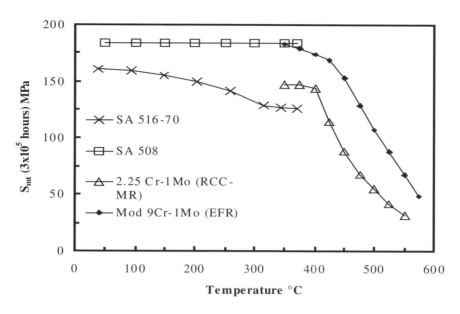

Figure 3. Materials temperature versus design code temperatures

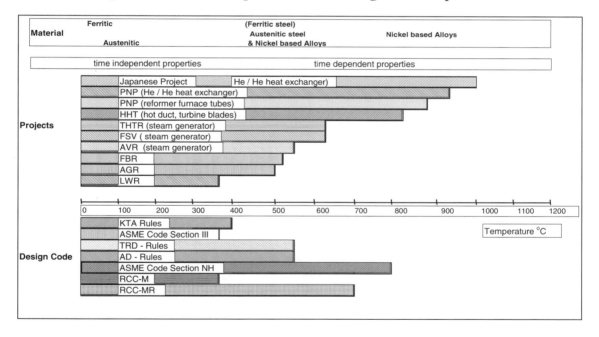

RADIALLY KEYED GRAPHITE MODERATOR CORES:
AN INVESTIGATION INTO THE STABILITY OF FINITE ELEMENT MODELS

W.R. Taylor[1], M.D. Warner[1], G.B. Neighbour[2], B. McEnaney, S.E. Clift[1]
Bath Nuclear Energy Group, Department of Engineering and Applied Science
University of Bath
Bath BA2 7AY, UK

Abstract

Finite element (FE) approaches are widely used to successfully predict the behaviour of engineering structures. However, the mechanical interactions between the keys and blocks in graphite moderator cores present particular computational problems due to the large number of contacting surfaces. Such models can often be computationally unstable. In this study, modelling strategies designed to improve solution stability have been explored. Results indicate that the incorporation of inter-layer friction can significantly enhance computational stability.

[1] Department of Mechanical Engineering, University of Bath, Bath BA2 7AY, UK
[2] Present address: Department of Engineering, University of Hull, Hull HU6 7RX, UK.

Introduction

The majority of the United Kingdom's gas-cooled thermal nuclear reactors have a graphite moderator core of the radially keyed design type. This core design is composed of large graphite bricks interspaced with interlocking moderator keys, geometrically configured in a manner which produces an overall negative Poisson's ratio under thermal and mechanical influences. It is critically important that the principal safety functions of control rod insertion and fuel cooling can be maintained throughout the reactor life. Based on existing evidence, it is presently believed that there is a large degree of redundancy in this functionality but this has yet to be quantified. Ultimately, the influence of factors such as cracked and broken keys on the relative displacement of the large number of core blocks needs to be evaluated in order to ensure safe operation.

The possibilities of assessing graphite moderator core behaviour in service are obviously extremely limited. One approachable way to proceed with an exploration of the mechanical behaviour of the core is to develop computer models, finite element (FE) analysis providing an obvious choice of technique [1,2]. There are, however, potential difficulties associated with this type of numerical approach. Graphite core behaviour can be highly non-linear due to the large number of independent bodies and contacting surfaces, and this can produce unstable or highly inefficient solutions.

The aim of this study was therefore to establish a stable modelling strategy which allowed the effect of contact conditions and restraints on key motion of a single key and a 3×3 multiple key-keyway model configuration to be investigated. While the work has been prompted by current UK requirements, the FE methodologies deployed are relevant to all graphite-moderated, high-temperature nuclear reactors.

Methodology

Investigations into key movement and analysis stability were first explored using a model of a single key and its surrounding keyway (ABAQUS FE software). The dimensions of the model were extracted from the geometry of a typical Magnox moderator core (Table 1), with an element height of 4.2 mm. For calculation efficiency eight noded elements were used with one internal integration point (CP8R: eight-node bilinear, with reduced integration). Contact surfaces were defined to allow relative motion between the key and the keyway bodies. The entire perimeter of the key was designated as the "master" surface, whilst "slave" surfaces were defined for both inner edges of the keyways.

Material properties for unirradiated Pile Grade "A" nuclear graphite show that the material exhibits high anisotropy (Table 2). The long axis of the moderator brick corresponds to the extrusion axis in manufacture and is denoted as the parallel direction. Since the finite element modelling was concerned with a horizontal slice of the moderator core, an isotropic model was used, with the perpendicular values taken from Table 2.

All entities of the model were constrained so that no out of plane motion could occur (z-direction). In addition, the lower keyway was completely restrained and the upper keyway was prevented from rotating. The model was loaded by applying a controlled displacement of 1 mm to the upper keyway in the y direction; this was deliberately applied at an oblique angle in order to produce a response involving multiple contacting surfaces. The behaviour of the model was then predicted.

Results and model developments

The model was run under the described conditions, and the upper keyway closed sufficiently to cause contact between the key and the upper keyway and the resulting displacement of the key.

Table 1. Geometrical data used in the model construction

Typical data, provided by BNFL

Model component	Dimensions (mm)
Diameter of all fuel channels	98.4
Square brick depth and width	170.7
Octagonal brick depth and width	221.0
Octagonal brick side face	170.7
Corner key depth	42.7
Corner key width	17.6
Side face key depth	30.0
Side face key width	17.6
All keyway widths	17.8
Keyway depth (side face, square brick)	16.7
Keyway depth (side face, octagonal brick)	18.5
Keyway depth (corner face, octagonal brick)	23.6
Key/keyway clearances	0.2
Wigner gap (side-faces)	1.1

Table 2. Pile Grade "A" materials properties

Typical data, provided by BNFL

Young's modulus		
Tension	11.0 ± 3.4 GPa	Parallel
Compression	7.6 ± 2.1 GPa	
Tension	4.8 ± 1.4 GPa	Perpendicular
Compression	4.1 ± 1.4 GPa	
Coefficient of friction		
Dynamic (>250°C)	0.03 to 0.08	Parallel and perpendicular
Static (>250°C) (add to dynamic values)	+0.05 to 0.10	Parallel and perpendicular
Poisson's ratio		
ν	≈ 0.1	Parallel and perpendicular

Figure 1. 3-D model of a single key-keyway

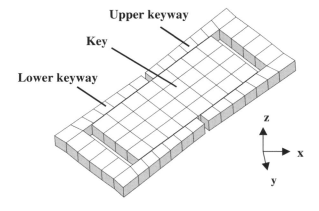

However, the solution failed to converge after only 15% of the prescribed displacement following severe cutbacks in time incrementation and excessive discontinuity iterations. This instability was attributed to a number of factors. Any change of direction in the key may cause numerical instabilities that the solution routine needs to deal with. Multiple changes in directions, which can be caused by multiple movements or rotations, cause "key rattle". The result of this is cutbacks in solution increments and possible failure of the solution routines to find a convergent solution. An additional instability of the system can be caused by unconstrained rigid body motion – this results from the lack of connection between the graphite blocks and keys that are not in contact. This unconstrained motion allows free movement of the keys within the keyways and hence allows key "bounce", a process that can continue *ad infinitum*. This causes the same numerical problems evident in key rattle.

As a result of this failure of the solution to converge, the model was run under a variety of modelling and contact conditions, each tested to improve the stability of the solution. These were run as follows.

Altering contact surface conditions

The construction of the contacting surfaces was altered from a single contacting key-keyway contact pair (for each of the upper and lower keyways) to sets of three contacting surfaces located around the edges of the key (Figure 2).

Figure 2. Location of master and slave surfaces in a key-keyway contact pair

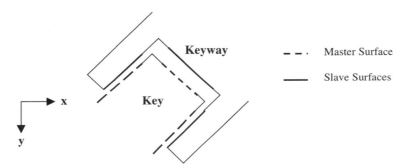

This approach produced improved solution efficiency as the master surface had a constant tangent direction for contact with each slave surface. The corner nodes of the key were not included in the master surfaces and were therefore able to pass through the slave surface. Due to the small key rotation however, this penetration never exceeded 0.08 mm.

This modification of the contact configuration produced a convergent solution, for the complete 1.0 mm displacement. The occurrence of "rattle" behaviour was confirmed and can be seen in Figure 3, which shows the displacement of the centre of the key. Whilst the solution finally converged, severe time increment cutbacks appeared in the solution, including discontinuity iterations.

Restriction of key separation after initial contact

In this investigation, the contact surfaces were restrained from re-separation once initial contact had occurred. The instability of the key motion was reduced, and a stable solution was reached. However, comparison of the key motion with previous solutions showed that the technique influenced the path of the displacement and the final resting positions of the key within the keyway. This procedure was considered inappropriate as it artificially influenced the final solution.

Figure 3. Displacement behaviour of the centre of the key

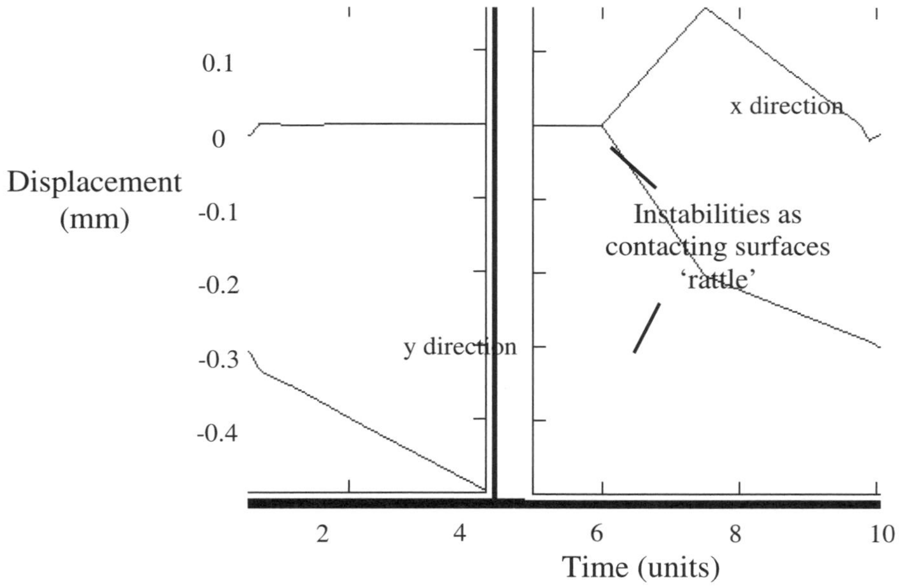

Modelling of key displacements with springs

Spring elements of very low stiffness were attached to the key in order to reduce the rattle effects of the key within the keyway. One end of the spring was situated in the centre of the key, the other was connected to ground. In addition, a small resistance to rotational motion was introduced to the system using a rotational spring. The concept was that forces may have been able to reduce any bouncing or rattle movements of the key and stop rigid body motion.

Despite the initial analysis stability after first contact with the keyway, however, further displacement caused the tendency for the key to slide and move in an attempt to return to its original position relative to the ground. As a result, this process presented a problem due to artificially imposing motion conditions on the solution. Although the problem was small in this case, since any significant contact pressure was enough to overcome these spring forces, solution stability was compromised since the direction of key displacements were often reversed in subsequent increments. Like many of the previous modelling techniques, this process caused inefficient solutions due to incremental cutbacks.

Investigation into the use of friction and mass

All springs were removed from the model. An underlying rigid plane surface was then defined against which inter-layer friction could be generated. The underside of the key was selected as a contact surface with this plane, whilst the base of the keyway elements was simply constrained in the z direction. In an initial process test, a normal force of 103N was applied to the centre of the key to simulate the weight of the key and that of the keys in the nine layers above. This provided the perpendicular force necessary for a frictional force between the key and the underlying surface to be generated.

219

Although the solution of each increment was slightly longer in comparison to previous models (on account of more contact surfaces), the full 1 mm displacement was achieved with this model, with minimal cutbacks in time incrementation. As a development of the most stable solution tested, mass effects replaced the force used to create friction, in an attempt to model this process more accurately. This was achieved by assigning a density of 1 731 kg/m^3 to the graphite key. A gravitational acceleration field was then applied to the system, perpendicular to the rigid layer (z direction). The two keyway components of the single key model were not assigned a value of density as they were fully constrained by the boundary conditions imposed. The result of these new conditions provided identical analysis stability. The modelling technique was therefore retained, as the conditions were considered more accurate than applying a single point force, which could locally distort elements and produce peaks in friction and hence artificial moments and forces.

Extrapolation of the single keyway model findings to larger models

A single generic unit cell, consisting of 1 226 solid elements, was designed to model a square section of the core surrounding either an interstitial channel, or, with an additional centre section, a corner key section (Figure 4).

Figure 4. Single generic unit cell

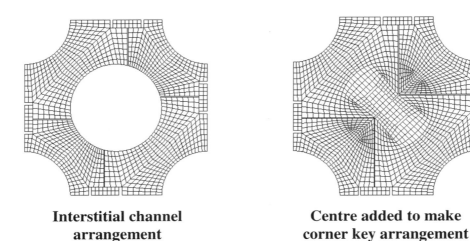

Interstitial channel arrangement	**Centre added to make corner key arrangement**

A 3 × 3 unit cell arrangement was created (Table 2/Figure 5) and a ramp uni-axial displacement of 5 mm was applied in order to assess the ability of this type of geometry to respond with a negative Poisson's ratio. The compression was applied over a period of 10 seconds. Simple boundary conditions were used. These were designed to produce multiple contacting surfaces. A much larger model, however, would be required to predict the effect of the actual core boundary conditions. The complete outer bricks enabled the displacements to be applied to the octagonal bricks only, in the manner of restraint rods at the core periphery.

Figure 6 shows the final position of the corner key and key-way that experienced the highest displacement. It can be seen that the key has experienced an anti-clockwise rotation and loaded in shear due to the relative movement of the two octagonal bricks that comprise the keyway. Transference of displacement occurring throughout the model is shown on an undeformed model in Figure 7.

Figure 5. Boundary conditions applied for uniaxial compressive loading

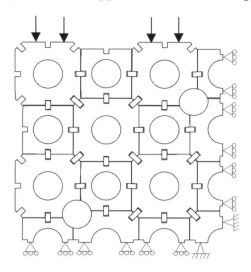

Figure 6. Key and keyway position after a uniaxial displacement of 5 mm was applied to the upper surface of the 3 × 3 model

upper brick
motion

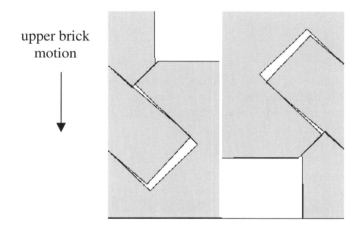

Figure 7. Displacement in the model (mm) under 5 mm uniaxial loading

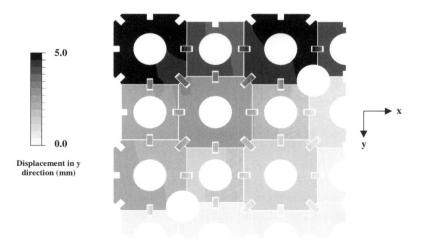

5.0

0.0

Displacement in y
direction (mm)

x

y

In a separate analysis, the displacement was allowed to continue until no more movement was possible. It was found that the outer surface could be displaced approximately 7 mm. At this point, all bricks and keys in the 3 × 3 model had come into contact with their immediate neighbours, i.e. all the gaps had been closed up. In addition, larger model sizes have been tested. The increasing model complexity significantly reduced the efficiency of the analyses. The dramatic increase in analysis time is considered a result of the increase in the number of contact surfaces required in the model. The resulting stress field is shown in Figure 8.

Figure 8. Distribution of von Mises stress under 5 mm uniaxial compressive displacement

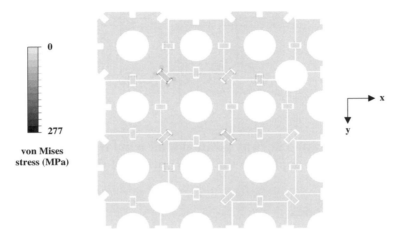

One possibility of improving model efficiency would be to reduce the level of detail at the centre of the graphite bricks. It is likely that no significant changes in the FE solutions would result from this simplification as the centre of the bricks experience very little stress or strain (Figure 7) as the problem is primarily one of contact between the bricks and keys.

Conclusions and recommendations for further work

- This study has shown that the graphite bricks themselves deform very little. The deformation applied to the model was accommodated by closing up the gaps between the layers of bricks.

- Finite element analyses of key-keyway interactions that use springs to represent the interactions between the contacting surfaces displayed unstable behaviour associated with keys rattling inside the keyways. This led to rapidly changing contact conditions and failure of the model to produce a convergent solution.

- The incorporation of inter-layer friction between the model and a rigid substrate produced a stable solution in the single key and multi-key models. However, the large number of contacting surfaces and the necessity to use three-dimensional elements significantly affected the efficiency of the model and produced impractical analysis times.

- Recommendation for further work: As the bricks themselves have been shown to undergo minimal deformation, the FE models could be made more efficient by reducing the numbers of elements used to represent them. This could convey considerable advantage in reducing the computer run times. This would allow the behaviour of larger models, with more realistic boundary conditions and materials properties, to be explored.

Acknowledgements

This work was funded by H M Nuclear Installations Inspectorate, Health and Safety Executive and is published with their permission. The views expressed in this paper are those of the authors and do not necessarily represent the views of the Inspectorate.

REFERENCES

[1]   N.P. Blackburn, P.J. Ford, "Impact Models for Nuclear Reactor Graphite Components under Seismic Loading", *Nuclear Energy*, 35, pp. 375-384 (1996).

[2]   M. Futakawa, S. Takada, H. Takeishi, T. Iyoku, "Evaluation of Aseismic Integrity in HTTR Core-bottom Structure V. On the Static and Dynamic Behaviour of Graphitic HTTR Key-keyway Structures", *Nuclear Energy and Design*, 166, pp. 47-54 (1996).

UNDERSTANDING OF MECHANICAL PROPERTIES OF GRAPHITE ON THE BASIS OF MESOSCOPIC MICROSTRUCTURE (REVIEW)

M. Ishihara, T. Shibata, T. Takahashi, S. Baba, T. Hoshiya
Japan Atomic Energy Research Institute
Oarai Research Institute
3607, Oarai-machi, Higashiibaraki-gun, Ibaraki-ken, 311-1394, Japan

Abstract

With the aim of nuclear application of ceramics in the high-temperature engineering field, the authors have investigated the mesoscopic microstructure related to the mechanical and thermal properties of ceramics. In this paper, recent activities concerning mechanical properties, strength and Young's modulus are presented.

In the strength research field, the brittle fracture model considering pore/grain mesoscopic microstructure was expanded so as to render possible an estimation of the strength under stress gradient conditions. Furthermore, the model was expanded to treat the pore/crack interaction effect. The performance of the developed model was investigated from a comparison with experimental data and the Weibull strength theory.

In the field of Young's modulus research, ultrasonic wave propagation was investigated using the pore/wave interaction model. Three kinds of interaction modes are treated in the model. The model was applied to the graphite, and its applicability was investigated through comparison with experimental data.

Introduction

In general, it is important to study macroscopic properties such as mechanical and thermal properties from a viewpoint of mesoscopic microstructure in order to develop microstructurally controlled new materials with the necessary material properties as well as to develop design and/or maintenance methods for structural materials. The authors have investigated the mesoscopic microstructure related mechanical properties as well as thermal properties of ceramics aiming at the nuclear application of ceramics in the high-temperature engineering field. In the paper, our recent activities with regard to the mechanical properties, strength and Young's modulus are presented on the basis of mesoscopic microstructure.

In the field of strength research, a brittle fracture model taking the mesoscopic microstructure into account was proposed [1]. Several researchers have expanded the model so as to treat the strength under stress gradient condition [2-5], e.g. bending strength, etc., since the model was applicable only to a uniform stress condition; specifically, the model can predict only tensile strength. Furthermore, we have modified the model so as to treat the interaction effect between crack and pore in order to predict strength over a wide range, including under oxidation conditions, etc. [6]. Moreover, the performance of the mesoscopic microstructure based fracture model was investigated from a comparison with the Weibull strength theory [4], which is generally applied in ceramics strength research.

In the field of the Young's modulus research, a prediction model has been developed [7-11]. In the model, the interaction between ultrasonic wave and mesoscopic pores is considered. Three kinds of wave/pore interaction modes are treated in the model:

1) *Direct wave mode*. The wave has no interaction with pores.

2) *Creeping wave mode*. The wave impinges on pores and then it propagates around the pore as a creeping wave.

3) *Direct wave mode*. The wave impinges on pores and scatters away.

The model can predict the ultrasonic wave propagation velocity as well as the height of echo signal due to pores, and the Young's modulus is obtained on the basis of wave propagation theory within a solid. The Monte Carlo simulation technique was also used so as to predict the non-uniformly distributed pore arrangements [12].

Generally speaking, the models presented are basically applicable to the other ceramic materials, such as the superplastic material 3Y-TZP zirconia (3 mol% yttria stabilised tetragonal zirconia polycrystal), silicon carbide, etc.

Strength-related activity [3-6]

Mesoscopic microstructure-based fracture model [4]

Uniform stress condition (Burchell model) [1]

The grain of graphite consists of a stack of parallel hexagonal net planes as schematically shown in Figure 1. Due to the weak van der Waais force operating in the *c* axial direction and the strong covalent bond connection in the *a* axial direction, cleavage within the grain occurs easily in the *c* axial

Figure 1. Microstructure-based fracture model

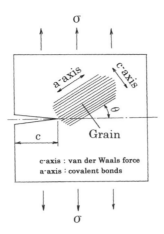

direction. If the inherent flaw in the graphite body faces a grain having inclination angle θ (as shown in Figure 1), the crack would deviate from its extension direction. When the uniform stress σ acts on inherent flaw size c, the probability that one grain will fracture, P_f, is:

$$P_f = \frac{4}{\pi} \cos^{-1} \left(\frac{K_{IC}}{\sigma \sqrt{\pi c}} \right)^{\frac{1}{3}} \tag{1}$$

where K_{IC} is a fracture toughness for grains, so-called particle K_{IC}. If there are n grains in the entire row, that probability that all grains will fail, P_n, is:

$$P_n = \left[\frac{4}{\pi} \cos^{-1} \left(\frac{K_{IC}}{\sigma \sqrt{\pi c}} \right)^{\frac{1}{3}} \right]^n \tag{2}$$

Here, it is assumed that the flaw will extend at a grain size, a, if the entire row with n grains fail. Therefore, Eq. (2) is thought to be the probability that the flaw size c will extend from length c to length $c + a$. The probability that the flaw will extend from c to $c + ia$, i.e. fracturing i rows of grains, is then expressed as:

$$P_n = \prod_{i=0}^{i} \left[\frac{4}{\pi} \cos^{-1} \left(\frac{K_{IC}}{\sigma \sqrt{\pi (c + ia)}} \right)^{\frac{1}{3}} \right]^n \tag{3}$$

which then may be approximated to:

$$\ln P_n = n \int_0^i \ln \left[\frac{4}{\pi} \cos^{-1} \left(\frac{K_{IC}}{\sigma \sqrt{\pi (c + ia)}} \right)^{\frac{1}{3}} \right] di \tag{4}$$

Now, when the probability $f(c)dc$ means that initial flaw size has a length between c and $c + dc$, the probability that one tip of a single flaw will fracture under applied stress σ may be written as:

$$\int_0^\infty f(c) \cdot p_n(\sigma, c) dc \qquad (5)$$

The survival probability P_s is then given as:

$$P_s = 1 - \int_0^\infty f(c) \cdot p_n(\sigma, c) dc \qquad (6)$$

Here, when N is the number of pores per unit volume and V is the specimen volume, the total survival probability of the volume V under stress σ, having $2NV$ flaw tips is $(P_s)^{2NV}$. Therefore, the total fracture probability of the specimen may be written as:

$$P_{ftot} = 1 - (P_s)^{2NV} = 1 - \left[1 - \int_0^\infty f(c) \cdot p_n(\sigma, c) dc \right]^{2NV} \qquad (7)$$

Stress gradient condition (expanded model with divided small elements) [2,3]

If the specimen has a large stress gradient, it would be necessary to divide into small elements so as to regard the uniform stress within the small elements as schematically shown in Figure 2. When the stress and volume of (k_k, j_j) element are $\sigma(k_k, j_j)$ and $V(k_k, j_j)$, the survival probability of the element is:

$$P_s(k_k, j_j) = \left[1 - \int_0^\infty f(c) \cdot p_n \{\sigma(k_k, j_j), c\} dc \right]^{2NV(k_k, j_j)} \qquad (8)$$

Therefore, the total fracture probability of the specimen may be written as:

$$P_{tot} = 1 - \prod_{k=k_1}^{k_n} \prod_{j=j_1}^{j_m} \left[1 - \int_0^\infty f(c) \cdot p_n \{\sigma(k_k, j_j), c\} dc \right]^{2NV(k_k, j_j)} \qquad (9)$$

Figure 2. Application of microstructure-based brittle fracture model to stress gradient condition

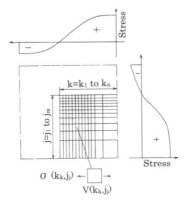

228

Application of the model

Tensile strength

The tensile strength of nuclear grade graphite, H-451, was estimated by Eq. (7) on the basis of the microstructure-based brittle fracture model. Input parameters are summarised in Table 1.

Table 1. Input parameters of H-451 graphite

Parameter	Value
Mean grain size (μm)	500*
Bulk density (g/cm^3)	1.79*
Mean pore size (μm)	42*
Standard deviation parameter of pore size	2.3
Mean pore area (μm^2)	700*
Number of pores per volume (m^{-3})	2.97×10^8*
Specimen volume (m^{-3})	4.65×10^{-7}**
Specimen breadth(mm)	10.29**
Grain fracture toughness (MN/m$^{3/2}$)	0.285*

* From Ref. [1].
** From Ref. [13].

In the table microstructural parameters were obtained by Burchell from an image analysis with microstructural observation [1], and specimen size was determined by the experimental data of the graphite [13]. Here, pore size distribution $f(c)$ in Eq. (7) is assumed to be a log-normal statistical distribution. The standard deviation parameter is determined so as to fit the mean tensile strength in the analysis. Figure 3 shows the prediction and experimental results. We can see from this figure that the microstructure-based brittle fracture model has a fairly good prediction of the tensile strength distribution.

**Figure 3. Prediction of tensile strength of H-451
graphite by microstructure-based brittle fracture model**

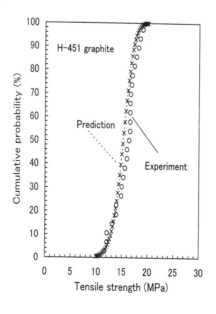

Tensile strength distribution

The experimental data are plotted with three kinds of probability plots; they are the Weibull, normal and log-normal probability plots. Figure 4 shows the Weibull probability plot of experimental data, and correlation coefficients for them are listed in Table 2. From them it is found that the tensile strength fits in well with the Weibull statistical distribution rather than the other two distributions. It is presumed, therefore, that the fracture of graphite might obey the so-called "weakest link theory", and the "weakest link" might correspond to the defect of the origin of fracture among many defects within the graphite body. The predicted result is also plotted with the Weibull probability plot in Figure 5. The prediction fit in well with the Weibull statistical distribution as opposed to the experimental data. Correlation coefficients of the prediction are also listed in Table 2. The prediction result shows the same tendency as the experimental data, i.e. both experimental and predicted data fit in well with the Weibull probability plot and next with the normal probability plot.

Figure 4. Weibull probability plot of experimental data

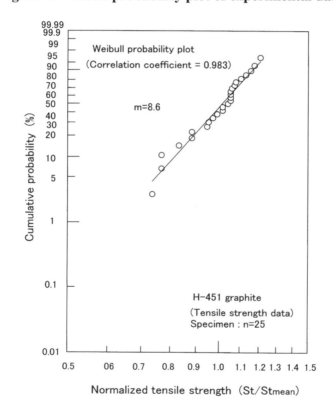

Table 2. Correlation coefficients for three kinds of statistical distributions

	Experiment*	Prediction
Weibull	0.983	0.998
Normal	0.974	0.997
Log-normal	0.960	0.987

* From Ref. [13].

Figure 5. Weibull probability plot of predicted data

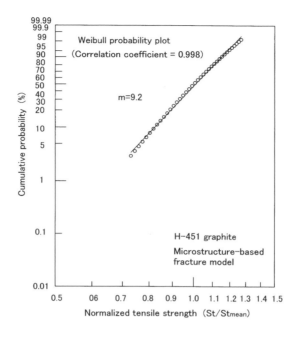

Bending strength

Microstructure-based brittle fracture model was applied to the four-point bending test in Eq. (9). The predicted analytical result is plotted with experimental data [13] in Figure 6. We can see from this figure that the model has a good bending strength prediction, although the predicted variance of the bending strength is somewhat smaller than the experiment; namely, the expanded model with divided small elements is applicable to the bending strength condition.

Figure 6. Prediction of four-point bending strength of H-451 graphite by microstructure-based brittle fracture model

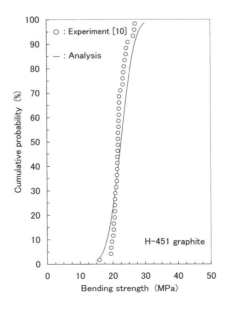

Comparison of bending strength with Weibull strength theory

The prediction by Weibull theory was also carried out to the bending strength condition. Here, the Weibull modulus, *m*, is determined by the tensile strength data [13] from a Weibull probability plot as shown in Figure 4. The predicted mean bending strength was 22.1 MPa, which is a fairly good prediction compared with experimental mean bending strength of 22.0 MPa. However, it is not possible to predict the strength distribution such as the microstructure-based brittle fracture model; this is a disadvantage of the theory from a viewpoint of an assessment of structural integrity.

Grain size effect on strength

Figure 7 shows the grain size effect on the tensile strength predicted by Eq. (7). In the prediction the pore size is assumed to be constant as listed in Table 1. We can see from this figure that the fine-grain material has high tensile strength. Figure 8 shows the Weibull probability plot of different grain size with constant pore size distribution. From this figure it is found that the material with fine grain size has large scatter of the strength. This is a reasonable trend if the relative pore size is taken into account; when the grain size becomes smaller, the pore size becomes relatively large, and at that time the pores would act well as the origin of fracture.

Pore distribution effect on strength

Figure 9 shows the pore size distribution as a function of standard deviation parameter in the log-normal statistical distribution. In this case the mean pore size is assumed to be constant at 42 μm. Figure 10 shows the pore distribution effect on the tensile strength predicted by Eq. (7). In the prediction the grain size is assumed to be constant at 500 μm. The material with large pore deviation parameter has low tensile strength, because there are many large-size pores in the material. Figure 11 shows the Weibull probability plot of different pore size distributions. It is understandable that the material with large pore deviation parameter shows large scatter of the strength, because there are also many large pores in the material.

Figure 7. Grain size effect on tensile strength

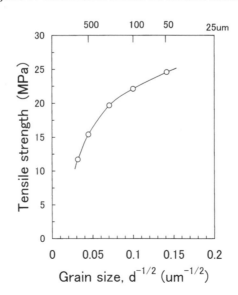

Figure 8. Weibull probability plot of different grain size

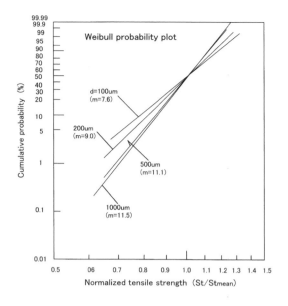

Figure 9. Pore size distribution

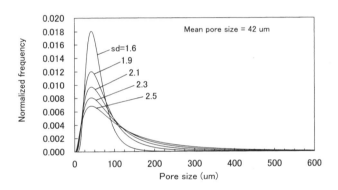

Figure 10. Strength change with pore deviation parameter

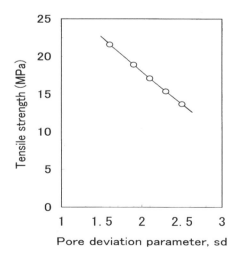

Figure 11. Weibull probability plot of different pore deviation parameter

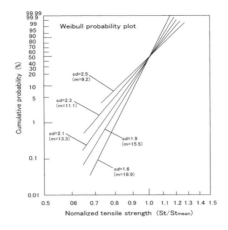

Young's modulus related activity [9-12,14-16]

Wave/pore interaction mode [7,8]

Pores with a radius r are assumed to be distributed homogeneously at a constant pitch p with N layers as shown in Figure 12 [7]. When the ultrasonic wave propagates from the left side to the right side, the waveform was changed by the pore-wave interaction. Three kinds of wave propagation mode due to pore-wave interactions were considered in the simulation analysis: the direct wave, creeping wave and scattering wave modes as shown in Figure 13 [7]. If we consider that the collision probability of incident ultrasonic wave to a pore is f, then the probability of direct wave is $1 - f$. Now, if c is the probability of the creeping wave after impinges on the pore at probability f, the probability of creeping and scattering waves are cf and $(1 - c) \times f$, respectively. Therefore the propagated wave becomes a mixed-mode wave in combination with both direct and creeping waves. Now, the time delay between the direct and creeping waves is:

$$t_d = \frac{l_p/4}{V_c} - \frac{r}{V_p} \tag{10}$$

where l_p is the circular length of the pore with radius r, and V_c and V_p are the sound velocities of creeping and direct waves, respectively.

Figure 12. Ultrasonic wave propagation within porous body [7]

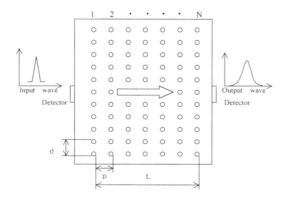

Figure 13. Pore-wave interaction model [7]

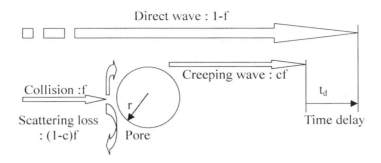

If the number of pore layers becomes large enough, the propagating ultrasonic waveform as a function of time, $h(t)$, becomes a Gaussian statistical distribution, and $h(t)$ is given by [7]:

$$h(t) = H \cdot e^{-\frac{(t-t_p)^2}{2t_w^2}}$$ (11)

where t_p, t_w and H are:

$$t_p = \{a + \sigma^2 \cdot \ln(c)\} \cdot t_d$$ (12)
$$t_w = \sigma \cdot t_d$$
$$H = \frac{c^{a+\frac{\sigma^2}{2} \cdot \ln(c)}}{t_d \cdot \sigma \sqrt{2\pi}}$$

where a and σ are given by:

$$a = N \cdot \phi$$ (13)
$$\sigma^2 = N \cdot \phi \cdot (1 - \phi)$$
$$\phi = \pi \cdot \sqrt[3]{\frac{4\pi}{3P_{or}}}^{-2}$$

Here, in the Gaussian statistical distribution given by Eq. (11), both mean value and maximum H is given by $t = tp$. In the case of the graphite, the porosity, P_{or} in Eq. (13), was determined from the true density of the graphite as follows:

$$P_{or} = \frac{2.26 - \gamma}{2.26}$$ (14)

where γ is a bulk density.

If the sound velocity of the longitudinal wave in the solid body is taken into account, the sound velocity is given by:

$$v_l = \sqrt{\frac{E(1-v)}{\gamma(1+v)(1-2v)}}$$ (15)

235

where E and ν are the Young's modulus and Poisson's ratio, respectively. Then the Young's modulus with pore is calculated by:

$$\frac{E}{E_0} = \frac{\left(1 - P_{or}\right)}{\left\{\frac{3}{4} \cdot P_{or} \cdot \frac{1}{r} \cdot \left(\frac{l_p}{4\alpha} - a\right) + 1\right\}^2} \tag{16}$$

where E_0 and E are Young's moduli without and with pore conditions, respectively. α is the ratio of sound velocity, and is given by:

$$\alpha = \frac{V_c}{V_p} \tag{17}$$

Analytical condition of graphite

Input parameter

The typical graphite properties of fine-grained nuclear grade graphite are listed in Table 3. Analysis input parameters include porosity P_{or}, pore radius r, probability of the creeping wave c and sound velocities V_c and V_p.

Table 3. Typical properties of fine-grained isotropic nuclear grade graphite (IG-110 graphite)

Bulk density (kg/m^3)	1.78×10^3
Tensile strength (MPa)	25.3
Compressive strength (MPa)	76.8
Young's modulus (GPa)	7.9
Thermal expansion coefficient (293 to 673 K) (10^{-6}/K)	4.06
Thermal conductivity (673 K) (W/(mK))	80

Since mean pore radius was estimated at 9.2 µm for fine-grained nuclear graphite, IG-110, by a quantitative image analysis technique [2], the pore radius was determined to be 9 µm in the analysis. The sound velocity of the creeping wave around the spherical pore was assumed to be $0.71 \times V_p$ [8] in this study. The input parameters used in the analysis are summarised in Table 4.

Table 4. Input parameters of simulation analysis

Porosity		0.2124
Pore radius (µm)		9
Wave velocity (m/s)	Direct wave	2 100
	Creeping wave	$2\ 100 \times 0.71$

Probability of creeping wave

In the pore-wave interaction model, the probability of the creeping wave is concerned with the attenuation characteristics of the ultrasonic wave. Therefore, at the beginning of the analysis, the probability of the creeping wave was determined from the attenuation data of the fine-grained nuclear graphite, IG-110 [14].

Figure 14 shows the noise echo height versus distance from the surface as a function of the probability of the creeping wave. The noise echo height decreased as distance increased, and also as the probability of the creeping wave decreased. This is an understandable change when taking into account the total number of the scattering wave, i.e. the total number increases when the distance from the surface increases due to the increasing of the pore-wave interaction chance as well as when the probability of the creeping wave decreases. From this figure the probability of the creeping wave is determined as $1 - c = 3 \times 10^{-4}$ in the analysis.

Figure 14. Noise echo height versus distance from the surface as a function of probability of creeping wave

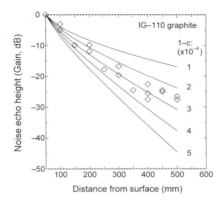

Application of the model

Estimation of Young's modulus

The ultrasonic wave/pore interaction model was applied under the oxidation condition, and the Young's modulus change with oxidation was estimated. The oxidation induced weight loss, burn-off, was considered as bulk density in the analysis in Eq. (14), and the Young's modulus was estimated by Eq. (16). Here, the Young's modulus without pore, E_0, was 17.5 GPa in this study, which is derived from the crystal Young's modulus in a and c axial directions with random grain orientation [16]. The analytical result is shown in Figure 15 with the experimental data. We can see from this figure that the interaction model predicts fairly well the Young's modulus under oxidation conditions.

Figure 15. Prediction of Young's modulus under oxidation condition (IG-11 graphite)

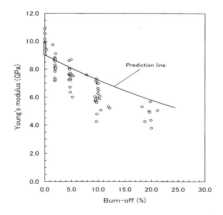

Now, the Young's modulus change with oxidation was analysed with several different kinds of graphites, fine-grained with small pores to coarse-grained with large pores, as listed in Table 5. Analytical results were plotted in Figure 16 in normalised value by each 0% burn-off value. We can see from this figure that the Young's modulus with the oxidation change can express the common line when the normalised value is applied.

Table 5. Input parameters in the oxidation effect analysis

Graphite	IG-11	PGX	H-451	AXF-5Q	AGX
Density (g/cm³)	1.72	1.74	1.79	1.8	1.648
Pore size (μm)	14	238	41	1	188

PGX from Ref. [2], all others from Ref. [1].

Figure 16. Young's modulus change with oxidation of several kinds of graphites

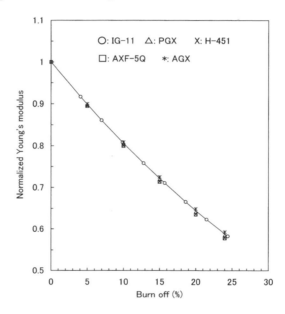

Estimation of oxidation damage

One of the important subjects using the UT is to detect the oxidation damage in the graphite component, since the oxidation greatly affects the strength of the graphite. If the graphite is oxidised by oxygen, water, etc., pores are produced within the graphite. Since the created pores would act as a crack initiation and/or extension, the strength of the graphite is reduced with oxidation. Therefore the wave propagation within the oxidised graphite components was investigated through simulation analysis. The pore radius was determined by the fracture toughness data under oxidation taking the fracture mechanics approach into account. Sato, *et al.* reported the mode-I fracture toughness under oxidised conditions of fine-grained nuclear graphite, IG-110, and they provide the following experimental equation:

$$\frac{K_{IC}}{K_{IC0}} = \exp(-5.27B) \tag{18}$$

where K_{IC} and K_{IC0} are the mode-I fracture toughness for oxidised and un-oxidised conditions, respectively. B is the so-called "burn-off" defined by the weights, W_0 and W, before and after oxidation, respectively.

$$B = \frac{W_0 - W}{W_0} \tag{19}$$

On the other hand, from tensile strength data under oxidation condition [15], the oxidation-induced strength degradation can be estimated as follows:

$$\frac{S_t}{S_{t0}} = \exp(-7.43B) \tag{20}$$

where S_t and S_{t0} are the tensile strength of oxidised and un-oxidised conditions.

From the fracture mechanics consideration:

$$K_I \propto S_t \sqrt{\pi \cdot c_r} \tag{21}$$

where c_r is a crack size. The crack size is then estimated by:

$$c_r \propto \left(\frac{K_I}{S_t}\right)^2 \propto \exp(4.32B) \tag{22}$$

Here, the crack size was assumed to be equal to the pore size in the analysis. The maximum waveform height in oxidation condition was normalised by that of un-oxidised condition, and plotted in Figure 17 as a function of burn-off. The ultrasonic signal echo height increases with increasing burn-off. For example the estimated echo height was about 1.5 times as large as that of the un-oxidised one at 5% burn-off. From this figure it is found that the UT is hardly applicable to the detection of the oxidation damage in graphite components under uniformly oxidised conditions, because normally the pass/fail judgement is done by the signal/noise ratio at 4, which corresponds to 12 dB. When the distributing oxidation condition, which would be caused by higher temperature oxidation at about 600 to 900°C and the near surface region have high burn-off, the UT might be an applicable method.

Figure 17. Ultrasonic signal echo height as a function of the burn-off rate

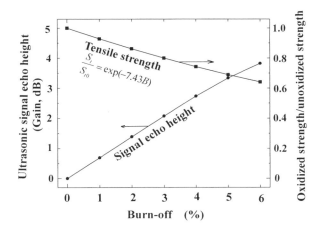

Summary

With the aim of nuclear application of ceramics in the high-temperature engineering field, the authors have investigated the mesoscopic microstructure-related mechanical properties as well as thermal properties of ceramics. The neutron irradiation study has been planned and proposed as an innovative basic research using the high-temperature engineering test reactor (HTTR). Currently, the preliminary study is being carried out using other research reactors such as the Japan Materials Testing Reactor (JMTR), etc., to determine the most effective irradiation test conditions using the HTTR.

In the paper, our recent activities with regard to the microstructure-related strength and Young's modulus are presented. To develop the mesoscopic microstructure-related material model is one of the key points in the research, and the model will be useful to develop microstructurally controlled new materials as well as to develop an advanced design and/or maintenance method for structural ceramics.

REFERENCES

[1] T.D. Burchell, *Carbon*, 34 [3], 297-316 (1996).

[2] K. Nakanishi, T. Arai and T.D. Burchell, International Symposium of Carbon, Tokyo, 332-333, November 1998.

[3] T. Takahashi, M. Ishihara, S. Baba and K. Hayashi, 1st World Conference on Carbon, Berlin, 397-398, 9-13 July 2000.

[4] M. Ishihara, T. Takahashi and S. Hanawa, 16th Int. Conf. on Structural Mechanics in Reactor Technology, Washington DC, August 2001, Paper No. 1920.

[5] M. Ishihara, T. Takahashi and S. Hanawa, presented at the Asian Pacific Conf. on Fracture and Strength and the Int. Conf. on Advanced Technology in Experimental Mechanics, Sendai, Japan October 2001.

[6] M. Ishihara and T. Takahashi, to be published.

[7] J. Takatsubo and S. Yamamoto, Trans. Jpn. Soc. Mech. Eng. (series A), 60 [577], 224-229 (1993) (in Japanese).

[8] J. Takatsubo and S. Yamamoto, Trans. Jpn. Soc. Mech. Eng. (series A), 60 [577], 230-235 (1993) (in Japanese).

[9] T. Shibata and M. Ishihara, *Nuclear Engineering and Design*, 203, 133-141 (2001).

[10] T. Shibata and M. Ishihara, Int. Symposium of Carbon, Tokyo, 642-643, November 1998.

[11] T. Shibata and M. Ishihara, 16th Int. Conf. on Structural Mechanics in Reactor Technology, Washington DC, Paper No.1114, August 2001.

[12]   M. Ishihara, T. Shibata and S. Hanawa, 8th Int. Conf. on Nuclear Engineering, Baltimore, USA, Paper No. ICONE-8517, April 2000.

[13]   General Atomic, "HTGR Generic Technology Program Fuels and Core Development", GA-A15093, pp. 3-1 to 3-17 and pp. 3-52 to 3-70 (1978).

[14]   M. Ishihara, *et al.*, Proc. 13th Int. Conf. on Structural Mechanics in Reactor Technology, Vol. 1 575-580 (1995).

[15]   M. Ishihara, *et al.*, JAERI-M 91-153 (1991).

[16]   T. Shibata and M. Ishihara, to be published.

ADVANCED GRAPHITE OXIDATION MODELS

Rainer Moormann, Hans-Klemens Hinssen
Forschungszentrum Jülich
ESS-FZJ, D 52425 Jülich
E-mail: R.Moormann@fz-juelich.de

Abstract

Graphite oxidation must be carefully considered in safety analyses for modern energy systems such as HTRs or fusion reactors. Restrictions of actual oxidation models and their improvement are discussed and some experimental data and calculation results concerning these subjects are presented. Advanced graphite oxidation models should solve the diffusion/reaction equation for in-pore diffusion controlled conditions (regime II), provided that burn-off dependent reactivities and diffusivities are known. It is suggested that a reasonable agreement between experimental rate data and calculation results is obtained only if graphite removal by erosion is considered in addition to gasification. Further, for high-temperature oxidation of graphite by oxygen, in regime III the increase of the mass transfer rate by a two-reaction mechanism should be taken into account. Finally, changes in geometry and configuration of carbon components induced by oxidation in regime III have to be considered.

Introduction

Graphite and other carbon-based materials are widely used in innovative energy systems like high-temperature reactors (HTRs) and fusion reactors. Safety considerations for these systems require a careful examination of accident scenarios with carbon oxidation (air ingress with carbon burning, water ingress). Accordingly, carbon oxidation codes were developed and used for these systems [1,2].

As described in more detail in [3], the chemical oxidation regime (regime I) is less important in safety examinations for accidents because of the comparatively low rates in this regime. Nevertheless, for oxidation under normal operation of HTR this has to be considered. In regime I volume-related rates r_I are given for small concentration ranges of the oxidising gas as:

$$r_I = k_{vo}\exp(-E_A/RT)c_e^n \cdot f_I(B_v) \qquad [mol \cdot m^{-3} \cdot s^{-1}] \qquad (1)$$

where n is the reaction order, k_{vo} is the rate constant, E_A is the activation energy, c_e is the educt gas concentration and B_v is the volume related burn-off, here used as a relative value normalised with the burn-off of the virgin material. For larger concentration ranges a Hinshelwood-Langmuir type equation has to be used, which also considers inhibition by reaction products [see right-hand side of Eq. (2)].

More relevant for safety analyses is regime II (in-pore diffusion controlled regime). At intermediate temperatures, consumption of oxidising gases within the pores becomes important and, accordingly, the oxidation attack becomes smaller in the depth of the material; this is due to the fact that the temperature dependence of the chemical process is much larger than that of the competing in-pore gas diffusion. Therefore, a concentration and – as a consequence of that – a burn-off (oxidation) gradient within the carbon is found. For integral isothermal rates an increase with (surface related) integral burn-off B_s up to a plateau value is usually observed [4]. Assuming carbon to be homogeneous with respect to gas transport and reactivity, Eq. (2), which is a superposition of second Ficks law and rate equation of regime I, has to be solved (written here for Hinshelwood-Langmuir kinetics [4,5], c_{pr} is the product gas concentration, x is material depth, ε is the transport porosity, D_{eff} is the effective gas diffusivity within the graphite and k are rate constants). The solution of Eq. (2) is substantially complicated by the fact that D_{eff} and k (particularly k_{vo} depend significantly on the local burn-off; this burn-off dependence is rarely known in a sufficient manner).

$$\varepsilon\frac{\partial c_e}{\partial t} = D_{eff} \cdot \frac{\partial^2 c_e}{\partial x^2} - \frac{k_{v0} \cdot \exp(-E_{A1}/RT) \cdot c_e}{1 + k_1 \cdot \exp(-E_{A2}/RT) \cdot c_e + k_3 \cdot \exp(-E_{A3}/RT) \cdot c_{pr}} \qquad (2)$$

The relation to the integral rates r_{II} of regime II is given by the concentration gradient on the geometrical surface:

$$r_{II} = -\left(D_{eff} \cdot \frac{\partial c_e}{\partial x}\right)_{x=0} \qquad [mol \cdot m^{-2} \cdot s^{-1}] \qquad (3)$$

Due to the aforementioned problems with the solution of Eq. (2), isothermally measured rates r_{II} have mainly been used up to now; this however has the disadvantage that non-isothermal processes cannot be exactly modelled and that information on the depth of oxidation attack is not obtained. Fitting equations for r_{II} use solutions of Eq. (2) for $B = 0$ containing an empirical burn-off dependence $f_{II}(B_s)$; as an example, the following expression for isothermal rates r_{II} is based on kinetics as in Eq. (1):

$$r_{II} = k_{II} \cdot \exp\left(-0.5 \cdot E_A/RT\right) \cdot c_e^{(n+1)/2} \cdot f_{II}\left(B_s\right) \qquad \left[mol \cdot m^{-2} \cdot s^{-1}\right] \tag{4}$$

with (for $B = 0$):

$$k_{II} = \sqrt{k_{v0} \cdot D_{eff}} \cdot 2/(n+1) \qquad \left[mol^{(1-n)/2}/\left(m^{0.5-1.5n} \cdot s\right)\right]$$

In contrast to this simplified approach, advanced carbon oxidation models require a complete solution for Eq. (2).

At high temperatures (regime III) external mass transfer to the outer surface is the rate-limiting step and the reaction is restricted to the geometrical surface only; mass transfer rules are usually applied here for rate calculations, meaning that kinetics are not influenced by the properties of the solid. It should be noted however, that mass transfer may be influenced under certain conditions by the chemical reaction, particularly in the case of air ingress. Carbon gasification can occur for air ingress conditions at high temperatures via the Boudouard reaction, whereas CO formed is oxidised by oxygen within the boundary layer to CO_2; this leads to a remarkable rate increase in comparison with mass transfer rules. Advanced oxidation models have to consider these influences. Further, graphite oxidation models have up to now been restricted more or less to conditions where the geometry and the configuration of the carbon components remain unchanged by oxidation. However for HTRs, in reality changing geometry will influence flow and mass transfer conditions; in case of large burn-off a pebble bed core may densify by movement of the debris of oxidised pebbles. Strength loss of block type HTR fuel or of reflectors may even lead to a break-down of the respective component combined with a complete change in flow conditions. Because the accident progress is substantially influenced by these geometry effects for later accident stages, these effects have to be carefully considered in safety analyses and should be modelled in advanced oxidation models.

For the approximate estimation of transition temperatures between different oxidation regimes Table 1 contains respective formula (valid strictly for $n = 1$ and $B_s = B_v = 0$; $l = V/Sexposed$, where $Sexposed$ = geometrical surface of the specimen exposed to the oxidising gas, V = specimen volume and β = mass transfer coefficient). Transition I-III – omitting regime II – occurs for $\beta = D_{eff}/l$ respectively $Sh/x_l < \psi/l$ (Sh = Sherwood number, x_l = characteristic specimen length as defined by analogy theory in flow/mass transfer, ψ = diffusion permeability D_{eff}/D).

Table 1. Approximate transition temperatures between oxidation ranges

Transition between oxidation regimes	Transition temperature [K]
I-II	$-E_A \Big/ \left(R \cdot \ln\left[\dfrac{D_{eff}}{k_{v0} \cdot l^2}\right]\right)$
II-III	$-E_A \Big/ \left(R \cdot \ln\left[\dfrac{\beta^2}{k_{v0} \cdot D_{eff}}\right]\right)$
I-III	$-E_A \Big/ \left(R \cdot \ln\left[\dfrac{\beta}{k_{v0} \cdot l}\right]\right)$

Numerical solution of the diffusion/reaction equation

Most relevant input data for models solving the diffusion/reaction Eq. (2) are burn-off dependent reactivities and diffusivities. In the next subsection, the data situation is briefly outlined for a typical nuclear grade graphite. Using these data, the computer model PROFIL is used for demonstration of the differences to the simplified approach using the example of reaction with air.

Burn-off dependent reactivities and effective diffusivities

Data on burn-off dependent reactivities of nuclear carbon are mainly available for oxidation in oxygen/air. For V483T fine-grain graphite (ash content: < 300 ppm, density: 1 810 kg/m^3) developed for use in the core support structure of HTRs, a continuous reactivity increase by a factor of 3.8 up to a burn-off of 0.37 is found at 7 73 K in air, followed by a nearly linear rate decrease [3]. This increase is due to enlargement of pores and the opening of blind pores, whereas the decrease is caused by reduction of the inner surface area due to oxidation-induced vanishing of pore walls; its activation energy is about 165 kJ/mol. For homogeneous graphites, no temperature dependence of the (normalised) reactivity versus the burn-off curve was found. Literature data on the regime I reaction order of other nuclear graphites in air (12-100 kPa) resulted in values of 0.54-0.60 [6]; sufficient data allowing for a fit on a Hinshelwood-Langmuir equation are not yet available. The more complicated behaviour of less homogeneous materials like the HTR fuel element matrix material A3-27 as reactivities of other types of nuclear carbons (CFCs, doped carbons) in regime I are given in [7]. Effective diffusion coefficients in porous carbons may be measured by static [8] or transient [9] methods; only transient methods are able to cover the effect of blind pores, which are also affected by oxidation. A systematic comparison of static and transient results is still to be done. Measurement of diffusion permeabilities at different pressures allow for the determination of the contribution of Knudsen diffusion and provides data on the average pore radius [6]. For V483T graphite it was shown through static measurements that the opening of closed micropores at low burn-off (air oxidation 673 K) led to a significant decrease of the average (transport) pore radius of A3-27 from 2.5 μm (virgin) to 0.45 μm (burn-off 0.15). Its diffusion permeability ψ increased in this burn-off range from $8 \cdot 10^{-3}$ to $5 \cdot 10^{-2}$. For burn-off dependence of diffusion permeability holds here: $\psi \sim B^{1.7}$. Data for the HTR fuel element matrix material are presented in [3].

Solutions of the transport/reaction equation with PROFIL

Numerical solutions of Eq. (2) are performed with the computer model PROFIL for V483T graphite because of its homogeneous character. The burn-off dependence of reactivity and permeability used are those used in the previous section. The oxidation medium is air at 0.1 MPa; a temperature-independent Hinshelwood-Langmuir term $k_{01} = k_1 \cdot \exp(-E_{A1}/RT) = 0.35\,[m^3/mol]$ is applied. No hindrance by external mass transfer (regime III) is taken into account. Geometry of a plane sheet (thickness = 0.02 m) is assumed with oxidation from both major surfaces.

At first, the relevance of instationary calculations were proven. Based on Crank [10], it was shown that quasi-stationary conditions ($dc_e/dt = 0$) are usually allowed in the solution of Eq. (2) because time constants for reaching diffusion/reaction equilibrium under constant burn-off are small compared with characteristic times, which are required for variation of burn-off dependent diffusion/reaction parameters by the chemical reaction.

Figure 1 contains numerical solutions of Eq. (2) for 1 073 K; r_{II} versus burn-off is given for different assumptions. Curve A represents a solution without consideration of any other weight loss than by chemical reaction. There is a strong rate increase, followed by slower decrease which accelerates in the final stage. Such behaviour is not observed under fast flow conditions, but may be found during oxidation in a thermogravimetric apparatus with small flow rates (low T). The rate increase is due to an increase of reactivity and porosity with burn-off at low burn-off values. The rate decrease is caused by the dominating decrease of reactivity at high burn-off, which also leads to an effect similar to a diffusion boundary layer. In the outer regions with high burn-off, the diffusivity is high and the reactivity is small and, accordingly, the mass transport resembles that in a boundary layer. Oxidation experience in regime II with flow, however, shows that there is a continuous size reduction due to erosion by flow-induced shear forces and some dust is usually found in these experiments (up to 20% of the total weight loss for turbulent flow in tubes, $Re = 2\,500$). Rate curves B model oxidation rates considering erosion, which is roughly approximated by an erosion rate r_e depending on the outer burn-off by $r_e = 1.4 \cdot 10^{-6} \cdot (B_{v,x=0} - 0.4)^2$ m/s $(B_{v,x=0} \geq 0.4)$. Erosion acts only on the outer surface. Curve B-1 contains the chemical oxidation rate and curve B-2 the total C-removal rate (chemical reaction plus erosion). Both curves lead to a plateau value, as observed in regime II under flow. A rate decrease as in curve A does not occur, because the aforementioned boundary-layer-like effect does not exist and an equilibrium between erosion, renewing of the outer surface, and chemical reactions with diffusion is established. This results in oxygen concentration profiles, which move nearly unchanged in shape through the porous carbon, as the density (oxidation) profiles do. Such oxygen concentration profiles are shown in Figure 2 for different time steps. For comparison one profile without calculation of erosion (H) is shown, too; the latter reveals the boundary-layer-like effect between 0-0.006 m. The depth of penetration of the oxidation effect is higher than estimated by usual formula, based on $B = 0$, even if erosion is considered.

Figure 1. Calculated regime II rates vs. integral burn-off

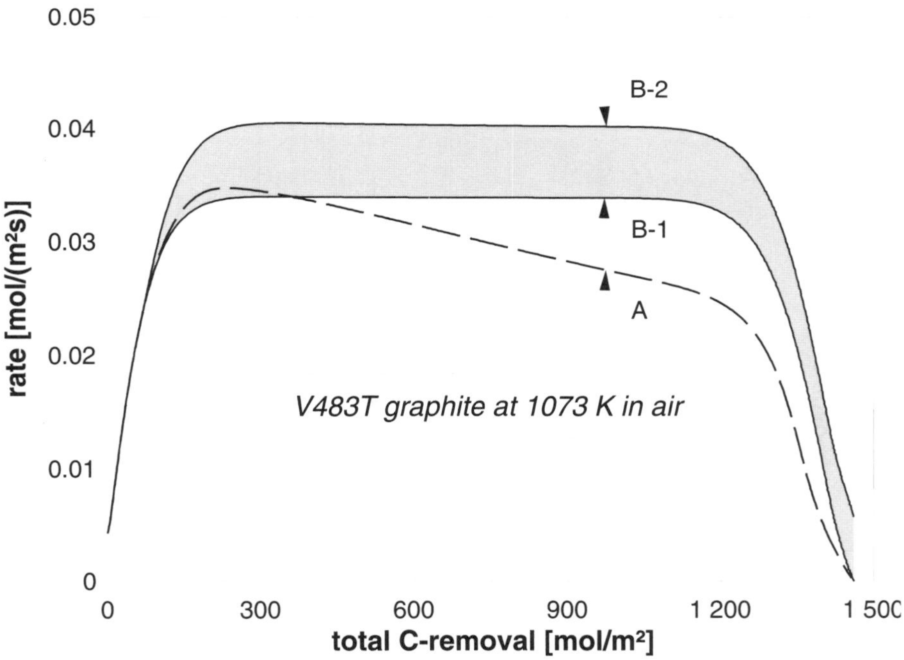

247

Figure 2. Oxygen profiles in V483T at 1 073 K in air

A: t = 2 000 s, Δt of subsequent curves A-G: 4 000 s; H: t = 30 000 s

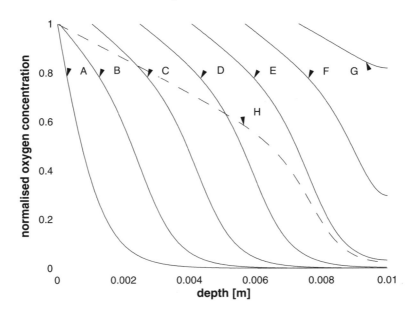

Additional calculations were performed for transient rate behaviour in regime II. The answer of r_{II} on jumps in temperature is demonstrated in Figure 3 (V483T in air, starting temperature 1 073 K, lower curve chemical reaction only, upper curve includes erosion). Rate r_{II} follows that jump, but because the aforementioned equilibrium is disturbed by temperature change, for a significant period of time resp. C-removal rate r_{II} changes its value until the equilibrium oxidation profile for the actual temperature is reached. This transient behaviour has to be considered in evaluation of experimental r_{II} data.

Figure 3. Regime II rates during fast temperature changes

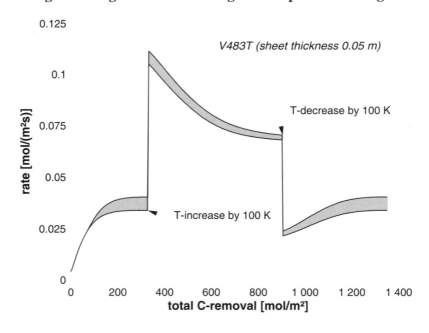

Comparison of these calculations with experiments in regime II, described resp. cited in a former paper [5], reveals sufficient agreement. The depth of penetration (density profiles) measured and calculated (considering different gas composition) are of the same size; for the increase of r_{II} with B_s a factor of 6 ± 3 was measured (this uncertainty is caused by the uncertainty of the measured starting value). Considering that ψ for low B used in calculations is probably too low due to neglecting dead end pores, the calculated increase factor of 9 might be overestimated; that can explain the disagreement. The burn-off resp. time range required to reach a stationary rate is in calculations larger than measured by up to a factor of 2, which might be due to uncertainties in the erosion function used. Agreement between measured and calculated stationary r_{II} values is still sufficient but suffers to some extent from the aforementioned differences between theoretical and measured apparent activation energies in regime II.

Continuation of these calculations is required to obtain a sufficient understanding of such reaction/transport processes: Differences of numerical solutions of Eq. (2) to results of approximate method are obviously too large to be omitted. Further, the described solutions of Eq. (2) are only valid for homogeneous materials; an extension of the model to inhomogeneous carbon materials (matrix graphites for HTR fuel elements, composites) has yet to be done.

Influence of chemical reactions on mass transfer

For the lower temperature range of regime III of the graphite/O_2 reaction it was experimentally verified that no influence of chemical reactions on mass transfer occurred [11]. The same holds for fast temperature transients as expected in fusion reactor accidents [12]. There are however several reports on rate measurements at higher temperatures in regime III, indicating an enhancement of rates compared to mass transfer rules by up to a factor of 4 [13]; an explanation of this effect is given in [13]. As schematically shown in Figure 4, the carbon oxidation by air may occur at high temperatures through a two-reaction process involving carbon gasification by Boudouard reaction and homogeneous CO-burning in the boundary layer. Steeper gradients and higher temperatures within the boundary layer can explain the above-mentioned rate enhancement.

Figure 4. Schematics of carbon oxidation in air by a two-reaction mechanism

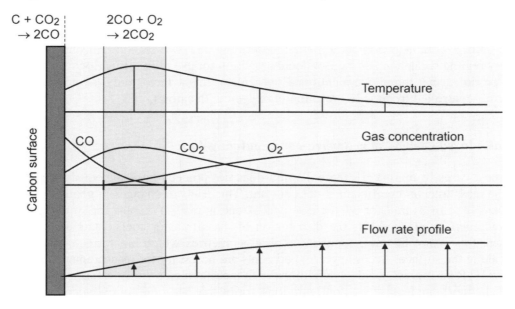

The rate enhancement is obviously coupled to the change in carbon gasification mechanism from C/O_2 to C/CO_2 (Boudouard) reaction. The conditions for this change will be discussed here in detail. The (regime I and regime II) rates of the Boudouard reaction are substantially smaller for a given temperature than those of the C/O_2 reaction; however, at high temperatures, a (common) rate limit exists by mass transfer conditions. Accordingly, CO_2, which is formed near the surface though a C/O_2 reaction, will with increasing temperatures compete with oxygen as gasifying agent. This is due to the fact that no mass transfer restrictions are acting under these conditions for the Boudouard reaction due to the lower chemical rates and due to the smaller diffusion distance to the carbon surface. The contribution of the Boudouard reaction to the total C gasification is roughly given by the quotient q:

$$q = \frac{\text{regime II rate of the Boudouard reaction}}{\text{mass transfer controlled rate of the } C/O_2 \text{ reaction}} \qquad \text{with } q \leq 1 \qquad (5)$$

One should bear in mind that not only the CO_2 formed near the carbon surface by C/O_2 reaction has to be taken into account in estimation of the Boudouard reaction rates, but also CO_2 already present in the bulk and diffusing to the surface. This quotient q describes the process fairly exactly for low q values; at high values some deviations may be found, because the homogeneous reaction moves from the surface into the boundary layer and with respect to boundary layer transport multi-component diffusion has to be considered. Nevertheless, $q = 1$ may be taken as a first approximation for conditions when the gasification has changed completely to the two-reaction mechanism.

A computer model that allows easy introduction of this two-reaction mechanism into graphite oxidation codes for calculation has not yet been developed. In addition it should be noted that in the majority of air ingress accident sequences to be considered, the two-reaction mechanism is not relevant, because the oxygen is already completely consumed in areas with relatively low temperatures, which correspond to low q values. Nevertheless, in case of loss of vacuum with air ingress into the vacuum vessel of fusion reactors, the two-reaction mechanism has to be considered in hot carbon areas for some minutes, until these areas are cooled down. Further, it should be noted that the regime II kinetics of the Boudouard reaction are not sufficiently known [14]. Additional measurements of Boudouard kinetics are however required, not only because of the two-reaction mechanism, but also because the Boudouard reaction is the main process attacking fuel elements for several important sequences of chimney draught air ingress accidents in innovative small-sized HTRs [14].

Altogether, we propose a rate enhancement factor of $4 \cdot q$ to be used in safety analyses for air ingress accidents until a complete computer model describing the two-reaction mechanism is available. Preliminary regime II rate data of the Boudouard reaction for use of in estimation of q are found in [5]. It should be noted that the above-mentioned rate enhancement factor only has to be applied on the mass-transfer-controlled (regime III) reaction of oxygen with carbon.

Oxidation-induced changes of geometries and configuration

For most cases in regime III, the oxidation does not proceed homogeneously along the surface, because the mass transfer coefficient is not constant. This leads to changes in geometry, which induce further changes of mass transfer coefficients and oxidation rates. Whereas for simple geometries like pebbles, cylinders or plane sheets the distribution of mass transfer coefficients along the surface is well known, this is not the case for the complex geometries which are formed by inhomogeneous oxidation along the surface. Accordingly, experiments are necessary for measurements of the respective effects. For HTR air ingress accident conditions such measurements were performed for (fuel element) pebbles in the most and least dense configurations, for pebble beds and for cylinders in crossflow.

Experiments on defined configuration were executed on a single graphite pebble surrounded by alumina pebble segments, simulating the surrounding in a bed [15]. In general it was found that the total burn-off rate for the least dense configuration is similar to that of a single pebble, but the most dense configurations lead to higher oxidation rates. Experiments on graphite pebble beds with the most stochastic configuration possible were first performed in the FZJ facility SUPERNOVA [16] at comparatively high air flow rates. Figure 5(a) shows the inner section of the furnace with the pebble bed; alumina pebbles are used in the bottom part of the pebble bed in order for the graphite pebbles to always have flow conditions as in the centre of the bed. Figure 5(b) contains a facility for measurement of the diameter distribution of oxidised pebbles. Figure 6 shows pebbles oxidised at about 1 000°C in air flow to 15 and 25% of burn-off in comparison with a virgin pebble. At contact points with other pebbles the oxidation remains small; no oxidation is found there up to a total pebble burn-off of 20%. The relation between the (local) maximum and the average burn-off of a pebble was measured to about 3.

**Figure 5. Inner section of SUPERNOVA with bottom part of test pebble bed (a)
and facility for measurement of diameter distribution of oxidised pebbles (b)**

Similar experiments were performed later on in the facility VELUNA (University of Duisburg) at low flow rates representative of air ingress by chimney draught in the HTR-Module. The results were in general similar to those of SUPERNOVA, but – due to the lower flow rates – the relation of maximum to average burn-off increased to up to 9. For both the SUPERNOVA and VELUNA experiments it was not possible to fit the results of the burn-off distribution along the pebble surface into a reasonable mass transfer model. The oxidation behaviour of graphitic cylinders in cross-flow is described in [11].

In recent safety analyses for air ingress in pebble bed HTRs with the code REACT/THERMIX we considered these geometry effects for the core as follows. We selected – depending on the total air flow rate within the HTR core – a suitable value for the relation between maximum and average burn-off between 3 and 9 with a corresponding (experimental) burn-off distribution. Further, these values were multiplied for each mesh point of the calculation grid by the relation of actual rate to mass transfer rate, i.e. weighted with the degree of mass transfer influence. In so doing, a more realistic answer to the question of at what time the first coated fuel particles are set free from surrounding graphite can be given.

Concerning changes in configuration of graphitic components by break down or debris movement, no reasonable model exists up to now. In safety analyses it was assumed [2], that the bottom reflector

Figure 6. Shape of oxidised pebbles compared with virgin ones at 15% (a) and 25% (b) burn-off

breaks down at a certain burn-off, but calculations were not continued beyond this point. Experimental data about the flow behaviour in the case of configuration changes may be obtained from the NACOK experiments [17], where a downsized HTR core can be oxidised in air.

Conclusion

The discussion of the status of advanced graphite oxidation models for oxidation regimes II and III leads to the conclusion that in safety analyses for air respectively steam ingress accidents in HTRs and fusion reactors the following topics should be considered:

- In regime II a solution of the combined reaction/in-pore diffusion equation is advisable for not too inhomogeneous carbons, provided sufficient reactivity and diffusivity data are available. This leads to more precise results for the transient behaviour and provides data on the depth of the oxidation attack. For inhomogeneous materials a sufficient model does not yet exist.

- For high-temperature oxidation by air in regime III an increase of the mass transfer rate by chemical reactions (two-reaction mechanism) has to be considered; a complete model for the complex process does not yet exist, but reasonable results are obtainable using a simplified approach.

- Data on changes in geometry by carbon oxidation in regime III are available for pebble beds, single pebbles and cylinders in cross-flow; this effect has to be taken into account, because deviations between maximum and average burn-off of a component may be large. A complete model for the interaction between geometry changes and mass transfer does not yet exist for pebble beds, nor for more simple geometries like single pebbles and cylinders; simplified approaches are available for the most safety-relevant issues. Concerning configurational changes of graphite components by oxidation no reliable data are available and additional measurements are required.

REFERENCES

[1]   A-K. Krüssenberg, Report Jül-3333 (1996).

[2]   R. Moormann, Report Jül-3062 (1995).

[3]   R. Moormann, H-K. Hinssen, Proc. Int. Conf. on Mass Transfer and Charge Transport in Inorganic Materials, 28 May-02 June 2000, Venice/Jesolo-Lido (in press).

[4]   R. Aris, "The Mathematical Theory of Diffusion and Reaction in Permeable Catalysts", Oxford (1975).

[5]   R. Moormann, S. Alberici, H-K. Hinssen, A-K. Krüssenberg, C.H. Wu, *Advances in Science and Technology*, 24, 331-338 (1999).

[6]   J.P. Lewis, P. Connor, R. Murdoch, *Carbon*, 2, 311-314 (1964).

[7]   H-K.Hinssen, A-K. Krüssenberg, R. Moormann, C.H. Wu, "High-temperature Materials Chemistry", Pt. II, Jülich, ISBN 3-89336-277-0, pp. 565-568 (2001).

[8]   H-K. Hinssen, R. Moormann, W. Katscher, Proc. Jahrestagung Kerntechnik'86, Aachen, ISSN 0720-9207, pp. 211-214 (1986).

[9]   J.D. Clark, C.S. Ghanthan, P.J. Robinson, *J. Mater. Sci.*, 14, 2937 (1979).

[10]   J.E. Crank, "The Mathematics of Diffusion", 2nd ed., Oxford, Chapter 14.2 (1975).

[11]   M. Ogawa, B. Stauch, R. Moormann, W. Katscher, Report Jül-Spez-336 (1985).

[12]   H-K. Hinssen, M. Hofmann, A-K. Krüssenberg, R. Moormann, C.H. Wu, *Fusion Technology*, 1996, pp. 335-338 (1997).

[13]   E. Specht, R. Jeschar, *Ber. Buns. Phys. Chem.*, 87, 1099 (1983).

[14]   R. Moormann, *et al.*, Basic Studies in High-temperature Engineering, Paris, ISBN 92-64-17695-0, pp. 161-172 (2000).

[15]   Report Nukem FuE – 83007, H. Huschka, ed. (1983).

[16]   W. Katscher, in Report Jül-Conf-53 (1985).

[17]   M. Kuhlmann, PhD thesis, RWTH Aachen (2001).

IRRADIATION CREEP IN GRAPHITE – A REVIEW

Barry J. Marsden
The University of Manchester School of Engineering, UK

Stephen D. Preston
SERCO Assurance, Risley, Warrington, Cheshire, UK

Abstract

In the presence of fast neutron radiation graphite creeps at a faster rate than it would without radiation. This is fortunate, as otherwise shrinkage and thermal stresses would lead to early graphite component failure in reactor. Work on irradiation creep carried out since the early 1950s is described in this paper.

Definition of irradiation creep

Irradiation creep in graphite is defined as "the difference in dimensions between a stressed sample and a sample having the same properties as that stressed sample when irradiated unstressed".

In unstressed graphite, fast neutron irradiation changes dimensions, modulus and the coefficient of thermal expansion (CTE). In addition, stressed graphite exhibits creep strain, additional changes to CTE and additional changes to modulus. It is also postulated that creep strain modifies the graphite dimensional change rate.

Therefore creep strain is the total strain observed in the stressed specimen minus:

- The strain due to dimensional change alone.

- The strain due to irradiation-induced change to CTE in an unstressed specimen.

- The strain due to irradiation-induced change to Young's modulus.

- The strain due to the additional change in CTE due to creep strain.

- The postulated dimensional change due the change in creep strain, due to the change in CTE.

The assessment of irradiation creep and creep rate, in the UK and other countries, is usually represented by the following relationship:

$$\frac{d\varepsilon_{cr}}{d\gamma} = \alpha(T)\frac{\sigma}{E_c}[\exp(-b\gamma)] + \beta(T)\frac{\sigma}{E_c}$$

where $\alpha(T)$ and $\beta(T)$ are temperature-dependent constants, b is a constant, σ is the applied stress, γ is the irradiation dose and E_c is the creep modulus, which is modified by structural changes or oxidation, but not the initial irradiation hardening (pinning) discussed later.

Early work

Simmons [1] observed that specimens irradiated under constant stress at 164°C exhibited a permanent set, which could not be thermally annealed out and was different in magnitude in the perpendicular and parallel extrusion directions.

Losty and Davison [2,3] carried out irradiation creep experiments on graphite and carbon springs. They showed that carbon crept at a much faster rate than graphite, but both exhibited non-linear primary irradiation creep. The constant strain experiments carried out at temperatures between 30 and 330°C showed that the creep rate in both graphite and carbon decreased with dose as a function of initial stress. From these results it appeared that the behaviour was not dependent on temperature but was a function of irradiation dose alone. The constant stress experiments showed that for carbon, creep increased at an ever-increasing rate with dose, whereas for graphite, the behaviour appeared to be linear. There was a partial recovery of creep strain in graphite on load removal.

Losty, et al. [4] carried out further experiments in one of the Calder Hall reactors. He demonstrated that irradiation creep strain was proportional to stress and that there was an initial non-linear rapid increase in strain with dose (primary creep) after which the creep rate was linear and appeared to be independent of irradiation temperature. Further experiments by Losty, et al. [4] were carried out in compression in HERALD (AWE). These experiments covered a range of stresses up to 18.75 MPa, to a dose of ~2.4×10^{19} n/cm^2 EDND at 70°C. The experiments showed a bi-linear dependence on creep strain with stress.

Gray, *et al.* [5] reported higher dose experiments for a number of graphites including Pile Grade A and some new improved candidate graphites. The graphites were irradiated to 50×10^{20} n/cm^2 EDND at a nominal temperature of 410°C in both compression and tension. In analysing these experiments he accounted for the irradiation induced changes in CTE and modulus.

Both PGA and the new isotropic graphites exhibited non-linear primary creep strain to ~2.5×10^{20} n/cm^2, followed by linear secondary creep strain. PGA graphite samples cut perpendicular to the extrusion direction crept at twice the rate of PGA samples cut parallel to the extrusion direction. There appeared to be little or no effect of irradiation temperature. The authors noted that there appeared to be a correlation between the linear (secondary) creep rate and the initial elastic deflection due to the loading. It was also noted that in a previously radiolytically oxidised specimen with 18% weight loss, a creep strain of 7.5% was attained, which was much greater than would have been expected for non-oxidised material. It was also noted that large strains were achieved in tension, that would have failed an unstressed sample irradiated to a similar dose. The observed change in CTE due to creep strain is also presented in the paper; compressive creep strain increasing CTE, tensile creep strain decreasing CTE.

Brocklehurst and Brown [6] later reported further data on the BR-2 experiments for semi-isotropic graphites with irradiation temperatures between 300-650°C. This work also included irradiation creep results on an HOPG graphite sample. Although originating from different sources, all of these semi-isotropic graphites had similar values for the Young's modulus. They fit the following relationship to the combined tensile and compressive data:

$$\varepsilon_{cr} = \alpha\sigma + \beta\sigma\gamma$$

where the constant α equals $(0.9 \pm 0.1) \times 10^{-6}$ per psi per 10^{20} n/cm^2 (or 6.2×10^{-6} MPa per 10^{20} n/cm^2) and the constant β equals $(0.2 \pm 0.1) \times 10^{-6}$ per psi per 10^{20} n/cm^2 (or 1.38×10^{-6} MPa per 10^{20} n/cm^2).

Upon conversion of the creep rate to elastic strain units this gives 1.03 for α and 0.23 for β. The primary creep coefficient has been taken to be unity and the secondary creep coefficient 0.23 for UK assessments in the range 300 to 650°C ever since.

The pyrolytic specimen irradiated in these experiments was loaded parallel to the basal plane ("a" direction). Before irradiation the modulus was measured as 75 GPa which was an order of magnitude lower than the theoretical value of ~1 000 GPa. Brocklehurst and Brown took this as an indication of the large amount of basal plane shear that was possible within the crystal. They plotted the creep data for this specimen along with the semi-isotropic creep data and the previously reported PGA creep data in elastic strain units [5]. The resulting data fitted the equation given above.

Measurements on other properties showed that there was an additional change in CTE due to creep strain as described above, but not a change in Young's modulus above that was expected due to fast neutron irradiation. The change in thermal expansion could be annealed out, but the creep strain could not be fully recovered.

Jenkins and Stephen [7] irradiated graphite springs at temperatures between 70 and 300°C. The creep rate was measured in reactor. Unfortunately the experiment was subject to several unexpected trips and shutdowns. However analysis of the results showed that transient creep accounted for about

0.31 elastic deflections (lower than that found by the other works above, but similar to that given by Kennedy, see below). The results clearly showed that all of the transient creep strain was recoverable on load removal in reactor, but only partially recoverable out of reactor. The steady state creep rate increased with irradiation temperature by a factor of about 1.6 between 100 and 300°C. This result is the opposite to that given by Kennedy, who predicted a higher creep rate at 100°C and is contrary to that found by other UK works, which measured no change in rate between these temperatures.

The early UK creep law is illustrated in Figure 1.

Early work in the USA and Russia

Kennedy [8] carried out irradiation creep experiments on various graphites between 150-370°C. His results indicated that graphite exhibited primary and secondary creep strain. He proposed that the primary creep behaviour could be related to a model proposed by Hesketh [9] and not the pinning/unpinning model discussed later. From the results he concluded that primary creep was independent of temperature and that it was proportional to the irradiated modulus. He also noted that the experiments showed that primary creep was recoverable in a reactor. He concluded that (even at these low temperatures) secondary creep did exhibit temperature dependence. Further results at low temperature and at 700°C are presented in a later ORNL report [10]. He concluded that the creep mechanism in graphite could be explained by a model for uranium, given by Roberts and Cottrell [11], which relates creep to the growth and yield of the crystals.

Similar results to these have been produced at the Kurchatov Institute in Moscow who proposed the creep law:

$$\varepsilon_c = \sigma K \ln\left(\frac{F}{Q_1} + 1\right) + B\sigma(F - Q_2)$$

where F is the irradiation fluence (or dose in units $E_n > 0.18$ MeV) and K, Q_1, Q_2 and B are temperature-dependent creep constants.

This Russian work [12] provides a fundamentally different creep law to the UK creep law, although the Russian creep values are in good agreement with US data from Oak Ridge, see Figure 2.

High-temperature creep experiments

As part of the DRAGON programme, irradiation creep experiments on Gilsocarbon graphite were carried out by Manzel, *et al.* [13] in the range 850 to 1 240°C to a dose of 11.9×10^{20} n/cm^2 EDND. These experiments consisted of three sets of 12 specimens connected together and loaded with three different loadings. Unfortunately there are insufficient data to confidently separate out primary and secondary creep coefficients, but the results clearly show that there is a temperature dependence on creep rate and probably primary creep.

Veringa and Blackstone [14] carried out irradiation creep experiments in HFR Petten. These restrained creep experiments were carried out over a large temperature range from 400 to 1 400°C. The specimens consisted of a dumbbell inside a restrainer. Both dumbbell and restrainer were manufactured from graphite, however the restrainer was manufactured from a different (more irradiation stable) graphite. The data appeared to indicate that creep rate was not only a function of temperature, but also a function of flux. Veringa and Blackstone [15] presented the creep coefficient multiplied by Young's modulus and the neutron flux. These data form a consistent set of data, which they fitted to the relationship:

$$\text{creep coefficient} = \text{cont.}\frac{\exp(-E_{im}/kT)}{E_{xx}\phi}$$

where E_{im} is an activation energy, k is Boltzman's constant, T is temperature, E_{xx} is the modulus and ϕ is the flux.

No other author has put forward the possibility of flux dependence on irradiation creep and results by Brocklehurst [16] for boronated graphite appear to show that there should be no rate dependence. However, the present authors consider that the results from this work give such a good fit to the above relationship, that it merits further analysis and comparison with creep data from other sources.

Two extruded coke graphites – AS2-500 and P$_3$JHAN – were irradiated between 850 and 1 100°C in graphite creep experiments in the Siloé reactor at Grenoble by Jouquet, $et\ al.$ [17]. The experiments were monitored continuously in reactor to a fast dose of $\sim3 \times 10^{21}$ n/cm^3 EDND. The results show primary and secondary creep strains. Once again it is demonstrated that the primary creep strain is recoverable. The primary creep strain was found to be between 0.74 and 1.45 elastic deflections with the exponential coefficient for primary creep between 2.8 and 4.0. The secondary creep rate $\beta(T)$ was between 0.4 and 0.72 compared with 0.23 as defined by the UK creep rule for temperatures between 300-650°C.

The effect of modulus on creep strain

It is reasonable to assume that in a constant stress creep experiment, the irradiation creep strain should be modified by changes to the modulus and it is known that fast neutron irradiation significantly changes the modulus in graphite. Furthermore in the AGR and Magnox reactors, there are significant changes to modulus due to radiolytic weight loss. Weight loss is also of interest in HTR designs incorporating a steam cycle and helium/water heat exchanger, since there is also the possibility of weight loss due to thermal oxidation in the event of leakage.

To study this effect Brocklehurst and Brown [18] irradiated various isotropic graphites, under load, in the BR-2 reactor to a dose of 60×10^{20} n/cm^2 EDND between 300-650°C. Some of the specimens had been pre-irradiated to 140×10^{20} n/cm^2 EDND and some had been pre-oxidised radiolytically to 28% weight loss. The experiments were carried out in compression and tension. In order to understand the findings it was then necessary to try to separate out the creep strains from the changes due to dimensional change, modulus changes and CTE changes.

The main difficulty was to sort out the change in modulus due to pinning, the change in modulus due to structure and the change in modulus due to oxidation. The method used was to anneal the specimens to remove the pinning effects.

When plotted as elastic strain, the pre-oxidised specimens, as predicted by the UK creep law, appear to creep at a lower creep rate than the controls (see Figure 3). However, the effect is lower than would be predicted by reduction in modulus alone.

In the case of the pre-irradiated specimens, these appear to show a reduced creep rate as expected (see Figure 3), but again the data are insufficient to arrive at any firm conclusion.

These findings, along with the theoretical consideration discussed later, were used as the basis for the UK creep rule.

Creep at high dose and high temperature

ORNL and KFA have carried out irradiation creep experiments to high dose at high temperatures. Data exist for H451 graphite and ATR-2E and ASR-1RS. The experiments were carried out at temperatures of 300, 500, 600 and 900°C for various tensile and compressive loads. Most of the experiments were carried out past "turnaround" in the dimensional change curve.

Data for ATR2-E and H451 were analysed by Kennedy, Cundy and Kliest [19]. They put forward a model to account for the non-linear secondary and tertiary creep rate as given below:

$$\varepsilon_{crs} = K \times \left(\frac{\sigma}{E_o}\right), \; K = K'\left[1 - \mu\left(\frac{\Delta V/V_o}{(\Delta V/V_o)_m}\right)\right]$$

where K' is the secondary creep coefficient (0.23 as discussed above), μ is a constant equal to 0.75 in the graphite assessed and the term in square brackets accounts for the structural changes to the graphite as a function of the irradiation-induced volume change normalised by the maximum volume change $(\Delta V/V)_m$.

This relationship was successfully used to predict the high dose creep data for ATR-2E at 300°C and 500°C and H451 at 900°C (see Figure 4).

Brocklehurst and Gilchrist [20] presented further high-temperature data to a relatively low dose. They assessed creep data for VNMC, SM2-24 and IM1-24 at 850°C and 1 050°C irradiated to a dose of ~12 × 10^20 n/cm² EDND. The data indicate that there is a significant increase in primary creep with irradiation temperature and also an increase in secondary creep coefficient.

These experiments also supplied data for Poisson's ratio in creep. These data were recently reassessed by Mitchell [21] using an averaging technique, giving a Poisson's ratio of 0.25 in creep, which is near to the elastic value of 0.2. The value is certainly less than 0.5 as used in assessments for creep behaviour in metals.

In support of the HTTR in Japan Oku, *et al.* [22] carried out irradiation creep experiments on IG-110. The experiment was carried out at temperatures of 756-984°C to a dose of 12 × 10^20 n/cm² EDND. The secondary creep rates they calculated from the results were 0.51-0.72 (MPa)^-1 (n/cm² × 10^20)^-1. This should be compared with the values above from the UK programme.

Theoretical methods

The UKAEA methodology

Kelly and Brocklehurst [23] outline in detail the so-called UK pinning/unpinning methodology. The theory predicts that the irradiation-induced polycrystalline creep rate, in a particular direction, taken as x in this case, can be described by:

$$\dot{\varepsilon}_{xx} = KT_{xx} \frac{C_{44}}{E_{xx}}$$

where K is a constant, C_{44} is the crystal shear modulus, E_{xx} is the polycrystalline shear modulus and T_{xx} is the applied stress.

The shear modulus for crystal basal shear is modified by irradiation due to dislocation pinning and the low dose change in polycrystalline modulus is also attributed to pinning. Therefore the equation above predicts, as observed, that there should be no change to creep strain due to low dose irradiation.

Furthermore, tests on single crystal and highly orientated pyrolytic graphite (HOPG) show that only loading in shear gives a non-linear stress-strain curve. However, polycrystalline graphite exhibits non-linear stress-strain behaviour in all loading modes. Irradiation tests on single crystal and HOPG show that the compliance S_{44} is greatly reduced whilst the other compliances do not change significantly (except where large lattice strains occur for irradiation temperatures <300°C).

Kelly and Forman [24], using a computer model, assessed this process and compared their results with experimental data. The results were compatible with observations, but they did not provide a definitive proof of the mechanism.

However, it should be noted that changes to the modulus caused by structural modifications at high dose, or those changes caused by oxidation, would be expected to change the secondary creep rate as these will affect the bulk modulus, E_{xx}, but not the crystal shear modulus, C_{44}.

Cottrell model

The model proposed by Roberts and Cottrell [11] to explain irradiation creep in uranium assumes that the crystals yield due to internal stress caused by dimensional change, allowing the graphite to creep under external load. The formulation is:

$$\varepsilon_{xx} = A_{xx} \left[\frac{[\dot{e}_{zz}]_r - [\dot{e}_{xx}]_r}{\sigma_{yc}} \right] T_{xx}$$

where A_{xx} is a constant and σ_{yc} is the yield stress of the crystal. This model was further developed by Williamson and Jenkins [25] but was never accepted by the UK nuclear industry, though it did find favour in the US.

Jenkins [26] also put forward a model to explain transient creep. This model was further developed by Green [27] who also made an attempt to visually observe the microscopic changes to the structure.

There is obviously more work required in this area to resolve these two approaches. At present the pinning/unpinning model is preferred.

The effect of creep strain on CTE and dimensional change rate

As previously discussed it had been observed in experiments that creep strain modified CTE. Kelly [28] observed that using the relationships derived by Simmons [29], that if creep strain modified

CTE it may also be expected to modify dimensional change. He related the modification to a correction term in the creep strain of:

$$\varepsilon_{cr} = \varepsilon'_{cr} - \int_0^\gamma \left(\frac{\alpha'_x - \alpha_x}{\alpha_c - \alpha_a} \right) \left(\frac{dX_T}{d\gamma} \right) d\gamma$$

where ε_{cr} is the true creep strain, ε'_{cr} is the apparent creep strain and the term in the integral is the correction term in which α'_x is the measured CTE in the crept specimen, α_x is the measured CTE in the irradiated control specimen, α_a and α_c are the crystallite CTEs which are unchanged by irradiation and $\left(\dfrac{dX_T}{d\gamma} \right) = \dfrac{1}{X_c} \dfrac{dX_c}{d\gamma} - \dfrac{1}{X_a} \dfrac{dX_a}{d\gamma}$ is the difference in graphite crystallographic dimensional change rates in the "c" and "a" directions.

Kelly also observed that there must be some change to the CTE in the direction perpendicular to the loading direction. He presented some limited data on this lateral change, but there were insufficient data to give a definitive lateral coefficient. Kelly and Burchell [30] took this analysis and applied it to the analysis of creep experiments on H451 irradiated at 900°C and ATR-2E irradiated at 500°C at 5 MPa in compression.

The analysis reconciled the previous discrepancy between predictions of creep strain at high dose and experimental measurements using the UK method as illustrated by Mitchell [22]. However no data are presented for the tensile creep experiment on ATR-2E at 500°C or the experiments on ATR-2E at 300°C.

To try to resolve the lateral changes to CTE due to creep strain, an experimental programme [31] was devised to assess the effect of large elastic strains on the CTE of unirradiated graphite. Specimens of Gilsocarbon graphite were loaded to large strains in tension and compression. The change in CTE was measured under load, hoping that the small change in CTE expected could be detected. However, the change in CTE observed was much larger than expected and was of a similar magnitude to the change observed in the irradiation creep experiments, which involved much larger strains. The CTE reverted to almost the original value after load removal, although there was a small residual effect left. The CTE was measured in both the loading direction and perpendicular direction. The lateral effect was approximately equal to the negative of the elastic Poisson's ratio.

The effect of creep strain on Young's modulus

Finally, some comment is required on the significant change to Young's modulus with creep strain for isotropic graphite reported by Oku, *et al.* [32] (~35% for 0.23% creep strain). This is much greater than the few per cent change to modulus reported by Price [33] and Brocklehurst and Kelly [20]. Clearly such a significant change requires further investigation.

Conclusions

- Irradiation creep in graphite has a primary, secondary and tertiary stage.

- The graphite creep strain, when normalised by multiplying by the modulus and dividing by the stress, appears to give a common law for all types of nuclear graphite.

- There is evidence that primary creep is equal to about one elastic deflection for temperatures up to about 800°C. It may be greater above this temperature.

- There is strong evidence that primary creep can completely recover on load removal in reactor and may partially recover on shutdown.

- Transient creep is probably of little interest for ongoing irradiation-induced stresses as at most it only accounts for a few elastic deflections which is small compared with the secondary and tertiary creep strains. In addition, ignoring it for ongoing stress is probably pessimistic.

- Russian data and US data predict high creep rates at low temperature, whereas most UK data predicts no change in creep rate between 150 to 650°C. However at temperatures greater than ~650°C all of the authors agree that there is a significant increase in secondary creep rate.

- The proposed flux dependence of secondary creep rate proposed by Veringa [34] requires further investigation, as the data appear to be convincing.

- A better understanding of the microstructural processes behind primary, secondary and tertiary creep is required to give confidence in the methodologies being used.

- The changes to CTE and Young's modulus due to creep strain require further investigation.

- Further creep experiments are required to high dose. Sufficient measurements should be made of CTE, modulus and dimensional change on both the creep specimens and the control specimens to assess the results in detail.

REFERENCES

[1] J.H.W. Simmons, "The Effect of Irradiation on the Mechanical Properties of Graphite", Proc. 3rd Carbon Conference, pp. 559 (1958).

[2] H.W. Davison, H.H.W. Losty, "Mechanical Properties of Non-mechanical Brittle Materials", Butterworths, London, p. 219 (1958).

[3] H.W. Davison, H.H.W. Losty, Proc. 2nd Intern. Conf. on the Peaceful Uses of Atomic Energy, Paper No. A/Conf 15/P/28.

[4] H.H.W. Losty, N.C. Fielder, I.P. Bell, G.M. Jenkins, "The Irradiation Plasticity of Graphite", 5th Conference on Carbon, Pennsylvania, pp. 266-273 (1962).

[5] B.S. Gray, J.E. Brocklehurst, A.A. McFarlane, The Irradiation-induced Plasticity in Graphite under Constant Stress, *Carbon*, Vol. 5, pp. 173-180 (1967).

[6] J.E. Brocklehurst, R.G. Brown, "Constant Stress Irradiation Creep Experiments on Graphite in BR-2", *Carbon*, Vol. 7, pp. 487-497 (1969).

[7]   G.M. Jenkins, D.R. Stephens, "The Temperature Dependence of the Irradiation-induced Creep in Graphite", *Carbon*, Vol. 4, pp. 67-72 (1966).

[8]   C.R. Kennedy, "Irradiation Creep in Graphite", Oak Ridge National Laboratory C/L M9, ORNL-DWG 65-2899.

[9]   R.V. Hesketh, "The Mechanism of Irradiation Creep in Graphite", CEGB RD/B/N.188 (November 1963).

[10]   C.R. Kennedy, ORNL-DWG 66-4748.

[11]   A.C. Roberts, A.H. Cottrell, "Creep of Alpha Uranium During Irradiation with Neutrons" (1956).

[12]   V.N. Manevski, P.A. Platonov, O.K. Chugunov, V.M. Alekseev, V.I. Karpuhin, "The Calculation of Stress-strain State of Graphite Bricks of Uranium-graphite Reactors and Prognosis of the Destruction of Graphite Stack RBMK-reactor SmiRT-12", Elsevier (1993).

[13]   R. Manzel, M.R. Everett, L.W. Graham, "A High-temperature Graphite Irradiation Creep Experiment in the DRAGON Reactor", D.P. Report 752, May 1971.

[14]   H.J. Veringa, R. Blackstone, "The Irradiation Creep in Reactor Graphites for HTR Applications", *Carbon*, Vol. 14, pp. 279-285 (1976).

[15]   H.J. Veringa, R. Blackstone, "High-temperature Enhanced Creep in Graphite for HTR Application", Proc. of the 12th Conf. on Carbon, Pittsburgh (1975).

[16]   J.E. Brocklehurst, "The Irradiation-induced Creep of Graphite under Accelerated Damage Produced by Boron Doping", *Carbon*, Vol. 13, pp. 421-424 (1975).

[17]   G. Jouquet, M. Masson, R. Schill, G. Kliest, D.F. Leushacke, H. Schuster, "Irradiation Creep of Two Reactor-grade Graphites Measured Continuously Between 850 and 1 100°C", *High Temperatures – High Pressures*, Vol. 9, pp. 151-162 (1977).

[18]   J.E. Brocklehurst, R.G. Brown, "The Effect of Radiolytic Oxidation and High Fast Neutron Dose on the Irradiation-induced Creep in Isotropic Graphite", TRG Report 2013(C) (1970).

[19]   C.R. Kennedy, M. Cundy, G. Kliest, "The Irradiation Creep Characteristics of Graphite to High Fluences", Proc. Int. Conf. Carbon, Newcastle, pp. 443-445 (1988).

[20]   J.E. Brocklehurst, B.T. Kelly, "A Review of Irradiation-induced Creep in Graphite under CAGR Conditions", ND-R-1406(S) 1989 (plus 2 supplements, June 1989, April 1990).

[21]   B. Mitchell, "An Irradiation Creep Law Evaluated for Nuclear Graphite at High Dose and Temperature", accepted for publication in Journal of the British Nuclear Energy Society.

[22]   T. Oku, M. Eto, S. Ishiyama, "Irradiation Creep Properties and Strength of Fine-grained Isotropic Graphite", *J. Nuc. Mat.*, 172, pp. 77-84 (1990).

[23]   B.T. Kelly, J.E. Brocklehurst, "Analysis of Irradiation Creep in Graphite", Proc. 3rd Conf. on Industrial Carbons and Graphite, Soc. Chem. Ind., London, pp. 363-368 (1971).

[24] B.T. Kelly, A.J.E. Foreman, "The Theory of Irradiation Creep in Reactor Graphite – The Dislocation Pinning/Unpinning Model", *Carbon*, Vol. 12, pp. 151-158 (1974).

[25] G.K. Williamson, G.M. Jenkins, "The Mechanism of Irradiation Creep in Graphite", Proc. 2nd Conf. on Industrial Carbon and Graphite, London, pp. 567-572 (1965).

[26] G.M. Jenkins, "Transient Creep and Recovery in Graphite", *Phil. Mag.*, Vol. 8, No. 90 pp. 903-910 (1963).

[27] W.V. Green, J. Weertman, E.G. Zukas, "High-temperature Creep of Polycrystalline Graphite", *Mat. Sci. Eng.*, Vol. 6, pp. 199-211 (1970).

[28] B.T. Kelly, "Irradiation Creep in Graphite – Some New Considerations and Observations", *Carbon*, Vol. 30, No. 3, pp. 397-383 (1992).

[29] J.H.W. Simmons, "Radiation Damage in Graphite", Pergamon Press Ltd. (1965).

[30] B.T. Kelly, T.D. Burchell, "The Analysis of Irradiation Creep Experiments on Nuclear Reactor Graphite", *Carbon*, Vol. 32, No. 1, pp. 119-125 (1994).

[31] B.J. Marsden, S.D. Preston, M.A. Davies and N. McLachlan, "The Interaction of Strain, the Coefficient of Thermal Expansion and Dimensional Changes in Graphite", IAEA Specialists Meeting on Graphite Moderator Lifecycle Technologies, 24-27 September 1995, University of Bath (UK).

[32] T. Oku, S. Ota, M. Shiraishi, M. Eto, Y. Gotoh, "The Effect of Compressive Pre-stress on Young's Modulus and Strength of Isotropic Graphite", IAEA-TECDOC-901 (1996).

[33] R.J. Price, "Irradiation-induced Creep in Graphite: A Review", GA-A16402, UC-77 (1981).

[34] H.J. Veringa, "The Irradiation Creep Coefficients of Reactor Graphites Determined by Restrained Shrinkage Method Between 400 and 1 400°C", RCN-75-047, April 1975.

Additional references

G. Kleist, M.F. O'Connor, "Irradiation Creep Performance of Graphite Relevant to the Pebble Bed HTRs", IAEA Specialists Meeting on Mechanical Behaviour of Graphite for HTRs, France, June 1976.

H.H.W. Losty, "The Problems Raised by the Influence of Neutron Irradiation on Stress-strain Relationships of Graphite", Proc. of the 4th Carbon Conference, Buffalo, pp. 593-597 (1960).

H.H.W. Losty, "The Interpretation of Irradiation Creep Experiments on Graphite Springs", Inst. Metals Symp. on Uranium and Graphite, Paper No. 12, pp. 81-85 (1962).

B.J. Marsden, S.D. Preston, M.A. Davies and N. McLachlan, "The Relationship Between Strain, the Coefficient of Thermal Expansion and Dimensional Changes in Polycrystalline Graphite. The European Carbon Conference "Carbon 96", Newcastle (UK), July 1996.

T. Oku, M. Eto, "The Effect of Compressive Pre-stressing on Mechanical Properties of some Nuclear-grade Graphites", *Carbon*, Vol 11, pp. 637-647 (1973).

Figure 1. UK creep data and rule

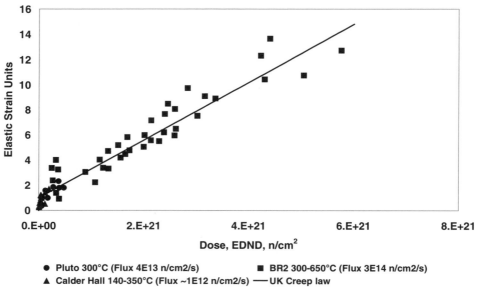

- Pluto 300°C (Flux 4E13 n/cm2/s)   ■ BR2 300-650°C (Flux 3E14 n/cm2/s)
- ▲ Calder Hall 140-350°C (Flux ~1E12 n/cm2/s) ── UK Creep law

Figure 2. Russian and USA creep/temperature dependence

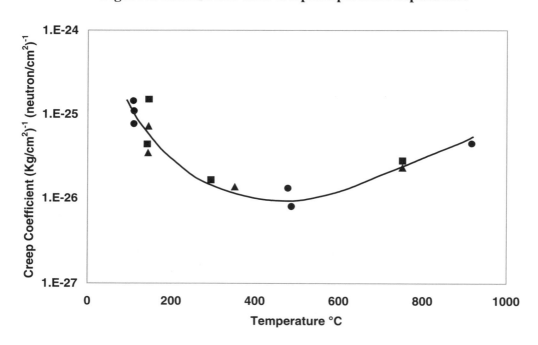

● Russian Graphite ■ EGCR (American Graphite) ▲ CGB (American Graphite)

Figure 3. UK high dose data and effect of radiolytic oxidation

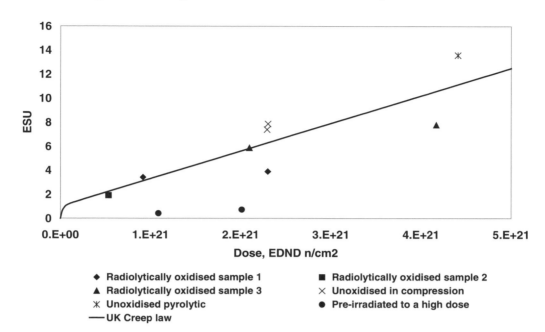

- ◆ Radiolytically oxidised sample 1
- ▲ Radiolytically oxidised sample 3
- ✳ Unoxidised pyrolytic
- —— UK Creep law
- ■ Radiolytically oxidised sample 2
- ✕ Unoxidised in compression
- ● Pre-irradiated to a high dose

Figure 4. German and USA high dose creep data

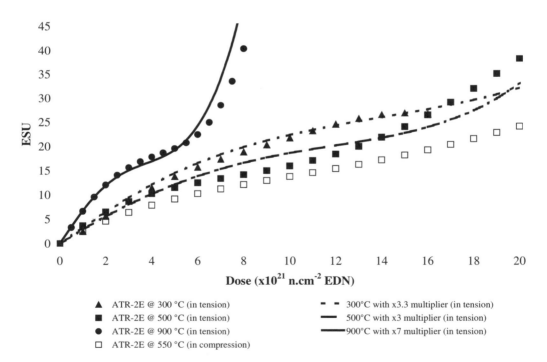

- ▲ ATR-2E @ 300 °C (in tension)
- ■ ATR-2E @ 500 °C (in tension)
- ● ATR-2E @ 900 °C (in tension)
- □ ATR-2E @ 550 °C (in compression)
- – – 300°C with x3.3 multiplier (in tension)
- —— 500°C with x3 multiplier (in tension)
- —— 900°C with x7 multiplier (in tension)

AN ANALYSIS OF IRRADIATION CREEP IN NUCLEAR GRAPHITES

Gareth B. Neighbour
Department of Engineering, University of Hull
Hull, HU6 7RX, United Kingdom

Paul J. Hacker
British Energy Generation Ltd
Barnett Way, Barnwood, Gloucester, GL4 7RS, United Kingdom

Abstract

Nuclear graphite under load shows remarkably high creep ductility with neutron irradiation, well in excess of any strain experienced in un-irradiated graphite (and additional to any dimensional changes that would occur without stress). As this behaviour compensates, to some extent, some other irradiation effects such as thermal shutdown stresses, it is an important property. This paper briefly reviews the approach to irradiation creep in the UK, described by the UK Creep Law. It then offers an alternative analysis of irradiation creep applicable to most situations, including HTR systems, using AGR moderator graphite as an example, to high values of neutron fluence, applied stress and radiolytic weight loss.

Introduction

Nuclear graphite under load shows remarkably high creep ductility with neutron irradiation, well in excess of any strain experienced in un-irradiated graphite (and additional to any dimensional changes that would occur without stress). As this behaviour compensates, to some extent, some other irradiation effects such as thermal shutdown stresses, it is an important property. This paper briefly reviews the approach to irradiation creep in the UK, described by the UK Creep Law. It then offers an alternative analysis of irradiation creep applicable to most situations, including HTR systems, using AGR moderator graphite as an example, to high values of neutron fluence, applied stress and radiolytic weight loss.[1]

Early irradiation creep experiments

There are various ways of measuring in-pile irradiation creep of graphite that vary in degree of sophistication, accuracy and the nature of the data generated. In early experiments, stress relaxation was studied in loaded graphite springs and thin slabs where it became evident that a large proportion of the stress relaxation took place during irradiation [1]. These experiments were limited to strains well below the failure strain, such that the stress relaxation effect could almost be attributed to the irradiation-induced increase in Young's modulus. Blackstone [2] attributes the first series of "real" irradiation creep experiments to C.R. Kennedy [3] in which large cantilever beams were placed under constant load in flexure and the strain measured at various temperatures between 150-1 000°C in the Oak Ridge Research Reactor. He found that there were two distinct stages in the irradiation creep process, a transient creep and a steady-state creep. The transient creep strain was reported to saturate very quickly with a saturation value close to the elastic strain under the same force. Kennedy [3] also established that steady-state creep strain was linearly proportional to stress and fluence (and as became evident later also the un-irradiated elastic compliance).

Three experimental techniques are used to study irradiation creep. The first is a "real" irradiation creep experiment in which a specimen is subjected to a constant tensile or compressive stress while under irradiation with continuous strain measurement. This gives an accurate measure of the irradiation and stressing conditions, but the complexity and cost of this method make an investigation of a large number of specimens impractical. The second technique can be found in either the "restrained growth" or the "restrained shrinkage" experiments, first used by Losty, *et al.* [4] and most commonly used in the irradiation temperature range 400-1 400°C. In this temperature range, all graphites undergo irradiation-induced shrinkage, but the shrinkage rates can differ markedly for different graphites. Using this knowledge, the shrinkage of a graphite tensile specimen can be restrained by pads or split sleeves made of a graphite that exhibits a smaller shrinkage rate or even expands under irradiation. By measuring the unconstrained dimensional change as well, the induced creep strain of samples within the restrained shrinkage assembly can be calculated knowing the various differences in shrinkage rates of the various components. The main disadvantage with this technique is that there is no possibility to make measurements during the irradiation. After irradiation, the stress at temperature deduced from established creep laws. Although this method is less accurate and precise than constant stress experiments, the advantage is that it is possible to obtain many results for a range of conditions and materials [5]. The final technique method can be regarded as a combination of a stress relaxation and

---

[1] Graphite oxidation in high-temperature reactors (HTRs) can occur due to either steam or air ingress into the cooling circuit (principally due to an accident). Although in future generating systems under development, concurrent oxidation is excluded, oxidation may still occur to small degrees in the conventional HTR as a consequence of in-leakage from the water circuit and in all HTR designs through incomplete removal of atmospheric air and moisture from the gas circuit. However, overall oxidation in all these systems will be low.

a restrained shrinkage experiment. A large block of graphite is irradiated under a temperature and neutron flux density gradient. Strips are then cut from the block and their curvatures are measured. These results are only meaningful if a stress calculation can be performed [6]. This method was primarily used in the Dragon Reactor Experiment. As Blackstone [2] remarks, the above three techniques may be regarded as complementary since real irradiation creep experiments are used to derive creep laws, the restrained shrinkage experiments provide information on different materials, temperature and flux gradients and the larger block experiment provides a test of the data and stress models under more realistic conditions. However, this brief history of early irradiation creep experiments has demonstrated the complexity involved and the difficulty in obtaining large data sets.

The stages of irradiation creep

There are two main stages in the irradiation creep process. The first stage, called transient creep, is characterised by a rapid (reversible) deformation that soon decreases in rate and can be represented by:

$$\varepsilon_I = \frac{P}{E} e^{-\alpha\gamma} \int_0^\gamma \sigma e^{\alpha t} \alpha dt \text{ or at constant stress } \varepsilon_I = P\frac{\sigma}{E}\left(1 - e^{-\alpha\gamma}\right) \tag{1}$$

where ε_I is the transient creep strain, α is the exponential transient creep strain rate, E is the Young's modulus of the material, P is a constant to relate the total transient creep strain to the elastic strain at stress σ, γ is the fluence and t is an integration parameter. The second stage, called steady-state creep, is characterised by a linear dependence of the creep strain, ε_{II}, on stress, σ, and fluence, γ, such that:

$$\varepsilon_{II} = k \int_0^\gamma \sigma dt \text{ or at constant stress } \varepsilon_{II} = k\sigma\gamma \tag{2}$$

where the constant k is called the irradiation creep coefficient. As transient creep saturates rapidly, the more technological important mechanism is the irreversible steady-state creep with most experiments being designed to measure the value of k. There is strong evidence to suggest that the irradiation creep coefficient, k, of various nuclear graphites is inversely proportional to the Young's modulus and also dependent upon the neutron flux [2]. Analogous to thermal creep, there is little information on the existence of a tertiary stage in irradiation creep, probably because creep experiments generally failed to reach sufficiently high fluence. Therefore, the total creep strain, ε_c, in irradiated graphite can be expressed by the sum of both transient and steady-state creep components such that:

$$\varepsilon_c = \varepsilon_I + \varepsilon_{II} \text{ or } \varepsilon_c = P\frac{\sigma}{E}\left(1 - e^{-\alpha\gamma}\right) + k\sigma\gamma \tag{3}$$

As can be seen from Eq. (3) above, creep strain is a linear function of stress. Various studies have shown Eq. (3) to be a good representation of creep data at low fluence. At high fluence, typically greater than 60×10^{20} n cm^{-2} EDN, the steady-state creep rate decreases with increasing fluence. Kelly (1992) comments that this observation was expected theoretically and to some extent demonstrated experimentally that changes in E_0 due to structure changes in the graphite would modify the creep rates, and has since been observed in high fluence HTR experiments. Kelly [7] remarks that the normal creep model is only strictly valid for small creep strains <0.5% and that changes in Young's modulus may not be significant in that range of strains.

The UK Creep Law

The UK Creep Law for AGR moderator graphites as elaborated by Method Statement C6/1 (GCDMC/P28) [8], differs slightly from Eq. (3). Brocklehurst [9] reports that the observed decreasing steady-state creep rate at high fluence can be attributed to structural changes that modify the internal stress distribution. He presents a theoretical approach that suggests the steady-state creep rate is directly proportional to the effective crystal shear modulus divided by the polygranular Young's modulus [9]. Since dislocation pinning increases both moduli in proportion, only the structural effects on Young's modulus will affect the steady-state creep strain. Thus, under constant stress, the steady-state creep is modified by substituting E_0 for the effective modulus, E_{eff}, such that:

$$E_{eff} = \left(\frac{E}{E_0}\right)_{Structure} \times E_0 = SE_0 \tag{4}$$

where S represents the proportional change in Young's modulus due to structural changes due to irradiation damage at high fluence.[2] Hence, at constant stress, the steady-state creep strain becomes:

$$\varepsilon_{II} = k\sigma \int_0^{\gamma} \frac{1}{S} d\gamma' \tag{5}$$

Brocklehurst [9] also comments that the effect of radiolytic oxidation is to reduce the structure factor, S, or more precisely the effective elastic modulus in irradiation creep, E_c, by $W(= (E/E_0)_{OX})$, the proportional change in Young's modulus due to radiolytic oxidation only, such that:

$$E_c = E_{0,S} SW \tag{6}$$

where $E_{0,S}$ is the unirradiated static Young's modulus. At present, for AGR moderator graphite, W is given by $\exp(-bx)$ where $b = 3.4$ and x is the fractional weight loss. From various studies, the constants in Eq. (3) for AGR moderator graphite can be shown to be $\alpha = 4.0$, $k = 0.23/E_c$ and P assumed to be unity. The coefficients of 0.23 and 4.0 are reported to be applicable for all polygranular graphites irradiated below ~650°C [10]. The coefficient of 0.23 is actually derived from the slope of creep strain per unit initial elastic strain versus fluence and can be shown to equal kE_0 [11]. Therefore, in summary, the current UK Creep Law for AGR moderator graphite (originally proposed by Brocklehurst and Kelly [12]) under uniaxial stress, for transient and steady-state creep respectively is as follows.

$$\varepsilon_I = 4.0\exp(-4\gamma)\int_0^{\gamma} \frac{\sigma}{E_c}\exp(4\gamma')d\gamma' \text{ or at constant stress } \varepsilon_I = \frac{\sigma}{E_c}\left(1-\exp(-4\gamma)\right) \tag{7}$$

$$\varepsilon_{II} = 0.23\int_0^{\gamma} \frac{\sigma}{E_c}d\gamma' \text{ or at constant stress } \varepsilon_{II} = 0.23\frac{\sigma}{E_c}\gamma \tag{8}$$

Alternatively, the UK Creep Law for constant stress can be expressed as:

$$\varepsilon_c = \frac{\sigma}{E_c}\left(1-\exp(-4\gamma)\right) + 0.23\frac{\sigma}{E_c}\gamma \tag{9}$$

---

[2] Kelly and Burchell [17] suggest the structure factor, S, is similar in free and crept samples.

Generally, the UK Creep Law represents data well for fluence up to $\sim 60 \times 10^{20}$ n/cm^2 EDN. At higher fluence, the UK Creep Law progressively underestimates the measured creep strain, ε_c'. Originally, the structure term was added to account for any discrepancy between the measured and predicted creep strain, however this term has limited success, no real physical meaning and cannot be measured directly.

Irradiation creep data

The bulk of all available irradiation creep data were obtained principally by the UKAEA, KFA Jülich, Reactor Centrum Nederland and CEN Saclay in co-operation with the DRAGON project. The results are concerned with the changes in creep strain or coefficient as a function of fluence or temperature. Notable exceptions include studies by Kelly and Brocklehurst [13] who discuss radiolytically oxidised specimens and Kennedy, *et al.* [14] and Brocklehurst and Kelly [15] who discuss changes in Young's modulus, CTE and Poisson's ratio as a function of creep strain. All these studies support the view that any nuclear graphite irradiated under constant stress, the tensile and compressive creep strains, plotted as creep strain/initial elastic strain, $\varepsilon_C E_0/\sigma$, are directly proportional to the fluence and independent of temperature in the range 140-650°C (for UK graphites see [16]). However, at high temperatures, i.e. >650°C, the creep strain/initial elastic strain is the same for all graphites over a wide range of stresses, but the gradient of the line is proportional to irradiation temperature. Only at very high stresses ($\sim$40 MN/m^2) is there a deviation from linearity.

Recent developments

Kelly [7] reviews the theory of irradiation creep and reiterates that the UK model is based upon plastic and elastic deformations that are dominated by basal plane shear. He states that this assumption explains the correlation between the initial secondary creep and the corresponding un-irradiated Young's modulus for all graphites irradiated at the same temperature and flux. He also explains that the UK model assumes that the creep rate of the component crystallites does not change with creep strain or fluence and more importantly the properties measured on a control sample exposed unstressed apply to the stressed sample. As Kelly states [7], the latter assumption is incorrect since CTE is known to change with irradiation creep and so other properties may also be affected.[3] For example, Kennedy, *et al.* [14] studied the dependence of CTE as a function of creep strain using a range of graphites all stressed in the extrusion direction. The CTE was found to increase linearly with creep strain. In addition, creep strain has been shown to affect the CTE of Gilsocarbon-based graphites. Creep under a tensile stress decreases CTE and creep under a compressive stress increases CTE (Figure 1), where the unstressed CTE values are taken from control samples irradiated to the same fluence as the stressed sample [17]. From this work, Kelly and Burchell [17] realised that the original definition of creep strain, i.e. the difference in length between a stressed specimen and a control, is only correct if the dimensional change component in the stressed specimen is the same as that of the control. However, as dimensional change is directly proportional to the CTE and since the CTE is modified by the creep strain, it was realised that the dimensional change component in the stressed specimen cannot be the

---

[3] It should be noted that work on un-irradiated graphite supports the observations found for the correlation of CTE with irradiation creep strain. For example, Matsuo and Sasaki [22] studied the CTE variation of H-327 grade nuclear graphite with temperature in the unstressed condition and at compressive stresses ranging from $\sim$45% to 95% σ_f. Their results showed that the thermal expansion increased linearly with applied compressive stress. Further, this work was supported by Matsuo and Sasaki [20] who found a linear relation between CTE of pre-stressed polygranular graphites and the residual strain. The change in CTE was found not to depend on the nature of the sample, the pre-stressing direction and the bulk porosity changes induced by pre-stressing.

same as that in the control specimen. That is, the true creep strain, ε_c, will be higher in tension and lower in compression than previously assumed. In the evaluation of creep strain, the following correction to the creep strain is required:

$$\varepsilon_c = \varepsilon'_c - \int_0^\gamma \left(\frac{\alpha'_x - \alpha_x}{\alpha_c - \alpha_a} \right) \left(\frac{dX_T}{d\gamma} \right) d\gamma' \tag{10}$$

where α_x, α'_x, α_a and α_c are the CTEs of the bulk unstressed and crept graphite in direction x, a and c axes of the graphite crystallites, respectively, and X_T is a crystal shape change parameter.

Analysis and discussion

The UK Creep Law, Eq. (9), has only been demonstrated to successfully apply at a fluence of less than $\sim 60 \times 10^{20}$ n cm^{-2} EDN and at stresses of up to 30 MN/m^2. Most of the existing creep data were obtained to relate creep strain to fluence, irradiation temperature or applied stress. However, the established experimental relationship between creep strain and Young's modulus allows for an intuitive argument for the likely behaviour of creep strain at high weight losses. Certainly, the UK Creep Law can be used to suggest that the effect of structural changes caused by radiolytic oxidation on creep can be treated independently of other factors such as irradiation temperature and fluence by incorporating such structural changes into the E_c term. At least, this seems to be supported by the experimental evidence in the fluence range studied by Kelly and Brocklehurst [16], i.e. up to $\sim 40 \times 10^{20}$ n cm^{-2} EDN. Young's modulus is known to decrease exponentially with thermal or radiolytic oxidation to high levels of weight loss. Therefore, it seems intuitive that irradiation creep can continue to be predicted by correcting E_c by E_{OX} in Eq. (9). At present, there is no evidence to suggest that these relationships will not continue to hold at high weight losses.

Emerging from the theory of irradiation creep (pinning/unpinning model) developed by Kelly and Foreman [18], the UK Creep Law is also expected to apply at applied stresses less than 30 MPa. From the limited amount of experimental data available on highly stressed samples (e.g. [16]), it appears that an increase in irradiation creep at a given fluence does not become apparent until samples experience stresses above ~ 30 MN/m^2. The theory of irradiation creep also predicts an increase in crystallite creep strain at high stresses although in this area the theory is less developed and only makes predictions that cannot be directly related to the externally applied stress.

The creep data at high fluence ($> 50 \times 10^{20}$ n cm^{-2} EDN) shows deviation from linearity. This is often attributed to changes in the elastic/plastic response of the bulk material, i.e. modification to the structure term, as it is difficult to explain a change in the underlying linear creep rate generated within individual crystals. Therefore, opinion in the UK has favoured the substitution of the appropriate structure term component of Young's modulus into Eq. (9) to improve the prediction of creep strains that would otherwise deviate from linearity at higher fluence. However, in the case of irradiation creep, it is the intra-crystalline movements of interstitial atoms that dictate the magnitude of the creep strain under irradiation. In this sense, the macroscopic structure does not change the fundamental mechanisms involved and therefore one should not expect a change in the nature of the relationship between creep strain and fluence at high fluence or as a function of oxidation.

As reported, the UK Creep Law has been demonstrated to represent well irradiation creep (up to 60×10^{20} n cm^{-2} EDN) for various nuclear graphites, a range of applied stresses (both tensile and compressive) in the temperature range 300-650°C. It follows, therefore, that nuclear graphites which have been oxidised prior to irradiation should also follow the UK Creep Law (in a similar manner to various grades of nuclear graphites with a range of densities). Examining the UK Creep Law it can be seen that as the Young's modulus decreases due to increasing oxidation, creep strain increases to compensate (assuming constant stress). However, the effects of simultaneous oxidation and irradiation appear not to have been addressed either experimentally or theoretically.

Irradiation creep – A new analysis

While the prediction of creep strain by the UK Creep Law appears to be valid for low fluence and low weight losses, it is not clear whether structural changes either through neutron damage (particularly post-turnaround) or radiolytic oxidation will render the UK Creep Law invalid. Separate from the UK Creep Law, the work of Kelly and Burchell [17] suggests that steady-state creep remains substantially linear to high fluence well beyond turnaround and that earlier discrepancies between measured (or apparent) creep and a linear extrapolation from low fluence are due to the incorrect determination of irradiation creep. Brocklehurst and Kelly [15] also show that German irradiation creep data on ATR-2E at 500°C under compression show behaviour in broad agreement with a linear creep law post-turnaround. On the balance of probabilities, a similar argument is likely to apply to high radiolytic weight losses. Therefore, using the correction given by Eq. (10), and without reference to a structure term, a new analysis may be undertaken to generate a relationship between the apparent creep strain, ε'_c, and fluence, γ, for high levels of fluence, applied stress and simultaneous radiolytic weight loss. From Eq. (10), the correction term, C_ε, to the apparent irradiation creep strain is:

$$C_\varepsilon = \int_0^\gamma \left(\frac{\alpha'_x - \alpha_x}{\alpha_c - \alpha_a} \right) \left(\frac{dX_T}{d\gamma} \right) d\gamma' \tag{11}$$

Strictly $(\alpha'_x - \alpha_x)$ is a function of applied stress, temperature and fluence, but because $(\alpha'_x - \alpha_x)$ can be described singularly by a function of apparent creep strain, ε'_c (c.f. Figure 1), the term $(\alpha'_x - \alpha_x)$ can be assumed to be a constant with respect to $d\gamma'$ in Eq. (11) such that:

$$C_\varepsilon = \left(\frac{\alpha'_x - \alpha_x}{\alpha_c - \alpha_a} \right) \int_{X_T(\gamma=0)}^{X_T(\gamma=\gamma)} dX_T \tag{12}$$

From an analysis of HAPG data extracted from Kelly and Brocklehurst [19], the term X_T fits the following empirical relationship:

$$X_T(\%) = 0.034 + 0.1864\gamma + 2.7803 \times 10^{-4} \gamma^2 \tag{13}$$

where γ is in units of $\times 10^{20}$ n cm^{-2} EDN. Therefore, Eq. (12) simplifies to:

$$C_\varepsilon = \left(\frac{\alpha'_x - \alpha_x}{\alpha_c - \alpha_a} \right) \left[0.1864 \times 10^{-2} \gamma + 2.7803 \times 10^{-6} \gamma^2 \right] \tag{14}$$

Analysis by Mobasheran for H-451 graphite as given in Brocklehurst and Kelly [15] give a linear relationship for $(\alpha'_x - \alpha_x)$ (compressive creep strains only) irradiated at 800°C such that:

$$\alpha'_x - \alpha_x = -0.504\varepsilon_c \times 10^{-4} \tag{15}$$

Brocklehurst and Kelly [15] also discuss a review of Price in which creep experiments at Hanford and ORNL are described. They report Price found that for most data for near-isotropic graphites, the ratio of CTE of a creep specimen, α_c, to that of an unstressed sample irradiated under identical conditions, α_c/α_0, may be related to creep strain, ε_c, by the expression:

$$\frac{\alpha_c}{\alpha_0} = 1 - 18\varepsilon_c \tag{16}$$

Therefore, based upon the above work of Mobasheran and Price and also un-irradiated studies of CTE versus applied stress, e.g. Matsuo and Sasaki [20], there is some evidence to suggest that CTE variance with irradiation creep strain is linear. A statistical analysis of UK AGR graphite data (Figure 2) seems to suggest that the most appropriate empirical fit (lowest R^2 value) be:

$$\left(\alpha'_x - \alpha_x\right) = -0.922\varepsilon'^2_c - 0.8919\varepsilon'_c \tag{17}$$

where $\left(\alpha'_x - \alpha_x\right)$ is in units of $\times 10^{-6}$ K^{-1} and ε'_c is per cent apparent creep strain. However, since the magnitude of creep is suggested to be independent of the mode of applied stress, then the relationship between $\left(\alpha'_x - \alpha_x\right)$ and ε'_c is more appropriate to be a linear fit over the tensile and compressive irradiation creep strains, as previously suggested by both Mobasheran and Price. The best-fit linear fit for AGR data[4] is:

$$\left(\alpha'_x - \alpha_x\right) = -0.6106\varepsilon'_c \tag{18}$$

which is between the prediction generated by Mobasheran and Price ($\alpha_0 = 4.81$), see Figure 3. Between 20-120°C, the CTEs for the c axis and a axis of the graphite single crystal can be estimated as 27.035×10^{-6} K^{-1} and -1.088×10^{-6} K^{-1}. Therefore, the term $(\alpha_c - \alpha_a)$ equals 28.123×10^{-6} K^{-1}. By combining Eqs. (14) and (18) and inserting the above values, the following relationship may be obtained for apparent creep strain, ε'_c, with increasing fluence, γ (in units of $\times 10^{20}$ n cm^{-2} EDN):

$$\varepsilon'_c(esu) = \frac{\left(1 - \exp(-4\gamma) + 0.23\gamma\right)}{\left(1 + \left(0.4043 \times 10^{-2}\gamma + 6.0316 \times 10^{-6}\gamma^2\right)\right)} \tag{19}$$

Figure 3 shows a plot of Eq. (19) and also the UK Creep Law. As can be seen, at low fluence the correction required is small, but at much higher fluences, say 200×10^{20} n cm^{-2} EDN, the apparent creep measured is approximately half the true creep strain. These results are in line with those of Kelly and Burchell [17] who found that significant corrections needed to be applied to the apparent irradiation creep of H451 graphite irradiated at 900°C at a fluence of 25×10^{20} n cm^{-2} EDN. Thus the reduction in apparent (measured) creep strain and the prediction given by the UK Creep Law can be attributed to the CTE corrections given in Eq. (10). Further, predictions can now be made that incorporate radiolytic oxidation into the model without reference to a structure factor.

Neighbour [21] in analysing dimensional changes for AGR graphites proposed a rough correlation between radiolytic fractional weight loss, x, and fluence γ (in units of $\times 10^{20}$ n cm^{-2} EDN) such that

---

[4] It is interesting to note that Marsden, *et al.* [23] found a single linear relationship between CTE and applied strain for AGR graphite in both tension and compression. Changes in CTE of ~3.5 were noted for compressive strains of ~-1%, much higher than that expected from irradiation creep strains.

$x = a\ \gamma$ where $a = 1.889 \times 10^{-3}$. By integrating Eq. (8) to determine the effects of simultaneous oxidation only then it can be shown that:

$$\frac{E_0 \varepsilon_{II}}{\sigma} = 0.23 \left[\frac{\exp(ab\gamma) - 1}{ab} \right] \tag{20}$$

Implementing the above equation, a prediction of both true and apparent creep strains versus fluence with simultaneous radiolytic oxidation can be made as shown in Figure 4 where radiolytic oxidation progressively increases the creep rate compared with neutron irradiation only. Figure 4 also shows for comparison a prediction for a sample with 15% weight loss from radiolytic oxidation prior to neutron irradiation and simultaneous enhanced oxidation (by a factor of 1.5 over that which would be expected in an AGR at a given dose) with neutron irradiation.

Conclusions

The UK Creep Law is a semi-empirical method of describing changes in irradiation creep strain and indicates that Young's modulus is the controlling parameter for relating creep strain with increasing fluence. Consequently, since the prediction of Young's modulus with radiolytic oxidation can be made with confidence, the effects of radiolytic oxidation to high weight losses on creep strain should also be predictable. However, its should be noted that the case of simultaneous oxidation and irradiation does not appear to have been addressed either experimentally or theoretically for the UK Creep Law.

At low fluence, the steady-state creep is observed to increase linearly with fluence, but it progressively deviates from linearity and an apparent reduction in creep strain is observed. The UK Creep Law accounts for this by changes in the "structure term", however this term has no satisfactory physical meaning. A more physically satisfying explanation of the apparent decrease in creep strain is the modification of CTE, and consequently dimensional change, by creep strain.

An analysis was presented which showed that the steady-state creep rate should remain constant without reference to a structure factor. The approach also allowed the prediction of irradiation creep of graphite subject to simultaneous irradiation and oxidation, using AGR moderator graphite as an example. The analysis provided a more physically satisfying explanation for the decrease in the apparent creep strain observed at high fluence than the UK Creep Law did. It is envisaged analysis presented can also be applied to HTRs systems. This review of irradiation creep experiments also identified the need to undertake new creep experiments under better and fully monitored conditions.

Acknowledgements

We wish to thank British Energy Generation Ltd for providing MTR data and financial support for this work.

REFERENCES

[1] H.W. Davidson and H.H.W. Losty, "The Effect of Neutron Irradiation on the Mechanical Properties of Graphite", Proc. 2nd Int. Conf. on Peaceful Uses of Atomic Energy, 7, 307 (1962).

[2] R. Blackstone, "Radiation Creep of Graphite – An Introduction", *Journal of Nuclear Materials*, 65, 72-78 (1977).

[3] C.R. Kennedy, "Gas-cooled Reactor Program Semi-Annual Progress Report", USEAC Report ORNL-3445, 221 (1963).

[4] H.H.W. Losty, N.C. Fielder, J.P. Bell and G.M. Jenkins, "The Irradiation-induced Plasticity of Graphite", Proc. 5th Biennial Conf. on Carbon, 1, 266 (1962).

[5] H.J. Veringa and R. Blackstone, "The Irradiation Creep in Reactor Graphites for HTR Applications", *Carbon*, 14, 279-285 (1976).

[6] A.N. Kinkaed, P. Barr and M.R. Everett, "The Thermal and Mechanical Performance of HTR Fuel Elements", Proc. BNES Conf. on Nuclear Fuel Performance, London, paper 41 (1973).

[7] B.T. Kelly, "Irradiation Creep in Graphite – Some New Consideration and Observations", *Carbon*, 30, 3, 379-383 (1992).

[8] GCDMC (1996)/P28, "The Compendium of CAGR Core and Sleeve Data and Methods".

[9] J.E. Brocklehurst, "Irradiation Damage in CAGR Moderator Graphite", UKAEA Report ND-R-117(S) (1984).

[10] B.C. Mitchell, B.J. Marsden, J. Smart and S.L. Fok, Evaluating an Irradiation Creep Law for Nuclear Graphite at High Dose and Temperature, private communication (2000).

[11] B.T. Kelly and J.E. Brocklehurst, "Irradiation-induced Creep in Graphite; A Review of UKAEA Reactor Group Studies", TRG Report 2878(S) (1976).

[12] J.E. Brocklehurst and B.T. Kelly, "Proposed Modifications to the Irradiation Creep Design Rules for AGR Graphite", ND-M-3785(S); AGR/GCWG/P(87)22 Rev. 2 (1987).

[13] B.T. Kelly and J.E. Brocklehurst, "Analysis of Irradiation Creep in Reactor Graphite", Proc. 3rd Conf. on Industrial Carbon and Graphite, London, 363-368 (1970).

[14] C.R. Kennedy, W.H. Cook, W.P. Eatherly, "Results of Irradiation Creep Testing at 900°C", Proc. 13th Biennial Conf. on Carbon, 342 (1977).

[15] J.E. Brocklehurst and B.T. Kelly, "A Review of Irradiation-induced Creep in Graphite under CAGR Conditions", UKAEA Report ND-R-1406(S) (1986).

[16]   B.T. Kelly and J.E. Brocklehurst, "UKAEA Reactor Group Studies of Irradiation-induced Creep in Graphite", *Journal of Nuclear Materials*, 65, 79-85 (1977).

[17]   B.T. Kelly and T.D. Burchell, "The Analysis of Irradiation Creep Experiments on Nuclear Reactor Graphite", *Carbon*, 32, [1], 119-125 (1994).

[18]   B.T. Kelly and A.J.E. Foreman, "The Theory of Irradiation Creep in Reactor Graphite – the Dislocation Pinning/Unpinning Model", *Carbon*, 12, 151-158 (1974).

[19]   B.T. Kelly and J.E. Brocklehurst, "High Dose Fast Neutron Irradiation of Highly Oriented Pyrolytic Graphite", *Carbon*, 9, 783-789 (1971).

[20]   H. Matsuo and Y. Sasaki, "Thermal Expansion of Some Pre-stressed Nuclear Grade Graphite", *Carbon*, 23, [1], 51-57 (1985).

[21]   G.B. Neighbour, "Modelling Dimensional Change with Radiolytic Oxidation in AGR Moderator Graphite", BNMG/REP/036; GCDMC/A/P(99)126 (1999).

[22]   H. Matsuo and Y. Sasaki, "Thermal Expansion of Nuclear-grade Graphite under Compressive Stress at High Temperatures", *Journal of Nuclear Materials*, 101, 232-234 (1981).

[23]   B.J. Marsden, S.D. Preston, N. McLachlan and M.A. Davies, "The Interaction of Strain, Coefficient of Thermal Cxpansion and Dimensional Changes in Graphite", in Proc. IAEA Specialists Meeting on Graphite Moderator Lifecycle Technologies, University of Bath, UK, 313 (1995).

Figure 1. Effect of "apparent" creep strain on the CTE measured parallel to the stress axis of Gilsocarbon graphite (data taken from [17])

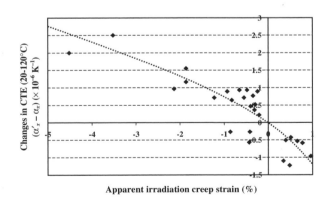

Apparent irradiation creep strain (%)

Figure 2. Effect of "apparent" creep strain on the CTE of Gilsocarbon graphite with linear and polynomial best-fits and predictions based on the work of Price and Mobasheran

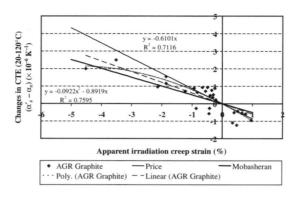

Figure 3. A plot of irradiation creep strain (esu) from both the UK Creep Law [Eq. (9)] and the prediction for equivalent apparent creep strain from Eq. (18)

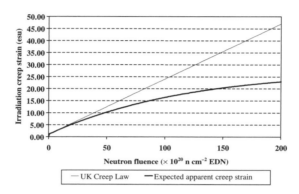

Figure 4. Predictions of (a) true creep and (b) apparent creep for expected and enhanced simultaneous oxidation with neutron irradiation, irradiation only and radiolytic oxidation (15% weight loss) prior to neutron irradiation

(a)   *(b)*

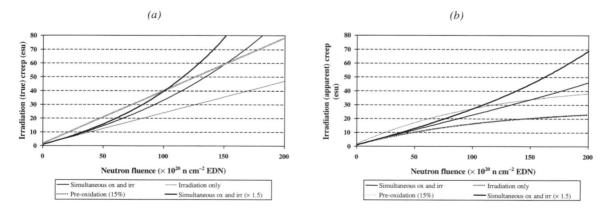

PLANNING FOR DISPOSAL OF IRRADIATED GRAPHITE: ISSUES FOR THE NEW GENERATION OF HTRs

B. McEnaney and A.J. Wickham
Bath Nuclear Energy Group
Department of Engineering and Applied Science
University of Bath
Claverton Down, Bath BA2 7AY, UK

M. Dubourg
Societé Carbone-14
2, rue Saint Cyran
Le Mesnil St. Denis
78320 France

Abstract

The disposal of radioactive wastes is an issue which preoccupies the authorities in numerous countries. Most plan to dispose of wastes in underground repositories, but the execution of such plans have been fraught with difficulty and delay, and no adequate analysis of comparative risks seems to have been applied to the issue.

In the specific cases of graphite, the argument for consideration of other disposal options is strong, especially when all relevant factors are assessed on a technical basis. One may conclude that it is entirely the political agenda, based upon inadequately analysed environmental perceptions, which has led to the intent to dispose of carbonaceous wastes by land burial.

In this paper, the numerous alternatives are discussed in the context of existing and potential HTR carbonaceous materials as intended for use in reflectors and fuel compacts, highlighting the issues which the designer needs to take into account in preparation for the eventual decommissioning of his plant and disposal of the irradiated material. Particular reference is made to the valuable experience gained from the successful dismantling of the Fort St. Vrain plant, and the value of applying this experience at the design stage is highlighted.

Introduction

The disposal of radioactive waste materials from the world's nuclear plants is probably the biggest issue facing the nuclear industry today, at least in the public and political perception. Yet, in our previous Information Exchange Meeting on the Basic Studies on High-temperature Engineering held in September 1999, there was not a single paper on this "back-end" issue for any newly-built reactors.

Nuclear regulators now require potential licensees to have strategies in place for all activities on their plant for the complete life cycle, including decommissioning. Specifically, this requires the identification of waste streams and the integration of the disposal of those wastes with the national programme for the creation of waste disposal processing sites.

It is therefore now increasingly important for reactor designers to look far ahead to the eventual destiny of plant components and waste products when selecting materials and designing the reactors. This paper concentrates more specifically upon the graphite and carbon components, evaluating the lessons which have been learned from the present experience of decommissioning activity on earlier graphite-moderated reactors.

A brief overview of decommissioning experience on graphite-moderated reactors

Magnox and production reactors

Three countries which have operated significant numbers of graphite-moderated plants are now engaged in practical decommissioning activity and/or have plans in place to evaluate some of the likely practical problems which such activity will face – these are the United Kingdom, France and the Russian Federation. Japan, although operator of only one graphite reactor at Tokai (discounting the current HTTR project), is also devoting considerable energies to this problem on the basis that the reactor site is required for further development. The USA has not contemplated serious decommissioning of its former production reactors at Hanford and Savannah River, but has successfully completed the dismantling of the Fort St. Vrain HTR, and this particular experience is of considerable value in the present context.

In the UK, active dismantling of the reactor vessel and core has begun for the Windscale Prototype Advanced Gas-cooled Reactor, and it is also intended to dismantle the former production reactor Windscale Pile No. 1 in which a serious fire occurred in 1957. Other graphite-moderated plants are to be maintained intact (from the pressure-vessel inwards) through periods of "care and maintenance" and "safe store" lasting up to 135 years in some cases. The objectives of such a delay are to await decay of the shorter-lived isotopes in order to diminish operator doses and to utilise improvements in dismantling technology which are expected to be developed in the intervening period. France also plans safe storage of its commercial Magnox reactors, although for a lesser period.

This strategy was recently compared with that planned for reactors of the former Soviet Union [1]. Under the Soviet system, little attention was paid to the eventual need to dispose prudently of the components from nuclear plants. These problems are now being addressed in Russia for the production reactors and for two prototype AMB reactors at Beloyarskaya NPP and also, along with colleagues from Lithuania and Ukraine, for the future disposal of the RBMK plant. In the case of the earlier reactors, certain fuel ponds and cores have been stabilised with epoxy material to prevent the spread of contamination and have been further sealed with welded steel plates pending future disposal actions. This is in contrast with the UK philosophy that no actions should be taken ahead of dismantling which make the operation more difficult to accomplish although, with the very high levels of fission-product contamination present in certain cases, such action may have been unavoidable.

Setting aside Fort St. Vrain for closer examination in a later section, the principal carbon and graphite issues which have been identified so far arising from these activities are as follows:

- During "care and maintenance" and "safe storage": (a) core stability, (b) activity transport by air movement or leaching, with particular attention being paid to releases of tritium, (c) increase in reactivity to air through catalytic contamination during vessel storage and (d) colonisation by organic material, possibly leading to the release of ^{14}C and other isotopes.

- Wigner energy release, especially during cementation of components and subsequent storage (hardly an issue for HTR).

- A potential for a carbon dust explosion where impure dusts exist.

- Fragmentation of graphite components through excessive neutron irradiation effects during reactor operation.

- Activation of the impurity content of the material, leading to radiation hazards during handling and storage and potentially introducing additional leaching problems.

- Contamination of the material both with fission products from failed fuel and also with activated corrosion products from other parts of the circuit.

- Specific activation of long-lived isotopes, augmented by coolant sources (for example, ^{14}C from nitrogenous coolants adding to the smaller amounts arising from ^{13}C).

- Premature short-term actions resulting in the complication of dismantling and treatment (for example, the impregnation with "stabilising" resins).

- Difficulty in removing components, including: (a) sticking caused by carbonaceous deposits in carbon-dioxide-cooled systems, (b) lack of features included at design stage to facilitate removal and (c) failure of such features as threaded holes due to distortions.

- Mixing of graphite and non-graphite materials in storage vaults, usually from components containing steel or Magnox wires or struts.

- No strategy or final destiny for graphite wastes in place, resulting in the need for temporary surface storage boxes and buildings and hence greatly inflated costs.

- Unwillingness to consider other options for disposal, despite credible technical solutions being available.

This final point is extremely important. Serious anomalies exist in the regulation of the nuclear industry compared with other industries. Whilst nuclear operators are being forced through political agreements to drive towards zero emissions of radioactive materials, there is no regulation on the fossil-fuel industry which discharges large quantities of ^{14}C and other isotopes; this has led to the farcical situation in the UK whereby a new fossil-fuel power plant intended to supply power to the Sellafield nuclear site has been built outside the perimeter fence to avoid compromising the site's permitted discharge limits. There is growing debate, viewed with a reluctant sympathy by some nuclear regulators, that the playing field should be levelled.

Fort St. Vrain

An excellent review of this HTR decommissioning project was given at an IAEA Technical Committee Meeting in 1998 [2]. Graphite components were removed using a grab working from a rotating platform above the core, which was flooded with water. The blocks had been hand-placed during construction and no means to retrieve many of them was included in the design; consequently it was necessary to employ specially designed equipment or to drill holes to facilitate the removal. A proportion of the blocks had been originally machined with threaded holes to facilitate lifting during decommissioning [3], but it was frequently not possible to engage with these threads, although the reason is thought to be mechanical problems rather than neutron damage of the graphite (M. Fisher, personal communication).

A major issue was the dose rate on some of the blocks, which exceeded 3 Sv (the distance of this reading from the surface is not specified) on one large permanent side block, and it is thought that the bulk of this activity arose from original impurities within the graphite since the extent of activity transport around the circuit was low. The blocks were transferred to a steel flask whilst underwater, and the flask was then used for transhipment to a hot cell where the components were transferred to a shipping cask and removed from the site as low-level waste.

This disposal route did not, of course, include the 1 482 hexagonal graphite fuel blocks for which a special storage facility was built, providing storage and cooling for steel cylinders each containing six elements. Whilst this aspect of the programme has deferred, rather than solved, the disposal problem for these components, it has allowed the remainder of the reactor to be completely removed and the site returned to free use.

A particular issue identified at Fort St. Vrain was the difficulty of quantifying the levels of tritium and ^{55}Fe isotopes which are particularly hard to detect.

Graphite materials and plant design

It is evident that the future requirements of the decommissioning programmes need to be taken into account in the selection or specification of graphite materials, in addition to the structural and operational needs for these components. In this paper we focus on the graphite or boronated carbon reflector material for the new HTRs.

It is pleasing to note that the designers of the two prototype HTRs currently being commissioned have given careful thought to future decommissioning needs. As examples, a brief survey of these issues for the Chinese HTR-10 is given in [4], whilst the question of the reliable recovery of fuel particles from the compacts from the Japanese HTTR, and the disposal of the graphitic debris through high-temperature oxidation in CO_2, is given in [5].

A great deal of work has also taken place to identify the most suitable graphite for the reflectors of the newly planned modular HTRs, concentrating on the ability of the chosen material to withstand thermal shock (good thermal conductivity equates with a high degree of graphitisation), to have a high degree of dimensional stability (frequently correlated with a relatively high coefficient of thermal expansion), and to have a good resistance to thermal oxidation from coolant impurities (particularly moisture). However, the published papers (e.g. [6]) make scant reference to potential problems from activation of impurities (other than the close attention to boron with respect to the neutron absorption cross-section). A particular problem is the formation of ^{60}Co, and it is especially ironic that the imposition of a specific target cobalt concentration part-way through the production of the Gilsocarbon graphite for the UK AGR programme actually allowed the manufacturers to relax the controls compared with the former manufacturer-controlled specification. This resulted in an *increase* in cobalt content, creating a severe handling problem in the later reactors even with just the very small quantities of graphite associated with the irradiation-monitoring programme. Thus, it is extremely important that the specifications for impurities in the new materials take the activation issue into account.

Equally, there should be an analysis of the rest of the design of the plant, firstly in order to attempt limitation of material transfer to the graphite from other parts of the circuit, such as oxidation products, which could become activated within the graphite pores, and secondly the individual blocks should be tailored with the need for the eventual removal taken into account. In the case of Fort St. Vrain, whilst the latter point received consideration, and was in the main a success, activation was a severe problem.

Finally, the reactor vessel itself needs to be of such a design that the eventual removal of the graphite and other internal components is made as simple as is practical (although there is one option for graphite disposal which would see the graphite disposed of *in situ*, see below).

Routes for graphite disposal

There is currently of the order of 200 000 tonnes of irradiated nuclear graphite in the world's reactors and associated nuclear facilities. Most of this consists of moderator blocks, but there is a sizeable quantity of fuel-sleeve debris, small-scale debris and items such as moulds from nuclear devices. Currently, most of this material appears to be destined for eventual disposal in land-burial sites, despite the innate potential of graphite to be converted to gaseous forms with consequent delivery of the isotopic content, principally ^{14}C, into the biosphere.

Such a methodology highlights three important factors:

- Volume of waste, and consequent capacity of repository facilities.

- Isotopic content (half-lives, leaching etc.).

- Future security, over extremely long times.

These factors combine into an overall and unavoidable matter – very high cost.

There are powerful political and environmental factors influencing this decision, including the Oslo-Paris Commission (OSPAR) Agreement which effectively commits European governments to work quickly towards a "near-zero" radioactive release philosophy by 2020,[1] and it does not appear that serious consideration is any longer being given to possible alternatives. With a new generation of graphite-moderated HTRs now becoming a realistic possibility, with the potential to add further large quantities of waste-irradiated graphite to the stockpile, it is the authors' belief that an objective and wholly independent comparative risk assessment of the full range of options is timely, if not overdue. Should such an analysis indicate that some of the alternatives do actually represent potentially more appropriate routes, then it is the scientists' responsibility to ensure that both the public, and political leaders, are given the full facts on which to review their judgements.

Before reviewing a range of alternative treatments, we should mention the possibility of recycling of existing graphite by using ground graphitic material as filler in the manufacture of new graphites. Whilst this suggestion has been made a number of times, the cost of setting up a dedicated manufacturing

---

[1] Not including releases from "non-nuclear" industry.

plant to handle active raw materials would be prohibitive, especially with different nations producing the initial waste and manufacturing the new graphite. A great deal of effort would also be necessary to ensure that a new graphite was produced which displayed appropriate properties, and there is an additional problem in that the irradiation properties of such new materials could not be obtained through MTR experiment in sufficient time to justify the necessary safety case or commercial lifetime assessments for the new reactors. One would also have to face the consequences of constructing a new reactor with already radioactive materials.

The merit of such a procedure would be a net volume reduction in the total quantities of waste, and it is the overall objective of volume reduction which provides the greatest incentive for a serious re-evaluation of alternatives to permanent deep burial of graphite wastes. An alternative form of recycling existing graphite waste would be as backfill in containers for other types of active waste.

Repository burial

A serious argument for pursuing alternatives to the deep repository approach is that in a number of countries, public opposition, shared in many cases by local government authorities, has meant that no repository with adequate capacity exists. The earliest that such a repository could now become available in the UK, for example, is 2020. The public appears now to prefer supervised shallow or surface storage, with consequent additional and ongoing personnel costs. Another public perception is that by following this route, the authorities are handing on the problem to future generations rather than dealing with it, although when suggested alternative ways of dealing with it are proposed, these obtain equal public opposition.

On technical grounds, the potential problems are groundwater in-leakage and leaching (more complex still in proximity to the coast), seismic stability and the high cost of construction. The long-term environmental risk clearly remains because material is not fully "removed" from the environment.

Sea disposal

Although there is now a moratorium on sea dumping, it is interesting to note that a review conducted by the European Commission in 1985 concluded that it was very favourable in comparison with land-disposal options for disposal of UK nuclear graphite [7]. Prior to the moratorium, a great deal of nuclear waste had been dumped in the oceans, following an assumption that if the immediate containers eventually failed, the collective population dose was protected by the "dilute and disperse" strategy. Public opinion now expects radioactivity to be "locked in" and controlled. Essentially there are two locations which might commend themselves for sea disposal – a deep ocean trench remote from land, giving maximum effect to the dilution factor, and a subduction zone in which the waste containers, within geological time scales, are conveyed into the mantle of the Earth.

Volume reduction

An obvious alternative to the "dilute and disperse" philosophy for radioactive materials is one of "concentrate and contain". We now turn attention to work in progress on a variety of techniques with the potential to significantly reduce the residual volume of active material, albeit introducing some additional complications in the process.

Disposal of irradiated graphite by burning raises a number of technical difficulties:

- It is impractical to burn graphite in a conventional furnace, at least without the presence of a significant catalyst.

- Burning releases not only CO_2 (environmentally sensitive, although compared with the fossil fuel burn the quantities are very small) but also radioactive gases and particulates which are dominated by [14]C (as CO_2 and also as CO if the combustion process is not wholly complete), [36]Cl and residual tritium (as water).

- Residual ashes, enriched in other isotopes such as ^{60}Co, remain to be disposed of by other methods although there is clearly a very large volume reduction.

One problem that is automatically solved through incineration, for reactors which have operated at low temperatures, is the risk of Wigner-energy release in the long term.

Numerous papers have appeared in the literature in recent years, offering a variety of solutions to increase the reactivity of the graphite to air prior to incineration. It seems clear that any industrial process would certainly be facilitated by the introduction of a catalyst. Simply soaking the graphite in a solution of lead acetate ahead of introduction to the furnace would increase its reactivity by orders of magnitude (A.J. Wickham, personal communication). Research in France has resulted in the successful operation of an industrial pilot plant (without catalytic assistance) using fluidised-bed incineration [8].

However, the incineration issue is dominated in Europe by concern about the release of ^{14}C, in contrast to the view apparent in the USA [9] where the arguments are based upon comparisons with the quantities released in A-bomb tests and the quantities produced by cosmic-ray bombardment of atmospheric nitrogen. It would be feasible, and indeed reasonable, to add the combustion of billions of tons of fossil fuels, releasing not only ^{14}C but also other isotopes into the environment.

Nair [10] developed a model in 1983 for the consequences of combustion of the graphite from all UK Magnox reactors[2] at the rate of one pile per annum commencing in 2000, taking into consideration the fate of the release of ^{14}C in terms of the removal through dissolution in the oceans and balance within the troposphere and phytomass growth and decay, and also taking into account projections on the extent of fossil-fuel burn, atomic-weapons testing and cosmic-ray production. His calculations indicated that this would result in a peak incremental dose to the individual of 10^{-2} $\mu Sv.y^{-1}$ followed by an exponential decay over about 50 years to 10^{-3} $\mu Sv.y^{-1}$. This should be compared with an individual dose from cosmic-ray bombardment alone of 10-14 $\mu Sv.y^{-1}$. The estimates also support the view that the *effective* half-life of ^{14}C in the accessible environment (to man) is about 35 years – very much less than the 5 760 years representing its true radiological half-life.

Of course, matters are not quite that simple, since the population dose close to the incineration plant would be very much higher, and there is a case for siting such plants in remote areas, which would mean transporting the graphite away from the original reactor.

Suggestions have been made in the past, in connection with graphite incineration, to avoid the release of carbon dioxide into the environment and to convert the CO_2 to carbonate. This results in a very large *increase* in the total volume of material compared with graphite – 8 times if converted to calcium carbonate and 16 times if as barium carbonate. Consequently, conversion to carbonate is only of interest if concentration of the ^{14}C component can be effected.

---

[2] Except for Calder Hall and Chapelcross which were separately owned at that time.

Fortunately there is a potential solution to both of these problems, which is to use a trapping technology to concentrate and retain the ^{14}C. There are several options: gas centrifuge, static Helikon process or a cryogenic separation technique first considered by Ontario Hydro in Canada to recover ^{14}C arising from irradiation of nitrogen annulus gas in the CANDU plant trapped on resin beds [11]. In applying this to graphite combustion, the CO_2 first formed would be converted into CO, the ^{14}CO fraction separated and reconverted to CO_2, and then to carbonates which would be re-used in the cementation of other radioactive wastes. There is also a Japanese patent for a similar process in which the trapping of both ^{14}C and tritium is claimed [12]. However, all of these options present a much-increased short-term cost.

Finally, we consider alternative technologies to direct combustion. Studsvik Inc. has developed a thermal organic reduction (THOR ^{SM}) process which utilises a pyrolysis and steam-reformation technology,[3] and also has a commercial low-level-waste processing facility operating in Erwin, Tennessee [13] which has been shown to have the capability to handle a wide range of solid and liquid LLW streams including graphite, charcoal and ion-exchange resins. Carbon-containing wastes are first reduced by pyrolysis in the absence of oxygen, or with limited oxygen, to volatile gases and a fixed carbon char. The char is then further processed with steam to gasify the carbon content. Essentially, wet-ground graphite may be introduced as a slurry directly to the second stage of the process since it contains only extremely small amounts of bound hydrogen. The product "water gas" ($CO + H_2$) can then be oxidised to give CO_2 and water. Again, there is the option to capture the CO_2 product and isolate the ^{14}C, as previously discussed.

It is proposed [13] that the steam-reformation process could, perhaps facilitated through suitable foresight at the design stage of the reactor vessel, be conducted *in situ*, obviating the need for complex handling procedures and thus contributing significantly to the minimisation of personnel dose. Of course, this does first demand a move away from a commitment to dispose of the graphite as stabilised solid waste.

This is not the only potential *in situ* treatment, since a Russian proposal [1] for a sintering procedure allied to the "Thermit" process could conceivably be conducted *in situ* although it would be necessary to have some method for mixing the chemical agents. The process involves the addition of aluminium and titanium dioxide to create a burning wave front which creates a carbonate matrix which might, again, be employed in some form as a suitable medium for the stabilisation of other types of radioactive waste.

Comparative risk

The consequence of the present drive towards minimising or eliminating and radioactivity release to the environment from the nuclear sector has been to retain the waste materials in interim surface storage, from which it can be retrieved for treatment by newer technologies as yet not identified. This results in larger radiation doses and increased risk to the operator and, indirectly, to the public at large. The lack of public trust in the procedures of the nuclear industry is currently having the opposite effect to that intended. However, it is not sufficient to rely wholly upon technical analysis: "… the assessment of risk has carried its own risk – namely, an undue reliance on logical quantitative techniques that fail to address the root causes of public concern and apprehension…" [14].

---

[3] Being developed in association with Bradtec Decon Technologies Ltd, Bristol, UK.

We believe that the time is now overdue for a major re-appraisal of the whole waste issue. The stages of this procedure would be similar for most materials, but we continue to concentrate here upon graphite and carbon wastes:

1) A thorough "state-of-the-art" technical appraisal of the alternative options for treatment/disposal.

2) An objective comparative risk assessment, ignoring social and socio-economic pressures; it is acknowledged that some parts of this will be extremely difficult to quantify, particularly the risks of penetration of deep repositories on geological time scales and the consequent threat to the population at that time.

3) A full and public evaluation of the technology and comparative risks, taking into account the expressed specific concerns of the public; it has to be recognised that whilst public perceptions of relative risk associated with the nuclear industry seem to the industry to be highly illogical compared with very high risks they are willing to tolerate,[4] these are deeply-held beliefs and a joint and tolerant approach to resolving the issue is essential.

On this last point, it is not sufficient for the industry to claim "transparency" [16,17]; without trust and confidence from the public, the exercise becomes pointless.

Résumé

It is evident that some basic precautions, such as limiting the presence of impurities in the graphite manufacturing process, were not as successful for the existing nuclear graphites as might have been hoped, and will as a consequence increase personnel doses during decommissioning as well as perhaps requiring the application of more complex procedures that would otherwise have been necessary. It is therefore most important that such mistakes are not repeated in the specification and quality control of new graphites and other carbon materials for the reflectors of the new generations of HTR.

Secondly, planning for eventual decommissioning is of paramount importance. The designers, potential operators and licensing authorities have a responsibility to engage in the public debate about the disposal of radioactive wastes now, and are encouraged to support the proposals for a detailed comparative risk analysis and open evaluation of these issues, which might most usefully be handled on an international basis. Finalised designs for the new reactors should allow for the accommodation of alternative strategies and should also have the eventual decommissioning needs in mind.

---

[4] Smoking, car travel, etc. It is considered that the risk of death from smoking 10 cigarettes per day is 1 in 200, an accident on the road is 1 in 8 000 and the risk from the UK nuclear industry, including waste disposal, is 1 in 40 000 000 [15].

REFERENCES

[1]   B.J. Marsden and A.J. Wickham, "Graphite Disposal Options: A Comparison of the Approaches proposed by UK and Russian Reactor Operators", Proc. International Conference on Nuclear Decommissioning '98, Professional Engineering Publishing, pp. 145-153 (1998).

[2]   M. Fisher, "Fort St. Vrain Decommissioning Project", Proceedings of a Technical Committee on Technologies for Gas-cooled Reactor Decommissioning, Fuel Storage and Waste Disposal, Jülich, Germany, September 1997, IAEA-TECDOC-1043, pp. 123-131 (1998).

[3]   V.F. Likar, "Decommissioning Progress at Fort St. Vrain", Proc. International Conference on Nuclear Decommissioning '95, Mechanical Engineering Publications, pp. 1-15 (1995).

[4]   Y.L. Sun and Y.H. Yu, "On the Issues of Fuel Storage and Decommissioning of the HTR-10 Test Reactor", Proceedings of a Technical Committee on Technologies for Gas-cooled Reactor Decommissioning, Fuel Storage and Waste Disposal, Jülich, Germany, September 1997, IAEA-TECDOC-1043, pp. 133-139 (1998).

[5]   K. Sawa, S. Yoshimuta, S. Shiozawa, S. Fujikawa, T. Tanaka, K. Watarumi, K. Deushi and F. Koya, "Study on Storage and Reprocessing Concept of the High-temperature Engineering Test Reactor Fuel", Proceedings of a Technical Committee on Technologies for Gas-cooled Reactor Decommissioning, Fuel Storage and Waste Disposal, Jülich, Germany, September 1997, IAEA-TECDOC-1043, pp. 177-189 (1998).

[6]   B.J. Marsden and S.D. Preston, "Graphite Selection for the PBMR Reflector", Proc. First International Exchange Meeting on Basic Studies on High-temperature Engineering, Paris, France, September 1998, OECD NEA, pp. 235-245 (1999).

[7]   I.F. White, G.M. Smith, L.J. Saunders, C.J. Kaye, T.J. Martin, G.H. Clarke and M.W. Wakerley, "Assessment of Management Modes for Graphite from Reactor Decommissioning", European Commission 1985, report EUR 9232.

[8]   J.J. Guiroy, "Graphite Waste Incineration in a Fluidised Bed", Proceedings of an IAEA Specialists Meeting on Graphite Reactor Lifecycle Behaviour, Bath, UK, September 1995, IAEA-TECDOC-901, pp. 193-203 (1996).

[9]   J. Carlos-Lopez, R.E. Lords and A. Patrick-Pinto, "Integrated Conditioning Process for Spent Graphite Fuels", Proceedings on the Conference Waste Management 1994: Working Towards a Cleaner Environment, Tucson USA, Feb-Mar 1994, 1, 571-574.

[10]   S. Nair, "A Model for Global Dispersion of ^{14}C Released to the Atmosphere as CO_2", *J. Soc. Radiological Protection*, 3, 16-22 (1983).

[11] S.A. Dias and J.P. Krasznai, "Selective Removal of ^{14}C from Ion-exchange Resins using Supercritical Carbon Dioxide", International Topical Meeting on Nuclear and Hazardous Waste Management, Seattle, WA USA, 18-23 August 1996 (American Nuclear Society); F.H. Chang *et al.*, "Producing ^{14}C Isotope from Spent Resin Waste", United States Patent No. 5286468, February 1994; see also J. Gotz, *et al.*, "^{13}C/^{12}C and ^{18}O/^{16} Liquid-vapour Isotopic Fractionation Factors in CO as a Function of Temperature", *J. Chem. Phys.*, 70, 5731 (1979).

[12] K. Tejima, "Processing Method for Radioactive Graphite Waste", Japanese Patent Document 6-94896/A/, 1994, application 4-244816 (1992).

[13] J.B. Mason and D. Bradbury, "Pyrolysis and its Potential Use in Nuclear Graphite Disposal", Paper Presented at IAEA Technical Committee Meeting on Nuclear Graphite Waste Management, Manchester, October 1999, IAEA-NGWM/CD 01-00120 (published as CD-ROM).

[14] A.M. Armour, "Rethinking the Role of Risk Assessment in Environmental Policy Making", in Environmental Policy: International Issues and National Trends, L.K. Caldwell and R.V. Bartlett, eds., Quorum Books, pp. 37-59 (1997).

[15] D.R. Williams, "What is Safe? – The Risks of Living in a Nuclear Age", Royal Society of Chemistry, London (1998).

[16] S. Frischman, "Transparency Doesn't Always Make it Right", Proc. VALDOR (Values in Decisions on Risk), Stockholm, Sweden, June 1999, European Commission, pp. 31-38.

[17] C.O. Wene and R. Espejo, "A Meaning for Transparency in Decision Processes", Proc. VALDOR (Values in Decisions on Risk), Stockholm, Sweden, June 1999, European Commission, pp. 404-421.

STATUS OF IAEA INTERNATIONAL DATABASE ON IRRADIATED GRAPHITE PROPERTIES WITH RESPECT TO HTR ENGINEERING ISSUES

P.J. Hacker[1], B. McEnaney, A.J. Wickham[2]
Bath Nuclear Energy Group
Department of Engineering & Applied Science
University of Bath
Bath BA2 7AY
United Kingdom

G. Haag
Institut für Sicherheitsforschung und Reaktortechnik-ISR 2
Forschungzentrum Jülich GmbH (FZJ)
Postfach 1913, D-52425 Jülich
Germany

Abstract

The International Database on Irradiated Nuclear Graphite Properties contains data on the physical, chemical, mechanical and other relevant properties of graphites. Its purpose is to provide a platform that makes these properties accessible to approved users in the fields of nuclear power, nuclear safety and other nuclear science and technology applications. The database is constructed using Microsoft Access 97 software and has a controlled distribution by CD ROM to approved users. This paper describes the organisation and management of the database through administrative arrangements approved by the IAEA. It also outlines the operation of the database. The paper concludes with some remarks upon and illustrations of the usefulness of the database for the design and operation of HTR.

---

[1] Present address: British Energy Generation Ltd., Barnett Way, Barnwood, Glos. GL4 3RS, UK.
[2] Current Chair, Steering Committee, IAEA International Database on Irradiated Graphite Properties.

Introduction

In September 1995, the International Atomic Energy Agency (IAEA) Specialists Meeting on Graphite Moderator Lifecycle Technologies was held at the University of Bath [1]. Delegates at the meeting drew attention to a number of factors:

- The decline in the number of graphite experts remaining in the nuclear power industry and research organisations.

- The possible withdrawal of certain countries from graphite-moderated reactor technology.

- A renewed interest in developments in high-temperature reactor technology (which has intensified considerably since 1995).

- The development of more sophisticated safety cases for continued operation and life extension of existing graphite-moderated plants.

- A new interest in certain types of graphite data as a result of decommissioning activity.

Taken together, these factors prompted a recommendation to establish a central archive for the storage of data on properties of irradiated graphite.

Following the meeting at Bath, a proposal to create a centralised archive of irradiated graphite properties was brought to the attention of the 13th meeting of the IAEA International Working Group on Gas-cooled Reactors (IWGGCR) in 1996. The Committee expressed support for the concept, but funding the development of the database was seen as an obstacle. After due consideration, the United Kingdom Health and Safety Executive (HSE), with the University of Bath acting as the main contractor, undertook the original development of the database. In 1999, after several meetings with representatives of interested Member States, a Working Arrangement was approved by IAEA. Within the IAEA, the Atomic and Molecular Data Unit in the Department of Nuclear Sciences and Applications maintain the Graphite Database.

Purposes and objectives of the Graphite Database

The Graphite Database consists of data on the physical, chemical, mechanical and other relevant properties of irradiated nuclear graphites contributed by Member States and organisations from the Graphite Database Members. The purpose of the Graphite Database is to preserve and further expand the existing scientific information on the physical, chemical, mechanical and other properties of irradiat graphites (including zero dose) relevant for nuclear power, nuclear safety and other nuclear science and technology applications, and to create a comprehensive international source for such information, including reference data.

The Graphite Database is intended to:

- Facilitate the development of national and international programmes on graphite-moderated reactors and fusion technologies.

- Assist the safety authorities in the assessment of safety aspects of graphite-moderated reactors, including the safety aspects of reactor decommissioning.

- Serve as a comprehensive source of scientific information for a broad range of material science applications, including the non-nuclear technology areas.

Membership of the Graphite Database

Full Membership of the Graphite Database is restricted to Member States of the IAEA that are willing and able to provide relevant data owned by them or non-governmental organisations within their territories for inclusion in the Graphite Database. To participate in the Graphite Database an official request to this effect must be made by the Member State to the Director General of the IAEA. A prospective member of the Graphite Database must provide the following information:

- The amount of data offered for immediate inclusion in the Graphite Database.

- A brief description of data offered for inclusion in the Graphite Database (including any access restriction).

- An estimate of the future data contributions to the Graphite Database.

- A statement of acceptance of the obligations stipulated in the International Agreement provided by the IAEA.

Member States that are not in possession of data relevant to the Graphite Database, but have a strong interest in the use of such data and in the development of the Graphite Database, may be accepted as Associate Members of the Graphite Database. A prospective Associate Member of the Graphite Database must make an official request to the Director General of the IAEA. Such a request shall include the following information:

- A description of the current and future needs of the Member State which motivate the application for Associate Graphite Database membership.

- A description of the expected non-data contribution of the applicant to the Graphite Database development and maintenance.

- A statement of acceptance of the obligations stipulated in the Database Arrangements provided by the IAEA.

Full database Member States

At present, the Full Member States of the Graphite Database are Germany (represented by Forschungzentrum, Jülich, FZJ) Japan, the United Kingdom and the United States of America. Lithuania is an Associate Member and it is expected that The Netherlands will be seeking membership very shortly. A number of other IAEA Member States are also considering participating in the project.

Database Sponsors

Non-governmental organisations making voluntary contributions to the operation and maintenance of the Graphite Database are Database Sponsors. A Database Sponsor may access restricted data with the approval of the Member who provided the data. Database Sponsors are informed about the Agenda of each forthcoming Steering Committee Meeting and may request to be invited to attend the meetings. Database Sponsors may on request be invited to all technical meetings that are organised by the IAEA in relation with the Graphite Database. At present, the Graphite Database is sponsored by Toyo Tanso Co. Ltd. (Japan) and Eskom Enterprises, Pebble Bed Modular Reactor (PBMR) Programme (South Africa).

The Steering Committee

The Steering Committee of the Graphite Database is made up of:

a) One representative from each Member State of the Graphite Database appointed by the appropriate national authority.

b) One representative of each Associate Database Member.

c) A Steering Committee Chairman.

The IWGGCR and the IAEA are each invited to send an observer to Steering Committee meetings. The Steering Committee meets once a year. Since its formation the Steering Committee has met in Vienna, Austria (1999), Oak Ridge National Laboratory (ORNL), USA (2000) and in Rain, Germany (2001).

The programme of the Steering Committee includes:

- All technical matters relating to data and Graphite Database operations.

- Consideration of annual progress report prepared by the IAEA in co-operation with the Liaison Officers of Graphite Database members.

- Evaluation of the Graphite Database status and operations, and the progress made in its development.

- Recommendations on the procedures regulating the operation of the Graphite Database.

- Consideration of financial aspects of the Graphite Database and recommending relevant measures and actions.

- Discussion of proposals from the Members and Associate Members of the Graphite Database relating to the operation and development of the Graphite Database.

The Chairman of the Steering Committee may invite representatives of Graphite Database Sponsors to attend the Steering Committee meetings as observers for obtaining information or advice on specific items of the agenda or following a specific request for attendance from the Sponsors. The invitation is extended through the IAEA.

Present members of the Steering Committee, who also act as Liaison Officers for their Member States, are as follows.

Germany	Dr Gerd Haag (FZJ)
Japan	Dr Kazuhiro Sawa (JAERI)
United Kingdom	Dr Anthony J. Wickham*
United States of America	Dr Timothy D. Burchell (ORNL)
Lithuania	Dr. Rimantas Levinskas, Lithuania Energy Institute
Together with IAEA representatives	

* Chair of Committee until February 2002.

Full details of the organisation and operation of the Graphite Database can be found at its web site [2].

Technical objectives of the database

The Graphite Database contains key physical properties data for various nuclear grade graphites under various reactor conditions; some of the data relate to high irradiation temperatures and high neutron fluences. As of September 2000, 32 400 records are in the database.

The development of the database had the following objectives:

- User friendly interaction with the database, based upon "point and click" activity and menus.

- No specialist IT knowledge.

- Clear presentation of data.

- Flexible methods of data retrieval.

- Flexible database structure to allow for different data types.

- Secure methods of data distribution.

At present the database concentrates on numerical data with limited textual and graphical material, although the Steering Committee may elect to extend the capabilities within the system for the inclusion of anecdotal references in due course.

Database platform

Microsoft Access 97 has been used as the main tool for storing data and for development of the database. The Microsoft Access relational database permits the rapid development of customised interfaces to data and it also facilitates rapid changes to data formats. Access 97 and the software developed using it can operate on any personal computer (PC) running Microsoft Windows 95 or higher. A CD ROM drive is also required for installation of the database.

Classification of data

At an early stage in the technical discussions it was realised that a number of Member States regarded irradiated nuclear graphite data as sensitive information or of commercial interest that should not be made available freely. This led to definitions of data in the database as follows.

- *Unrestricted data* are data published in the open literature and those offered by Database Members on behalf of the data supplier for free dissemination.

- *Restricted data* are all other data that are nonetheless made available to other Database Members for non-commercial purposes. Restricted data will be disseminated to Members only by CD ROM. Restricted data will also be made available to current Database Sponsors at the discretion of the Steering Committee and the representatives of the Member States that provided such data.

- *L2 restricted data* is another category of restricted data from operating plants that are provided by the commercial utilities in the United Kingdom and potentially for other commercially sensitive data from other States. The listings of the L2 restricted data are made available to all restricted data users. Potential users of these data are required to sign a confidentiality agreement directly with the utilities providing the data before achieving access to the detail. These data are also distributed by CD ROM.

Types of data in the database

The current content of the database consists principally of data recording changes in physical and mechanical properties of nuclear graphites upon fast neutron irradiation. The list of properties is set out in Table 1. The selection and prioritisation of data for input into the database is a responsibility of the Steering Committee, which has taken into account several factors. These include the development of safety cases for current plant and for new HTR designs.

Table 1. Classification of materials property data in the Graphite Database

Group 1. Sample Characterisation

* Irradiation Temperature
* Fluence
* Graphite Grade
* Burn-up
* Direction of Measurement (with respect to extrusion direction)
* Experiment

Group 2. Crystalline Parameters (a and c are lattice parameters and La and Lc are the crystallite dimensions) in Angstroms

* a Initial
* a Final
* c Initial
* c Final
* La Initial
* La Final
* Lc Initial
* Lc Final

Group 3. Elastic Moduli

* Initial E
* Final E
* Initial G
* Final G

Group 4. Electrical Resistivity

* Initial r
* Final r

Group 5. Coefficient of Thermal Expansion, CTE

* Initial CTE
* Final CTE
* CTE Temp Range

Group 6. Sample Properties

* Initial Density
* Final Density
* Initial Open Pore Volume 1

* Final Pore Volume
* Weight Loss, %

Group 7. Thermal Conductivity

* Initial TH-COND
* Final TH-COND
* Experiment Temperature

Group 8. Strength

* Initial S (Tensile)
* Final S (Tensile)
* Initial S (3 pt Bending)
* Final S (3 pt Bending
* Initial S (Compressive)
* Final S (Compressive)

Group 9. Poisson's Ratio

* Initial u
* Final u

Group 10. Dimensional Changes

* Length Change, %
* Diameter Change, %
* Volume Change, %

Group 11. Irradiation Creep Properties

* Appld Stress
* Creep Strain
* Creep Coefficient

Group 12. Acoustic Properties

* Initial Sonic Velocity
* Final Sonic Velocity
* Initial Attenuation
* Final Attenuation

Group 13. Stored Energy

* Total S
* dS/dt
* Release Temperature

An important factor that has been considered by the Steering Committee is the problem of different standards and definitions employed by the data suppliers concerning, for example, absorbed neutron dose. In order to preserve traceability of the data, the Steering Committee has decided to enter data in the units employed by the original data supplier without conversion or rationalisation. However, careful attention is paid to ensure that users are not confused by different units, and the Steering Committee regularly reviews appropriate conversion factors as well as any technical issues affecting these factors. For example, there has recently been a thorough discussion about the conversion of neutron dose units into terms of "displacements per atom", which has led to an energetic debate about the appropriate value of the displacement energy to be used for the conversion, on which the literature is inconsistent.

Operation of the database

Data input and formatting

Nuclear graphite data are contributed to the Graphite Database through its Members, either upon initial joining or through continuing data contributions. Database Members are responsible for the collection, validation, categorisation, specification and appropriate formatting of nuclear graphite properties data relevant to the Graphite Database. At present most data is being supplied to the IAEA in the original format as specified by the data contributors. Data is mostly submitted as technical reports that are formatted and arranged according to each particular authors' or institutions' requirements. Some data have also been submitted in electronic form. The flexible database structure ensures that the differing arrangements of data contributions can be entered easily into the database. However, expert appraisal is required to ensure that data entered are meaningful. It is preferable if each data set is submitted using a predefined template, as this minimises additional formatting; details of the template are given in Graphite Database web site [2] and the User Manual [3].

To ensure traceability of the data, each submitted data set is identified as a "Volume" in the database, which corresponds to a uniquely identifiable submission, referenced by a title, authors, source, report number and country of origin. The submitted data are stored and a copy is formatted for inclusion in the database as a "Table". Data received in paper form must be converted to electronic form before formatting into a "Table" in the database. Currently, this is done manually, as it was found that using optical character recognition methods introduced errors. Several verification methods have been employed to reduce errors during data logging and formatting. Data input rules have been used to restrict the range and/or type of data that can be entered under a particular property. After data entry and formatting random selections of records are produced and these are compared with the original data.

Using the database

Full details of the operation of the database are presented in the User Manual [3], so only a brief summary is given here. The relationships between the principal elements of the database are shown in Figure 1. Tables are similar to conventional spreadsheets with each property being identified by column heading. Tables are stored in a single Access 97 file together with other database components. Relationships between different properties are created from program code at run time, thus ensuring a flexible method of data retrieval.

Query screens ask the user for particular information relating to the final data set they wish to retrieve. A succession of query screens enables the user to specify the data set that is retrieved and displayed as a report. The user can also identify and, if needed, change the units in which the data are listed. Various filtering and sorting functions are available in the database to enable the data set to be

Figure 1. Relationships between the principal elements of the database

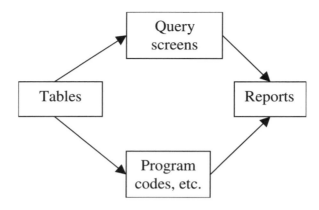

refined to suit the requirements of the database user. In addition to the numerical data, the database also contains small amounts of both graphical and textual information, for example experimental details that may be relevant to the data. Finally, the customised report can be exported to other applications for further analysis.

Relevance of the Graphite Database to new designs of HTR

The effects of irradiation upon the properties of nuclear graphites in HTR will depend upon the irradiation temperature neutron fluence domain during its lifetime. These domains will vary both from one design to another and, for a given design, with location within the reactor. As an example, Table 2 summarises these domains for various locations of graphites within the HTR-500 plant that was developed in the 1980s [4], but was never built.

Table 2. Operating temperatures and lifetime neutron fluences for graphite components in the HTR-500 (adapted from [4])

Component	Depth/m	Operating temperature/°C	Lifetime fluence/ 10^{22} n cm^{-2} (EDN)
Top reflector	–	250	1.5
Core	2.0	450	3.4
Core	3.0	475	1.9
Core	4.0	650	1.2
Core	5.0	675	0.8
Bottom reflector	–	700	0.4

In more modern designs, particularly those coupled with gas turbines, the irradiation temperatures are higher but the neutron fluences can be lower if the reactors are designed so that the inner reflector blocks are exchangeable. For fuel element graphites the temperatures are much higher (up to about 1 100°C) but the lifetime neutron dose is rather small [~0.5×10^{22} n cm^{-2} (EDN)].

Figures 2-4 show examples of the effect of neutron irradiation on several properties of a nuclear graphite that has been used in the development of an HTR. The property data are recorded in the Graphite Database. Figure 2 shows an increase in Young's modulus as a result of neutron irradiation at 720-770°C. For the neutron fluence irradiation temperature domain in this study Young's modulus also increases slightly over the range of fluences used. Figure 3 shows the dimensional changes for the

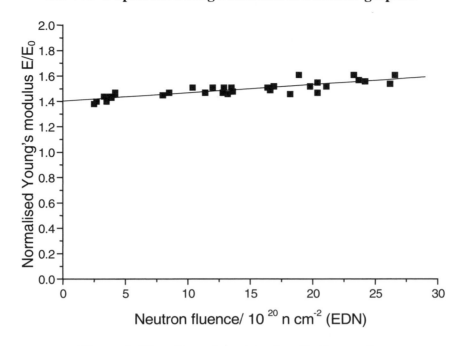

Figure 2. The effect of neutron irradiation in the range 720-770°C upon the Young's modulus of a nuclear graphite

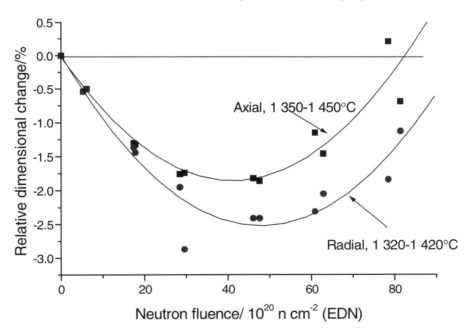

Figure 3. The effect of neutron irradiation on the relative dimensional changes of a nuclear graphite

Figure 4. Effect of neutron irradiation at 1 115-1 165°C upon the CTE of a nuclear graphite

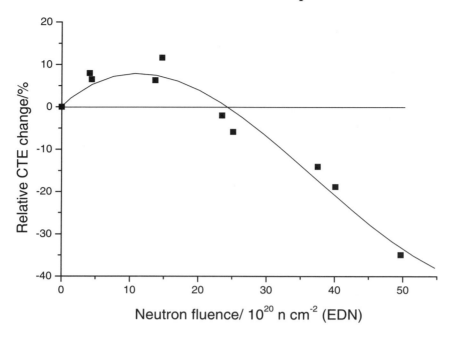

graphite for higher irradiation temperatures and a wider range of neutron fluences that are found in Figure 2. As is observed for many graphites under a range of irradiation conditions, the nuclear graphite used in this study shows initial shrinkage before turn around at about 50×10^{20} n cm^{-2} (EDN). Figure 4 shows the effect of neutron irradiation at 1 115-1 165°C upon the coefficient of thermal expansion (CTE) of the graphite. As is observed for most nuclear graphites under various irradiation conditions, there is an initial increase in CTE upon irradiation that reaches a broad maximum at about 10-15 $\times$ 10^{20} n cm^{-2} (EDN). Thereafter, CTE decreases to reach the un-irradiated value at about 28×10^{20} n cm^{-2} (EDN). At higher fluences the CTE continues to decrease until it reaches a value that is about 65% of the un-irradiated value at 50×10^{20} n cm^{-2} (EDN).

After a long period of low interest in the HTR technologies developed in the 1970s and 1980s, in recent years there has been a substantial increase in interest from several countries in developing new designs of HTR containing graphite components. It is likely that some new designs of HTR may require the development of new grades of nuclear graphite and so new irradiation test programmes will be desirable. In such cases, the Graphite Database may be used for scoping studies using data for graphites that are similar in structure to those being developed and that have been exposed to neutron fluence temperature domains that are close to those in the new designs. The value of the Graphite Database will increase if, in due course, the irradiation properties data for the newly developed graphites can be incorporated. Also, there are undoubtedly other relevant, extant data in files, reports and publications that could be incorporated into the database if they are made available.

Conclusions

The International Database on Irradiated Nuclear Graphite Properties contains data on the physical, chemical, mechanical and other relevant properties of graphites subject to fast neutron irradiation. Its purpose is to provide a platform that makes these properties accessible to approved users in the fields of nuclear power, nuclear safety and other nuclear science and technology applications. A flexible and user-friendly platform for the database has been constructed using Microsoft Access 97 software. This paper has described the organisation and management of the database through administrative arrangements approved by the IAEA. It also outlines the operation of the database in order to extract required data. The database contains much information that is useful for the design and operation of HTR. Its usefulness in this area will be increased if additional, extant data can be incorporated into the database. In the future, it will also be useful if results from any new materials testing experiments made to support the design of new HTR can be added to the database.

REFERENCES

[1]   "Graphite Moderator Life Cycle Behaviour", Proceedings of a Specialists Meeting held in Bath, UK, 24-27 September 1995, IAEA-TECDOC-901, August 1996.

[2]   International Database on Irradiated Nuclear Graphite Properties, to be found at the site: http://www-andis.iaea.org/graphite.html.

[3]   T.D. Burchell, R.E.H. Clark, M. Eto, G. Haag, P. Hacker, R.K. Janev, G.B. Neighbour, J.A. Stephens and A.J. Wickham, "IAEA International Database on Irradiated Graphite Properties", International Atomic Energy Agency Report (Nuclear Data Section) INDC (NDS)-413, February 2000.

[4]   W. Theymann W. Delle, H. Nickel, "Auslegung des Inneren Graphitereflektors eines Hochtemperaturreaktors", Report No. Jül-1906, March 1984, ISSN 0366-0885.

SESSION V

Basic Studies on HTGR Fuel Fabrication and Performance

Chairs: J. Kuijper, R. Moormann

EVIDENCE OF DRAMATIC FISSION PRODUCT RELEASE FROM IRRADIATED NUCLEAR CERAMICS

L. Thomé, A. Gentils, F. Garrido
Centre de Spectrométrie Nucléaire et de Spectrométrie de Masse
Bât. 108, 91405 Orsay, France

J. Jagielski
Institute of Electronic Materials Technology, 01-919 Warsaw, Poland
Andrzej Soltan Institute for Nuclear Studies, 05-400 Swierk/Otwock, Poland

Abstract

Crystalline oxide ceramics are promising matrices for plutonium and minor actinide transmutation in specifically devoted nuclear reactors. An important issue concerning these materials is the investigation of their ability to confine radiotoxic elements resulting from the fission of actinides. Here, we study the diffusion and release upon annealing or irradiation at high temperature of a toxic fission product (Cs) in a typical oxide ceramic (zirconia). The foreign species are introduced by ion implantation and the diffusion is studied by Rutherford backscattering experiments. The results show the influence of the fission-product concentration on the diffusion properties. The Cs mobility is strongly increased when the impurity concentration exceeds a threshold of the order of a few atomic per cent. Irradiation with medium-energy heavy ions is shown to enhance Cs diffusion with respect to annealing at the same temperature. The effects described in this paper have to be taken into account when developing innovative nuclear fuels in order to decrease the risk of accidental release of radiotoxic elements into the geosphere.

Introduction

The reduction of the excess amount of plutonium and other actinides arising from the nuclear fuel cycle or nuclear weapon dismantling is an important challenge for the industrialised world in the next decades. A solution to this issue is the incorporation of these radiotoxic species in non-fertile matrices and their burning in nuclear reactors specifically devoted to the task [1-3]. Crystalline oxide ceramics, more particularly zirconia (ZrO_2), were identified as very promising matrices for actinide transmutation due to their high melting point, their reasonable thermal conductivity and their strong resistance against irradiation [4-6]. One of the most serious problems to be solved for the final qualification of new nuclear fuels with respect to safety criteria is the evaluation of their ability to confine radiotoxic elements resulting from the fission of actinides. Such an investigation relies on the study of the mechanisms of fission-product (FP) diffusion and release, which depend on parameters such as the FP concentration and lattice location, the substrate temperature and the nature, amount and mobility of radiation damage produced in the host material.

Ion beams provide very efficient tools for diffusion studies since they can be used for irradiation, doping and characterisation of solids [7]. First of all, foreign species may be introduced into the host matrix by ion implantation in a wide range of concentrations (from a few ppm up to several atomic per cent). Then, the radiation damage produced in the nuclear fuel can be easily simulated by inert gas irradiation at various energies (from a few hundreds of keV up to several hundreds of MeV). Finally, the depth distribution of fission products following various treatments (annealing or high-temperature irradiation) may be performed using the Rutherford backscattering (RBS) technique [8] implemented with a MeV ion accelerator. The present paper reports the case study of caesium in zirconia, which can be generalised to other fission species and nuclear matrices.

Experimental procedure

The samples used in the experiments are cubic {100}-oriented zirconia single crystals, fully stabilised with with 9.5 mol.% yttria. They contain 1.5 wt.% hafnium as a contaminant. The surface was covered with a thin carbon layer (~15 nm) in order to avoid charging of the samples during ion beam analyses.

Caesium ions were introduced into zirconia single crystals by 300 keV room-temperature ion implantation at the IRMA implanter of the CSNSM Orsay facility (see Figure 1) [9]. Ion fluences ranged from 5×10^{15} to 5×10^{16} cm^{-2} in order to vary the Cs concentration from ~1 to ~10 at.%. The current density was always kept lower than 1 μA cm^{-2} to prevent excessive target heating during the process of implantation. The energy used ensures a good separation of the high-energy peaks of Cs from the Zr front edge in RBS spectra (see next section).

Implanted samples were then either annealed in vacuum during one hour or irradiated at the IRMA machine (also during one hour) with 360 keV Ar ions (total fluence: ~10^{15} cm^{-2}) at temperatures ranging from 300°C to 1 000°C. Irradiations with Ar ions were performed in order to simulate the damage created in a fuel matrix by the various nuclear reactions. Only half of the crystal surface was exposed to the Ar ion beam and both irradiated and un-irradiated regions were subsequently analysed.

The modifications of the Cs ion depth distribution due to thermal annealing or high-temperature Ar ion irradiation were investigated by RBS experiments at the ARAMIS accelerator of the CSNSM Orsay facility (see Figure 1) [9]. The RBS analysis was made using 3.06 MeV ^{4}He ions. The energy resolution of the RBS set-up was ~12 keV, which corresponds to a depth resolution of the order of 10 nm.

Results and discussion

Figure 2 presents RBS spectra recorded in a random direction on a ZrO_2 single crystal implanted with 300 keV Cs ions at a fluence of 3×10^{16} cm^{-2}, before and after thermal treatments. The data can be separated in two parts: (i) the low-energy plateaus (below channel 410) due to the backscattering of analysing particles from the Zr atoms of the target and (ii) the high-energy peaks (between channel 420 and channel 435) arising from the backscattering of analysing particles from implanted Cs atoms. It should be noted that the signals generated by the backscattering from O atoms of the target, appearing at an energy much lower than the Zr edge, are not displayed in the figure for the sake of clarity. The normalised yield versus channel scales can be easily transformed in atomic concentration versus depth scales (see Figure 2) in order to provide the depth distribution of implanted Cs ions.

The Cs depth profile before any heat treatment (filled squares in Figure 2) is well reproduced with a Gaussian distribution. The values found for the projected range, $R_p \sim 45$ nm, and the range straggling, $\Delta R_p \sim 30$ nm, are in reasonable agreement with the results of Monte Carlo simulations for implanted ions performed with the TRIM code [10].

The FP diffusion and release are monitored by the modifications occurring in the depth distribution of implanted elements upon annealing or high-temperature irradiation. The data of Figure 2 clearly show that annealing at 500°C (filled triangles) induces only a little modification of the Cs profile, whereas irradiation at this temperature (open triangles) or annealing at 600°C (filled circles) lead to a drastic modification of the Cs peak with a strong decrease of the Cs concentration (from ~4.5 down to ~2 at.%). The latter effect is the manifestation of the diffusion and subsequent desorption of implanted species.

The summary of the results obtained on Cs-implanted ZrO_2 samples submitted to thermal treatments is presented in Figure 3. It shows the variation of the Cs concentration at the maximum of the distribution as a function of the annealing or irradiation temperature for various implantation fluences. For crystals implanted at low atomic concentrations (i.e. below 2 at.%), no diffusion of Cs atoms occurs up to the highest temperature investigated in this study. In this concentration range, high-temperature Ar irradiation has no influence on the mobility of implanted species. For crystals implanted at high atomic concentrations (i.e. above 4 at.%), diffusion and release of Cs atoms occur upon annealing at ~550°C. The result is a decrease of the Cs concentration down to a critical value (c_c) of the order of 1.5 at.%. In this concentration range, Ar ion irradiation increases the mobility of Cs atoms, since they start to diffuse at ~450°C to reach the same final concentration as upon annealing. Furthermore, once the Cs concentration has dropped down to the critical concentration, the Cs depth profile remains almost unaffected up to the highest annealing temperature, revealing a high stability of the phases formed below c_c.

The determination of the local atomic configurations of foreign species in the host matrix may help to understand the diffusion results summarised in Figure 3. Previous RBS and channelling data recorded on ZrO_2 single crystals implanted with FP showed that a large fraction of Cs ions occupy substitutional lattice sites at low atomic concentrations (i.e. below ~2 at.%) [11-12]. In ZrO_2, Zr atoms are located in the centre of a cube formed by O atoms (ZrO_8 dodecahedron); the distance between Zr and O atoms is ~2.4 Å and the Zr atoms are in the Zr^{4+} oxidation state with a radius of 0.87 Å [13]. Since the radius of Cs atoms ranges from 0.174 to 0.202 nm [14-15], the substitution of a Zr atom by a Cs one creates a large stress, which can only be relaxed by the formation of anionic vacancies in the vicinity of the substituted atom. In this atomic configuration, Cs ions occupy positions only slightly displaced from the original locations of Zr atoms, leading to the value of the substitutional fraction measured by channelling.

When the Cs concentration exceeds c_c, the fraction of Cs ions in substitutional lattice sites falls down to zero, and a severe damage is induced into the zirconia lattice by ion implantation [11-12]. The solubility limit of Cs in ZrO_2 (at 2 000 K) was estimated to be ~1.5 at.% [16-17]. The decrease of the substitutional fraction at high Cs concentration is thus very likely due to either the precipitation of Cs atoms in metallic inclusions or the formation of ternary compounds (such as Cs_2ZrO_3 [16-18]).

Annealing above 550°C of crystals implanted at high ion fluences leads to a drop of the Cs concentration down to 1.5 at.% and to a release of excess Cs atoms from the ZrO_2 matrix. This result indicates that the atomic configurations built up at high Cs concentrations are not thermodynamically stable. The substitutional fraction measured in RBS and channelling experiments after annealing implanted ceramics up to 800°C remains essentially unchanged for all the samples investigated in this study. This result reveals that Cs atoms are randomly located in the case of heavily-doped crystals even when these crystals are annealed at temperatures above 550°C. Thus, despite the fact that the final concentration is the same after annealing for samples implanted at low or at high fluences, the local environment of implanted species is obviously different. The atomic configurations most probably formed are thus: isolated atoms surrounded by vacancies in the case of weakly-doped samples and precipitates in the case of highly-doped samples (in both cases before and after annealing). Therefore, the annealing of ZrO_2 crystals doped with high Cs concentrations likely induces a rearrangement of the clusters formed during implantation, leading to the diffusion and release of a part of implanted species.

Irradiation with rare gas ions of heavily-doped single crystals leads to an enhanced diffusion of Cs atoms as compared with sole annealing. Since 360 keV Ar ions essentially slow down via elastic collisions with the nuclei of the target atoms, it can be assumed that nuclear collisions assist the fission-product diffusion by shaking up the Cs clusters formed during implantation. The result is a modification of the number and size of these clusters accompanied by the release of excess FP from the host matrix at a temperature ~100°C lower than upon annealing.

It should be pointed out that very similar results were obtained in $MgAl_2O_4$ crystals implanted with Cs ions at similar fluences, with the difference that the Cs migration occurs in heavily-doped spinel samples at higher temperatures than in ZrO_2 (~800°C for annealing and ~700°C for high-temperature Ar ion irradiation). The same conclusion can be drawn from these results since previous RBS and channelling experiments performed in spinel also showed that a large fraction of Cs atoms occupies substitutional lattice sites at low concentration and random positions at high concentration [19].

Conclusion

The experiments presented in this paper demonstrate that ion beams provide very useful tools for the evaluation of oxide ceramics as potential matrices for nuclear waste incineration. The main issue addressed in this study is the diffusion and release of fission products upon annealing or irradiation at high temperatures (up to 1 000°C). Ion implantation was selected to dope the material with the species to be confined. Ion irradiation was used for the simulation of the damage created in the matrix by a radiative environment. The modifications of the depth distribution of fission products upon heat treatments were monitored by nuclear microanalysis techniques (RBS).

The results show that the diffusion and release of fission products in oxide ceramics is strongly dependent on the concentration of foreign species. The fission-product mobility increases when the concentration exceeds a threshold of the order of a few atomic per cent. Further enhancement of the fission-product mobility is observed when the temperature rise is accompanied by radiation damage production. It is of great importance to consider the dramatic fission-product release described in this paper when developing innovative nuclear fuels, in order to decrease the risk of accidental contamination of the geosphere with radiotoxic elements.

Acknowledgements

We would like to acknowledge the ARAMIS staff of the CSNSM in Orsay for assistance during implantation and RBS experiments. This work was partially supported by the GDR NOMADE and the NATO Collaborative Linkage Grant N° PST/CLG 974800.

REFERENCES

[1]   *Scientific Basis for Nuclear Waste Management XIX*, W.M. Murphy and D.A. Knecht, eds., Mater. Res. Soc. Symp. Proc., 412, 15 (1996).

[2]   Proceedings of the 4th Workshop on Inert Matrix Fuel, *J. Nucl. Mater.*, 274 (1999).

[3]   C. Degueldre and J.M. Paratte, *Nucl. Technol.*, 123, 21 (1998).

[4]   W.J. Weber, R.C. Ewing, C.R.A. Catlow, T. Diaz de la Rubia, L.W. Hobbs, C. Kinoshita, Hj. Matzke, A.T. Motta, M. Nastasi, E.K.H. Salje, E.R. Vance and S.J. Zinkle, *J. Mater. Res.*, 13, 1434 (1998).

[5]   Hj. Matzke, Proceedings of the International Workshop on Advanced Reactors with Innovative Fuels, OECD Publications, Paris, p. 187 (1999).

[6]   K.E. Sickafus, Hj. Matzke, Th. Hartmann, K. Yasuda, J.A. Valdez, P. Chodak III, M. Nastasi and R.A. Verrall, *J. Nucl. Mater.*, 274, 66 (1999).

[7]   L. Thomé and F. Garrido, Vacuum 63, 619 (2001).

[8]   *Handbook of Modern Ion Beam Materials Analysis*, J.R. Tesmer and M. Nastasi, eds., Materials Research Society (1995).

[9]   E. Cottereau, J. Camplan, J. Chaumont, R. Meunier and H. Bernas, *Nucl. Instrum. Meth.*, B 45, 293 (1990).

[10]   J.F. Ziegler, J.P. Biersack and U. Littmark, *The Stopping and Range of Ions in Solids*, Vol. 1, J.F. Ziegler, ed., Pergamon, New York (1985).

[11]   L. Thomé, J. Jagielski and F. Garrido, *Europhys. Lett.*, 47, 203 (1999).

[12]   L. Thomé, J. Jagielski, A. Gentils and F. Garrido, *Nucl. Instrum. Meth.*, B 175-177, 453 (2001).

[13]   J. Emsley, *The Elements*, J. Emsley, ed., Oxford University Press (1998).

[14]   R.D. Shannon and C.T. Prewitt, *Acta Crystal.*, B 25, 925 (1969).

[15]   R.D. Shannon, *Acta Crystal.*, A 32, 751 (1976).

[16] M.A. Pouchon, M. Döbeli, C. Degueldre and M. Burghartz, *J. Nucl. Mater.*, 274, 61 (1999).

[17] M.A. Pouchon, Ph.D. Thesis, University of Geneva (1999).

[18] L.M. Wang, S.X. Wang and R.C. Ewing, *Phil. Mag. Lett.*, 80, 341 (2000).

[19] L. Thomé, J. Jagielski, C. Binet and F. Garrido, *Nucl. Instrum. Meth.*, B 166-167, 258 (2000).

Figure 1. Schematic representation of the ion beam facility of the CSNSM Orsay [9]

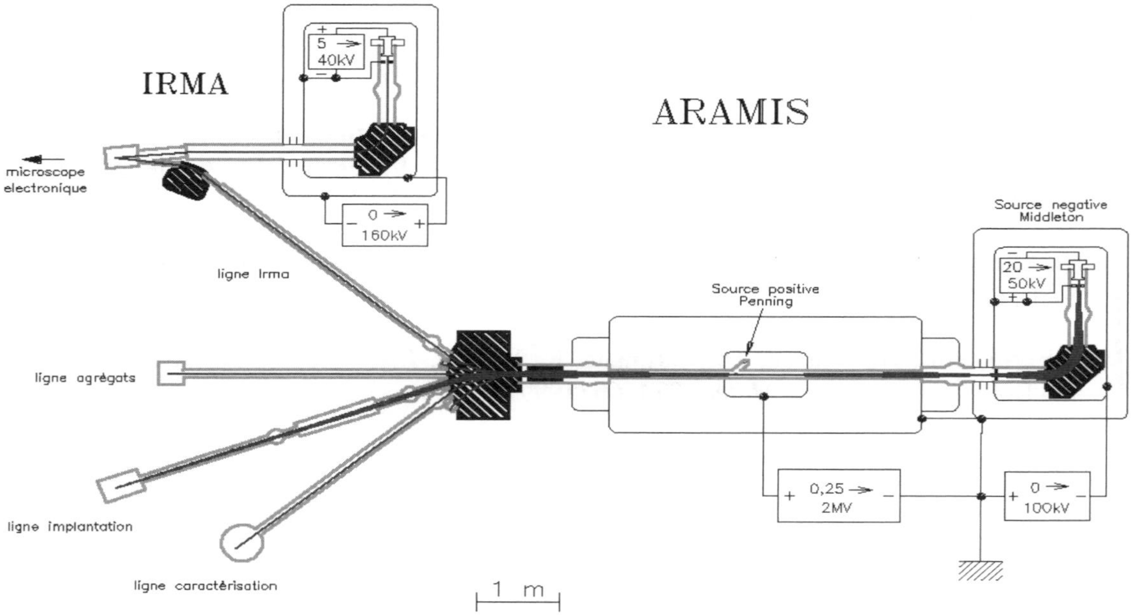

Figure 2. RBS spectra recorded in a random direction on a ZrO$_2$ single crystal implanted with Cs ions at a fluence of 3×10^{16} cm^{-2}, before and after annealing or irradiation with Ar ions at the indicated temperatures

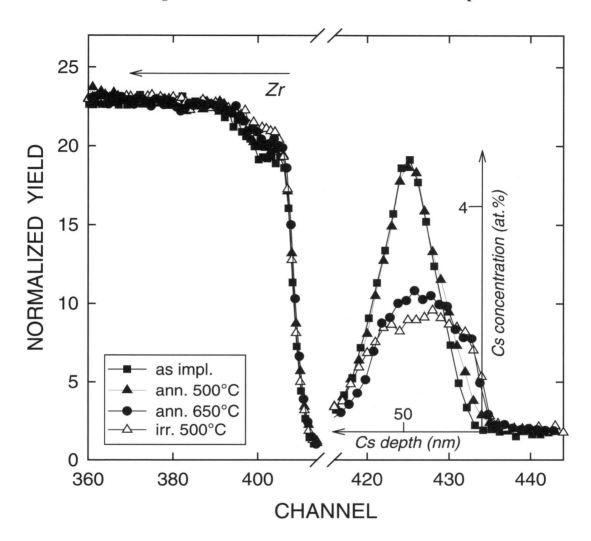

Figure 3. Maximum Cs concentration vs. annealing (filled symbols) or irradiation (open symbols) temperatures for ZrO₂ single crystals implanted with Cs ions at fluences of 5×10^{16} at.cm⁻² (triangles), 3×10^{16} at.cm⁻² (squares) and 10^{16} at.cm⁻² (circles)

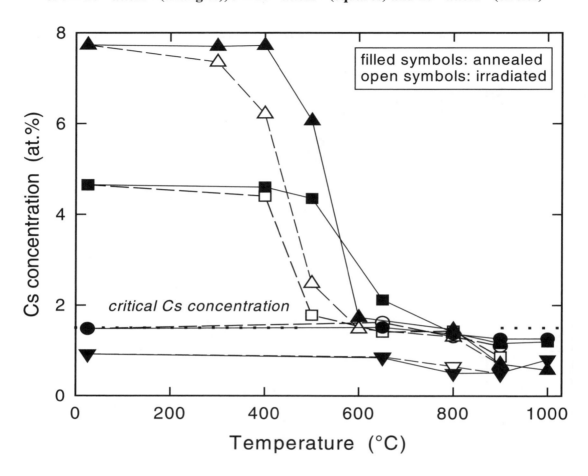

METHOD OF REACTOR TESTS OF HTGR FUEL ELEMENTS AT AN IMPULSE LOADING

A. Chernikov, V. Eremeev, V. Ivanov
Scientific and Industrial Association Luch
24, Zheleznodorozhnaya Str., Podolsk, Moscow Region, 142100, Russia

Abstract

The method employed for testing HTGR fuel elements under impulse loading (during several tens seconds) in the impulse graphite reactor (IGR) is described. Experiments with spherical fuel elements have shown reliability of the control system, providing maintenance and measurement of the specified level of temperature and energy release.

A program for computation of temperature fields in fuel elements and coated fuel particles and temperature stresses distribution along a section of fuel elements was developed.

As an illustration of the progress made, the results of fuel element tests are presented at an energy release of 22 kW.

Introduction

Uranium-graphite fuel elements based on coated fuel particles and a graphite matrix are efficient under the normal conditions of HTGR operation (temperature 1 000-1 600 K). In some accidents [1], for example, after spontaneous removal of control rods of the accident protective system, temperatures in the reactor core have risen up to 2 000 K and higher. At this time power energy release will increase tenfold and greater. The estimated time of reactor operation under conditions of increased power in such a situation can be from several fractions of a second up to several tens of seconds and more.

The necessity of investigating fuel element behaviour under short-term, high-energy loading is dictated by the fact that significant temperature gradients arise in this case in coated fuel particles dispersed in matrix graphite.

The paper considers the method of uranium-graphite fuel element reactor tests under impulse loading conditions and analyses temperature fields and stresses which occur during testing.

Experimental details

Tests were conducted in the impulse graphite reactor (IGR). The main physical characteristics of this reactor are presented in Table 1 [2].

Table 1. The main characteristics of the impulse graphite reactor [2]

Reactor core	Graphite impregnated with uranium (90% ^{235}U), dimension 1 400 × 1 400 × 1 400 mm, moving part 800 × 800 × 1 400 mm, block dimension 100 × 100 × 200 mm
Reflector	Graphite, thickness everywhere is not less than 500 mm, size of a brickwork 2 400 × 2 400 × 4 200 mm
Nuclear ratio, U/C	1:10 000
^{235}U load, kg	7.46
Excess reactivity, %	22 ± 2
Effective fraction of latent neutrons	0.00685
Lifetime of instantaneous neutrons, s	$(1.07\text{-}1.03) \times 10^{-3}$
Fluence of neutrons in the impulse after rods removal, n cm^{-2}	1.1×10^{17}
Maximum possible density of neutron flux in the flash mode, n cm^{-2} s^{-1}	1×10^{18}

Five spherical graphite elements 60 mm in diameter were exposed to tests. One of them was made of graphite and had no fuel, four others were models of spherical fuel elements consisting of the fuelled centre (core) and the fuel–free graphite shell. TRISO-type coated fuel particles ~1-1.1 mm in diameter (average radius of UO$_2$ fuel kernels was 0.25 mm) were uniformly dispersed in a matrix graphite of the core [3]. ^{235}U content in fuel elements No. 1 and No. 3 was 1 g, and in No. 2 and No. 4 were 1.5 g.

Reactor experiments were carried out in a stainless steel ampoule with wall thickness of 2 mm (Figure 1).

Figure 1. Test ampoule scheme

1-4 – Spherical fuel elements, 5 – Graphite ball, 6 – Ampoule walls,
7,8 – Graphite sleeves, 9,10 – Graphite fixtures and cover, T_{1-4} – Thermocouple

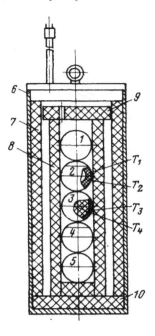

The ampoule was placed in a steel casing filled with helium at a pressure of 0.1 MPa; casing was in the central channel of the reactor. Tested samples were arranged along a vertical axis of a graphite sleeve with an internal diameter of 65 mm. The clearance fit of graphite spacers excluded the possibility of a stress occurrence in the fuel elements, resulting in their contact interaction after thermal expansion. Fission product release was measured by the radiometric analysis of a gas sample from the casing after each start-up.

The ampoule with the fuel elements underwent three reactor tests under impulse mode. Reactor power in each start-up was changed according to a trapezoid profile. At the first stage of loading heat release in the spherical fuel elements was increased linearly during 10-17 s up to some stationary value N_{st}, which was kept constant during 10-30 s. The power was then decreased linearly during 10-12 s. The first start-up at low power was used for developing the mode of reactor loading and to verify control and measuring devices. A stationary mode at power N_{st} was maintained during 30 s. The N_{st} value was 3.4 kW for fuel elements No. 1 and No. 3 and 5.1 kW for fuel elements No. 2 and No. 4 (hereinafter heat release is identified per fuel element). Methods of the determination of temperature along the fuel element section with use of experimental and calculated data were developed in the second start-up. The maximum power in this case was 13.6 kW in samples No. 1 and No. 3 and 20.4 kW in samples No. 2 and No. 4; duration of a stationary mode loading was 20 s. The computation model of a temperature field was checked in the third start-up. The N_{st} value in this case reached 22.1 kW for fuel elements No. 1 and No. 3 and 33.5 kW for No. 2 and No. 4 (Figures 2, 3).

Computation models

It was accepted that the thermal scheme of experiment realisation (Figure 1) is well described by the two-dimensional equation of thermal conductivity. The solution of a similar task requires simplification of boundary conditions. If the temperature of spherical fuel element surface is changed

Figure 2. Change of heat release and thermocouples
indications for fuel element No. 2 in the second start-up

1,2 – Temperature of fuel element surface and core respectively, 3 – Density of an energy release

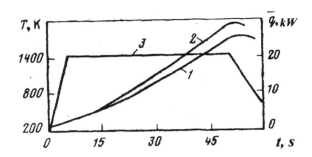

Figure 3. Change of heat release and thermocouples
indications for fuel element No. 3 in the third start-up

1 – Energy release density in the fuel element core,
2,3 – Temperature of fuel element surface and in the centre of the core respectively

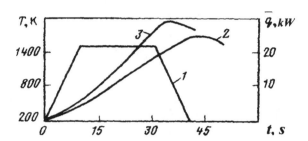

slightly in dependence on azimuthal co-ordinates at the specified moment of time, the real temperature distribution can conveniently be replaced by one-dimensional and to reduce a task to the solution of the thermal conductivity equation of a ball:

$$\partial T/\partial t = (a/r^2)(\partial/\partial r)[r^2(\partial T/\partial r)] + \overline{q}/c\rho \tag{1}$$

where r is the current radius, R_o is the fuel element radius, t is time, T is temperature; $\overline{q}$ is the energy release density in the core with radius $R_c < R_o$ and $a = \lambda/c\rho$ is temperature conductivity (λ is thermal conductivity, c is heat capacity and ρ is density).

The density of an energy release was simulated by continuous function of r and t taking into account the effect of neutron field "depression". Correction on the "depression" in experiments was insignificant.

Eq. (1) was solved by a method of running with the use of a three-dot pattern and an implicit scheme under boundary conditions of the first kind:

$$T(Ro,t) = f(t) \tag{2}$$

where the function $f(t)$ describes the temperature of a spherical fuel element surface which was measured during reactor start-ups.

Boundary condition (2) and the temperature dependencies of thermophysical properties a, λ, c and ρ were specified by seventh power degree polynomial.

Stresses in the spherical fuel elements during reactor tests in an ampoule depend on non-uniform graphite shrinkage during irradiation, temperature stresses due to interaction between them and walls of internal graphite sleeve and stresses due to non-uniform temperature fields in the tested samples. Stresses due to radiation were neglected because of the small value of the integrated neutron flux. The presence of backlashes between the fuel elements and the graphite sleeve, as well as the possibility of free temperature expansion of samples along the ampoule axis excluded the occurrence of contact loadings. Therefore the only source of stress and deformation was the non-uniform temperature field. According to the theory of thermoelasticity, radial σ_r and circumferential σ_Θ components of stresses are calculated with use of the formulas [4]:

$$\sigma_r = \frac{2aE}{1-\nu}\left[A(R_o) - A(r)\right] \text{ and } \sigma_\theta = \frac{aE}{1-\nu}\left[2A(R_o) - T(r)\right]$$

where $A(r) = 1/r^3 \int_o^r T(r) r^3 dr$, α is the linear thermal expansion coefficient, E is the Young's modulus and ν is the Poisson's ratio.

Temperature and stresses in the fuel elements were calculated with use of the program SHAR written on Fortran-4. The program was verified by comparison of calculated and known exact data for continuous sphere with a constant heat release. Calculated with use of the program SHAR, temperature differed by 0.2-0.3%, and stresses not greater than 0.5%. The temperature field in coated fuel particles was determined with the use of the same program as follows. At the first stage distribution of temperature $T(r,t)$ in the fuel elements was calculated. Then one or several characteristic coated fuel particles were chosen, their centre co-ordinates being $r_i < R_c$. If $T(r,t)$ is known, it is easy to find temperature in points $T(r_i - \delta_i, t)$ and $T(r_i + \delta_i, t)$, where δ_i is the radius of i-th coated fuel particle, and the average temperature on a coated fuel particle surface is equal to:

$$T_{avr} = \left[T(r_i - \delta_i, t) + T(r_i + \delta_i, t)\right]/2$$

At the second stage the equation of thermal conductivity (1) was solved for multi-layer sphere with the specified thermophysical properties of each layer, density of heat release in a fuel particle and the first kind boundary conditions on the external surface of coated fuel particle $T_{srf}(t) = T_{avr}(r_i, t)$.

Discussion of experimental results

The maximum temperature of matrix graphite in the fuel elements during the first methodical start-up did not exceed 800-1 000 K. Calculated temperature stresses were not greater than 3 MPa, that is less than the graphite bend strength of $\sigma_b = 22$-45 MPa. The temperature of fuel particles was less than 1 050 K, therefore fuel element characteristics were not changed, and coatings of coated fuel particles maintained their barrier properties with regard to fission products. Heat release in fuel elements No. 2 and No. 4 was 1.5 times higher, however it could not result in an increase of temperature and stresses up to a level which is dangerous as regards the of depressurisation of coated fuel particles.

The second start-up, when reactor power was 10 times higher then in nominal conditions, has confirmed serviceability of all measuring systems. Heat release in the fuel elements and temperature mode of tests were maintained in the specified intervals. As was expected, the temperature of the fuel element surface differed from point to point, this being the main source of a deviation of calculated values from experimental data. All calculations were conducted with use of a boundary condition (2), surface temperature was accepted equal to indications of the thermocouple, the location of which is shown by label T_2 in Figure 1.

The stress σ_r has a compressing character. Its maximum value was achieved in the ball centre, it was equal to zero on a surface. The circumferential stress σ_Θ at the ball centre was also compressing. With radius increased it became subject to tensile stress, achieving maximum value at the fuel element surface. σ_Θ^{max} estimated in one-dimensional model was 23-25 MPa; for fuel elements No. 1 and No. 3 it was 15-17 MPa.

The goal of the third start-up was to test fuel elements No. 1 and No. 3 at stresses close to matrix graphite ultimate strength (Figure 3). According to calculations similar stresses appeared when energy release density is $\overline{q}$ = 20-25 kW. In these conditions heat release in fuel elements No. 2 and No. 4 will be in the range of 33-35 kW, which should result in the destruction of their shells. Thermocouples fixed at the centre T_4 and on a surface T_3 of fuel element No. 3 gave information about temperature conditions of the test. At the stationary mode power was 22 kW. Calculated temperature at the centre of the fuel element was below the experimental temperature by 50-150 K. After correction of the boundary condition, an agreement between experimental and calculated data was improved (Figure 4).

Figure 4. Dependencies of calculated temperature T_{clc} at the centre (1) and of circumferential stresses (2) on the surface of fuel element No. 3 vs. the test time in the third start-up

o – Indication of thermocouple

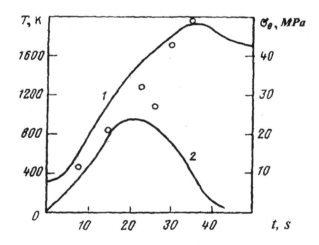

At the moment the reactor achieved a stationary power at t = 10 s, the maximum temperature of the matrix graphite in the centre of ball was equal to 800 K. Temperature continued to increase during the stationary stage of the test, and then during 5 s after power reduction begun at t = 30 s. Maximum temperature of graphite at the centre equal to 1 940 K was achieved after 35 s. Maximum temperature difference 740 K was fixed earlier at t = 30 s. Maximum stress σ_Θ^{max} on a fuel element surface outrun T change and was 24 MPa at 25 s (Figure 4).

The temperature of fuel particles was higher during the entire start-up period. The temperature difference in coated fuel particles increased smoothly from 0 up to 100 K at 10 s. Over the interval of 10 to 30 s it changed weakly, remaining on a level of 100-120 K, then began to fall as power was reduced. The maximum calculated temperature of fuel in fuel elements No. 1 and No. 3 was 2 050 K.

Fuel elements No. 2 and No. 4 in the third start-up were under heavier conditions. Circumferential stresses obviously exceeded ultimate strength. Shell destruction and partial destruction of these fuel element cores were found after opening of the ampoule.

It was the reason of radiation power growth of a gas sample taken out of an ampoule after of the termination of the start-up. Fuel elements No. 1 and No. 3 were intact. Their appearance was not changed, cracks or other defects were not been found after visual inspection of a surface once the series of experiments was at its end.

Conclusion

Tests have shown the capability of the IGR to investigate spherical fuel element behaviour under accident conditions. The control and measuring complex provides high accuracy of measurement and maintenance of the main parameters of a loading under an impulse mode at an energy release of ~20-30 kW and more at temperatures above 2 000 K. Experiments using fuel element models and appropriate calculated data allow to obtain information on distribution of temperatures in fuel elements and coated fuel particles, as well as about stresses in a graphite matrix.

REFERENCES

[1]   A.S. Chernikov, V.S. Eremeev, *et al.*, "Behaviour of HTGR Spherical Fuel Elements at Short-term Exposure in Impulse Graphite Reactor", Proc. of the 2nd JAERI Symposium on HTGR Technologies, 21-23 October 1992, Oarai, Japan.

[2]   G.A. Bat, A.S. Kochenov, L.P. Kabanov, "Test Nuclear Reactors", M.: Energoatomizdat, 1985, 280 p. (in Russian).

[3]   A.S. Chernikov, "Fuel and Fuel Elements of HTGR", *Atomic Energy*, Vol. 65, No. 1. pp. 32-38 (1988) (in Russian).

[4]   Bruno A. Boley, Jerome H. Weiner, "Theory of Thermal Stresses", Trans. from English. M.: Mir, 1964, 518 p.

GAS REACTOR TRISO-COATED PARTICLE FUEL MODELLING ACTIVITIES AT THE IDAHO NATIONAL ENGINEERING AND ENVIRONMENTAL LABORATORY

David Petti, Gregory Miller, John Maki, Dominic Varacalle and Jacopo Buongiorno
Idaho National Engineering and Environmental Laboratory
Idaho Falls, Idaho, United States

Abstract

The development of an integrated mechanistic fuel performance model for TRISO-coated gas-reactor particle fuel termed particle fuel model (PARFUME) has begun at the Idaho National Engineering and Environmental Laboratory. The objective of PARFUME is to physically describe the behaviour of the fuel particle under irradiation. Both the mechanical and physico-chemical behaviour of the particle under irradiation are being considered. PARFUME is based on multi-dimensional finite element modelling of TRISO-coated gas-reactor fuel. The goal is to develop a performance model for particle fuel that has the proper dimensionality and still captures the statistical nature of the fuel. The statistical variation of key properties of the particle associated with the production process requires Monte Carlo analysis of a very large number of particles to understand the aggregate behaviour. Thus, state-of-the-art statistical techniques are being used to incorporate the results of the detailed multi-dimension stress calculations and the fission-product chemical interactions into PARFUME. The model is currently focusing on carbide, oxide and oxycarbide uranium fuel kernels. The coating layers are classical TRISO type (IPyC/SiC/OPyC). Extensions to other fissile and fertile materials and other coating materials (e.g. ZrC) are currently under consideration. This paper reviews the current status of the model, discusses calculations of TRISO-coated fuel performance for cracking observed in recent US gas reactor irradiations, and presents predictions of the behaviour of TRISO-coated fuel at high burn-up that is currently under consideration in Europe.

Objectives and goals

The INEEL has begun the development of an integrated mechanistic fuel performance model for TRISO-coated gas-reactor particle fuel termed PARFUME (PARticle FUel ModEl). Compared to light water reactor and liquid metal reactor fuel forms, the behaviour of coated-particle fuel is inherently more multi-dimensional. Moreover, modelling of fuel behaviour is made more difficult because of the statistical variations in fuel physical dimensions and/or component properties, from particle to particle due to the nature of the fabrication process. Previous attempts to model this fuel form have attacked different pieces of the problem. Simplified one-dimensional models exist to describe the structural response of the fuel particle. Models or correlations exist to describe the fission-product behaviour in the fuel, though the database is not complete owing to the changes in fuel design that have occurred over the last 25 years. Significant effort has gone into modelling the statistical nature of fuel particles. However, under pressure to perform over one million simulations with the computing power available in the 1970s and 1980s, the structural response of the particle was simplified to improve the speed of the calculations. No publicly available *integrated* mechanistic model for coated-particle fuel exists in the United States as an accurate tool for nuclear design. The advent of powerful personal computers and the tremendous advancements in fundamental modelling of materials science processes now make more accurate simulations of particle fuel behaviour possible.

Thus, our objective in developing PARFUME is to physically describe both the mechanical and physico-chemical behaviour of the fuel particle under irradiation. Our goal is to develop a performance model for particle fuel that has the proper dimensionality and still captures the statistical nature of the fuel. The statistical variation of key properties of the particle associated with the production process requires Monte Carlo analysis of a very large number of particles to understand the aggregate behaviour. Thus, state-of-the-art statistical techniques are being used to incorporate the results of the detailed multi-dimension stress calculations and the fission-product chemical interactions into PARFUME. Furthermore, we want to verify PARFUME using data from historical TRISO-coated particle irradiations so that the code can be used to design advanced coated-particle fuel for the gas reactor with greater confidence and other particle fuel applications (Pu and minor actinide burning, fast reactors).

Key phenomena

Our mechanistic model for coated-particle fuel will consider both the structural and physico-chemical behaviour of particle-coated fuel system during irradiation. The following important phenomena will be included:

- Fission gas release from the kernel as a function of burn-up, temperature and kernel type (oxide, carbide, oxycarbide).

- Anisotropic response of the pyrolytic carbon layers to irradiation (shrinkage, swelling, and creep that are functions of temperature, fluence and orientation/direction in the carbon).

- Failure of the pyrolytic carbon and SiC layers based on the classic Weibull formulation for a brittle material either by traditional pressure vessel failure criteria or by mechanisms such as asphericity, layer debonding or cracking.

- Fission product inventory generation as a function of burn-up and enrichment of the particle.

- Chemical changes of the fuel kernel during irradiation (changes in carbon/oxygen, carbon/metal and/or oxygen/metal ratio depending on the kernel fuel type, production of CO/CO_2 gas) and its influence on fission-product and/or kernel attack on the particle coatings.

- Kernel migration.

- Fission-product diffusion, migration and segregation.

- Statistical variations of key properties of the particle associated with the production process, requiring Monte Carlo analysis of a very large number of particles to understand the aggregate behaviour.

The status of PARFUME with respect to modelling these phenomena is discussed in in the following sections. Later, the model is applied to predict the behaviour of recent US TRISO-coated particle irradiations and to define trends in the performance of EU particles at high burn-up. A summary and our future plans are presented in conclusion.

Key material properties

A typical TRISO-coated particle is shown in Figure 1. Fission gas pressure builds up in the kernel and buffer regions, while the IPyC, SiC, and OPyC act as structural layers to retain this pressure. The basic behaviour modelled in PARFUME is shown schematically in Figure 2. The IPyC and OPyC layers both shrink and creep during irradiation of the particle while the SiC exhibits only elastic response. A portion of the gas pressure is transmitted through the IPyC layer to the SiC. This pressure continually increases as irradiation of the particle progresses, thereby contributing to a tensile hoop stress in the SiC layer. Countering the effect of the pressure load is the shrinkage of the IPyC during irradiation, which pulls inward on the SiC. Likewise, shrinkage of the OPyC causes it to push inward on the SiC. Failure of the particle is normally expected to occur if the stress in the SiC layer reaches the fracture strength of the SiC. Failure of the SiC results in an instantaneous release of elastic energy that should be sufficient to cause simultaneous failure of the pyrocarbon layers.

Numerous material properties are needed to represent fuel particle behaviour in the performance model. These include irradiation-induced strain rates used to account for shrinkage (or swelling) of the pyrocarbon layers, creep coefficients to represent irradiation-induced creep in the pyrocarbon layers and elastic properties to represent elastic behaviour for the pyrocarbons and silicon carbide. The properties used in the model were obtained from data that was compiled in a report by the CEGA Corporation in July 1993. This data was based on a review and evaluation of material properties published in the literature to that date [1-7].

Figure 1. Typical TRISO-coated fuel particle geometry

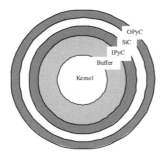

Figure 2. Behaviour of coating layers in fuel particle

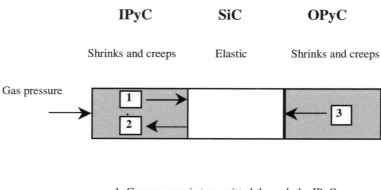

1 Gas pressure is transmitted through the IPyC

2 IPyC shrinks, pulling away from the SiC

3 OPyC shrinks, pushing in on SiC

Creep

Irradiation-induced creep in the pyrocarbon layers is treated as secondary creep, i.e. the creep strain rate was proportional to the level of stress in the pyrocarbon. The creep coefficient is applied as a function of pyrocarbon density and irradiation temperature. Because variations in pyrocarbon density are small, the creep is primarily a function of temperature, increasing significantly with increases in temperature. The creep coefficients used in the analysis ranged from 0.7 to $2 \times 10^{-29} (\text{MPa-n/m}^2)^{-1}$ over a temperature range of 600 to 1 200°C. There is considerable variation among values reported for the creep coefficient throughout the literature, making this a major source of uncertainty in the analysis. The remaining creep property is Poisson's ratio for creep of the pyrocarbon layers. In accordance with CEGA's recommendations, a value of 0.5 is used for secondary creep of the pyrocarbons.

Shrinkage

Due to anisotropy in the swelling behaviour of the pyrocarbon layers, the strains are different for the radial and tangential directions. The swelling strains are reported in the CEGA data to be functions of four variables, i.e. fluence level, pyrocarbon density, degree of anisotropy (as measured by the Bacon Anisotropy Factor, BAF) and irradiation temperature. Figure 3 shows swelling strains as a function of fluence and BAF for the radial and tangential directions. The plots presented correspond to a pyrocarbon density of 1.96×10^6 g/m^3 and an irradiation temperature of 1 032°C, and cover a range of BAF from 1.02 to 1.28. In the radial direction, the pyrocarbon shrinks at low fluences but swells at higher fluences for all but the lowest BAF values. In the tangential direction, the pyrocarbon continually shrinks at all levels of fluence and the magnitude of the shrinkage increases as the BAF increases. Figure 4 shows swelling strains as a function of fluence and temperature for the radial and tangential directions. The plots presented correspond to a pyrocarbon density of 1.96×10^6 g/m^3 and a BAF value of 1.08, and cover a range of temperatures from 600 to 1 350°C. Similar trends are seen in these curves, wherein the magnitude of shrinkage increases as the temperature increases.

In the ABAQUS model, the swelling data is input in the form of strain rates that are functions of four field variables, i.e. fluence level, pyrocarbon density, BAF and irradiation temperature.

Figure 3. Radial and tangential swelling (shrinkage) of pyrocarbon for variations in BAF

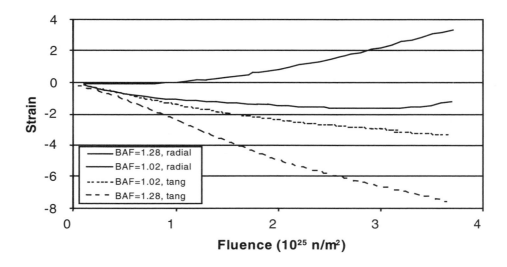

Figure 4. Radial and tangential swelling (shrinkage) of pyrocarbon for variations in temperature

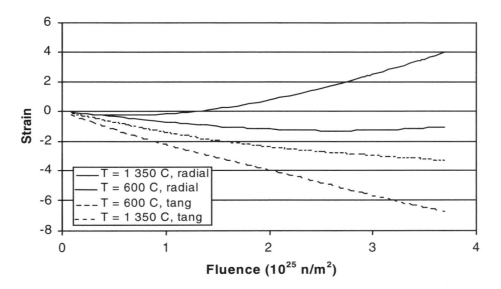

Weibull parameters

Because of the brittle nature of pyrolytic carbon and silicon carbide, the PyC layers and the SiC layer are expected to fail in a probabilistic manner according to the Weibull statistical theory [8]. As such, the failure probability for a PyC or SiC layer in a batch of particles is given by:

$$P_f = 1 - e^{-\int_V (\sigma/\sigma_0)^m dV}$$

where P_f is the probability of failure for the IPyC or SiC layer, m is the Weibull modulus for IPyC or SiC layer, V is the volume of the IPyC or SiC layer in μm^3, σ is the stress in the IPyC or SiC layer in MPa, and σ_0 is the Weibull characteristic strength for the IPyC or SiC layer in MPa-$\mu m^{3/m}$.

The Weibull parameters used in the PARFUME code are based on CEGA's recommendations. For the pyrocarbons, the Weibull modulus m is assumed to have a value of 9.5. CEGA's data indicates that the characteristic strength σ_0 increases with increasing values of BAF. For isotropic PyC (BAF = 1), the recommended value for σ_0 is 13.36 MPa-m$^{3/9.5}$. For a BAF of 1.06, which may typically be expected, the strength increases to 23.99 MPa-m$^{3/9.5}$. In accordance with CEGA's data, the Weibull modulus m for the silicon carbide layer is assumed to have a value of 6 and the corresponding characteristic strength σ_0 is assumed to be 9.64 MPa-m$^{3/6}$.

Elastic

The Young's modulus for the pyrocarbon layers was applied as a function of four variables (the same variables as used for swelling), while the Young's modulus for the silicon carbide layer was applied only as a function of temperature. A typical Young's modulus for the pyrocarbons is about 30 GPa, while that of the silicon carbide is about 370 GPa. Values of 0.33 and 0.13 were used for Poisson's ratio in the pyrocarbon and SiC layers, respectively. The stresses in the coating layers are not highly sensitive to variations in the elastic properties.

Models

Structural

The ABAQUS program [9] is used in the performance model to perform finite element stress analysis on coated fuel particles. This program is capable of simulating the complex behaviour of the coating layers, and can be used to evaluate multi-dimensional effects, such as shrinkage cracks in the IPyC, partial debonding between layers and asphericity. ABAQUS analyses are also used as a benchmark for validating simplified solutions that may be employed in PARFUME. We have shown (see Ref. [10]) that shrinkage cracks in the IPyC could make a significant contribution to fuel particle failures. Therefore, the condition of a cracked IPyC is included as a potential failure mechanism in PARFUME, and has been evaluated using ABAQUS analyses.

ABAQUS models for both normal and cracked three-layer geometries have been developed as shown in Figure 5. These are axisymmetric models that allow for non-symmetry in the plane, thus enabling an evaluation of multi-dimensional effects on stress behaviour of the coating layers. The model of the normal spherical particle has no cracks or defects in the layers of the particle. The IPyC and OPyC layers are assumed to remain fully bonded to the SiC layer throughout irradiation. This model is used to demonstrate behaviour of a normal particle in expected reactor conditions, as well as to determine stresses in the various layers throughout irradiation. The model consists of quadrilateral axisymmetric elements, giving the effect of a full sphere. Only the three structural layers (i.e. the IPyC, SiC and OPyC) of the particle are included in the model. The layer thicknesses for the IPyC, SiC and OPyC are nominally set at 40, 35 and 43 μm, respectively. Any of these dimensions can be varied as desired. The internal pressure applied in this analysis was ramped linearly from zero at the beginning of irradiation to a final value of 23.7 MPa, to simulate the build-up of fission gas pressure. A constant external pressure was also applied to represent either reactor or test conditions. Particles are analysed in a viscoelastic time-integration analysis that progresses until the fluence reaches 3×10^{25} n/m^2, occurring at a time of 1.2×10^7 s in the analysis. These are representative conditions that can be varied as desired.

The model for a cracked particle is identical in all respects to that of the normal particle except that it has a radial crack through the thickness of the IPyC layer. As discussed later, the crack is typical of those observed in post-irradiation examinations of the New Production Modular High-temperature

Figure 5. Finite element models for normal and cracked configurations

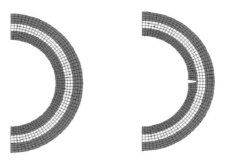

Gas Reactor (NP-MHTGR) fuel particles. During irradiation, shrinkage of the initially intact IPyC layer induces a significant tensile stress in that layer. If the tensile strength of the IPyC layer is exceeded, then a radial crack develops in the IPyC layer. This crack is included in the model from the beginning of the solution since it is not feasible to initiate the crack later in the ABAQUS analysis. Because the shrinkage in the pyrocarbons dominates the particle behaviour early during irradiation, large tensile stresses in the IPyC occur early. Therefore, the assumption of the presence of a crack at the beginning of the solution should be a reasonable approximation. The analysis does not include dynamic effects associated with a sudden failure of the IPyC, which could increase the magnitude of the stresses calculated.

Figure 6 plots the calculated tangential stress history for the SiC layer of a normal (uncracked) particle. As shown, the SiC remains in compression largely because of the shrinkage in the pyrocarbon layers (the IPyC pulls while the OPyC pushes on the SiC). Figure 6 also plots the maximum principal stress in the SiC layer near the crack tip of a particle with a cracked IPyC. In the particle analysed, the crack leads to a calculated tensile stress in the SiC layer of about 440 MPa. It can be seen that a cracked IPyC greatly changes the stress condition in the SiC, which significantly increases the probability of SiC failure.

Figure 6. Time histories for stress in SiC layer for normal and cracked particles

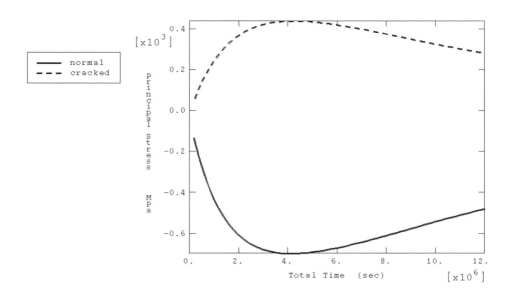

Statistical

We have also investigated the impact of the statistical variations in fuel particle design parameters on the structural response of the fuel particle. In the case of a normal particle, statistical variations in design parameters are treated with simplified solutions built into the PARFUME code, rendering finite element analysis unnecessary. In the case of a cracked particle, however, finite element analyses are performed to capture the multi-dimensional behaviour and thereby characterise the effects of variations in these parameters. Based on the results of analyses on cracked particles, the following six variables are judged to be important in describing the behaviour of the cracked particle and thus merited a detailed statistical evaluation: IPyC thickness, SiC thickness, OPyC thickness, IPyC density, BAF of the IPyC and irradiation temperature. For a particle with a cracked IPyC, statistical variations from the nominal value for other parameters such as kernel diameter and buffer thickness are less important, and thus have not yet been addressed in these studies because of the size of the statistical base. These parameters were held constant throughout the analyses at values typical of TRISO particles.

Three values for each of five factors were chosen for analysis in a statistical study of the cracked particle, as shown in Table 1. The sixth factor, irradiation temperature, was analysed at the four values shown. A full-factorial statistical analysis, involving 243 stress calculations for each irradiation temperature (972 runs total), allowed an evaluation of all six factors (i.e. A = IPyC thickness, B = SiC thickness, C = OPyC thickness, D = IPyC density, E = BAF (of IPyC), F = irradiation temperature) and their interactions (e.g. AB, ABF, BCDF, AB^2CD, BC^2D^2EF, ABCDEF).

Table 1. Range of parameters selected for ABAQUS analyses

Factor	Low	Nominal	High
A (μm)	30	40	50
B (μm)	25	35	45
C (μm)	33	43	53
D (10^6g/m^3)	1.8	1.9	2.0
E	1.0	1.16	1.32
F ($^\circ$K)	873	1 073, 1 273	1 473

The Design Expert program [11] was used to perform both an effects analysis and a regression analysis on the data obtained from the ABAQUS calculations on the cracked particle. The effects analysis showed the relative significance of varying each parameter, while the regression analysis produced an algorithm that can be used to predict the stress level in the SiC layer of a cracked particle. The program used response surface analysis to develop a sixth order polynomial that statistically fit the stress data to within 0.5% accuracy. This algorithm has been incorporated in the PARFUME code to calculate failure probabilities utilising a Monte Carlo sampling approach (within the range of parametric variations considered in Table 1). PARFUME then uses the Weibull statistical approach to estimate the potential for fracture of the SiC layer in a particle that has a cracked IPyC. A fracture mechanics approach was deemed to be impractical because the material discontinuity at the interface of the IPyC and SiC layers greatly complicates calculation of the stress intensity at the crack tip. In the failure probability calculations, the stress (in the SiC layer of a sampled particle) calculated by the algorithm is compared to a strength value to determine whether the particle fails. The mean strength for these comparisons is derived from the Weibull characteristic strength (σ_0) and accounts for the intensification of stresses that occurs in the region surrounding the crack tip.

When the PARFUME code samples a particle it first uses a closed form solution (Ref. [12]) to calculate stresses in the IPyC layer and thereby determines (with Weibull statistics) whether the particle has a cracked or un-cracked IPyC layer. If the IPyC layer is cracked, then the code uses the approach described above to determine whether the particle fails. If the IPyC is un-cracked, then the code uses the closed form solution to determine the SiC stress, and uses this stress in a Weibull statistical evaluation to determine whether the particle fails. In its Monte Carlo sampling, the code performs statistical variations on any number of input parameters (such as IPyC, SiC, OPyC thicknesses, IPyC BAF, etc.) by applying Gaussian distributions to these parameters.

Fission gas release model

The fission gas release model calculates the amount of CO and noble fission product gases released to the void volume of the fuel particle. This quantity is used to determine the internal gas pressure of the fuel particle according to the Ideal Gas Law. For each gas species, i, the amount released is determined by:

$$\text{(moles gas)}_i = \text{(release fraction)}_i \text{ (fission yield)}_i \text{ (burn-up) (moles fuel)}$$

Fission yields for the significant noble fission product gases, xenon and krypton, are taken from the ORIGEN-2 computer code database [13]. For uranium-based fuels, the production of krypton decreases with time, while the production of xenon increases with time due to the increasing yield contribution from conversion plutonium fission. Within reasonable accuracy, the sum of the xenon and krypton yields may be assumed to be a constant throughout the life of the fuel where a value of 0.259 is currently used.

The release fraction for noble fission-product gases considers recoil from the outer shell of the fuel kernel into the buffer and diffusive transport to free surfaces. Double counting is avoided whereby atoms released by recoil are not available for diffusive release. Recoil release is based upon standard geometrical considerations and average fission fragment ranges [14]. These ranges are calculated for a given fuel composition from elemental data [15]. Diffusive release is calculated according to the Booth equivalent sphere diffusion model [16]. A value of 10 μm is used for the radius of the equivalent sphere which is representative of uranium-based fuels. An effective diffusion coefficient is used which is the sum of the contributions from intrinsic diffusion, irradiation enhanced vacancy diffusion and irradiation induced athermal diffusion [17].

For UCO and UC_2 fuels, it is assumed that there is no free oxygen available to form CO or CO_2. It is also assumed that when free oxygen is available, as in UO_2 and ThO_2 fuels, only CO and not CO_2 forms at typical particle fuel temperatures [18]. For UO_2 fuel, the CO yield is determined from the correlation developed by General Atomics [19]. The CO fractional release is assumed to be 1.

Representative results from the fission gas release model are displayed in Figure 7 for typical German TRISO fuel (8% enriched UO_2 fuel, 500 μm diameter kernel, 900°C irradiation temperature, 8.5% FIMA). The end of life internal particle pressure of about 5 MPa for the German fuel compares well with an end of life internal pressure of 8.5 MPa for proposed GT-MHTGR TRISO fuel (19.7% enriched UCO fuel, 350 μm diameter kernel, 1 200°C irradiation temperature, 21% FIMA and other parameters based upon the fuel irradiated in the HRB-21 experiment [20]). Based upon future chemistry work, the fission gas release model will be enhanced with greater functionality in calculating fission product yields and CO gas formation.

Figure 7. Internal gas pressure for typical German TRISO fuel

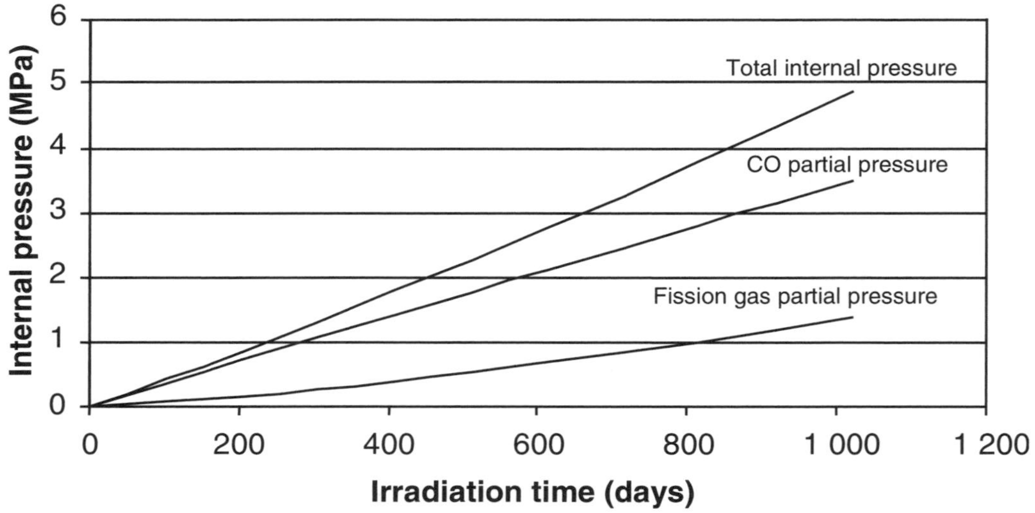

Fission-product chemistry, CO generation and fission-product transport

The goal of the fission-product chemistry module of PARFUME is to calculate the chemical state of the important fission products in the particle as a function of temperature and burn-up. Fission product inventories in the particle are calculated using the ORIGEN-2 computer code and statistically fit as a function of initial enrichment (between 4 and 19%) and burn-up of the particle. In the future, we plan to use a chemical thermodynamics code to calculate the chemical state of the fission products (i.e. oxide form, carbide form, metal form) for a range of fuel forms including uranium oxide, carbide and oxycarbide. Any excess oxygen generated from uranium fission that is not tied up with the fission products will be assumed to react with carbon from the buffer to form CO. This estimate of CO production can be used to compare with correlations that have been used in the past. The chemical state and concentrations of the fission products will be used as initial conditions for a fission-product diffusion module to track the transport of fission products through the layers of the TRISO coating. We expect to complete this work in the next two years.

Cracked model and results for NPR experiments

The capabilities of the PARFUME code to predict failure probabilities for fuel particles having a cracked IPyC were used in predicting failure probabilities for three irradiation experiments conducted as part of the NP-MHTGR program in the early 1990s. Fuel compacts were irradiated at the High Flux Isotope Reactor (HFIR) and the Advanced Test Reactor (ATR) in the United States. TRISO-coated particles containing high enriched uranium were irradiated at temperatures between 750 and 1 250°C, burn-ups between 65 and 80% FIMA and fluences between 2 and 3.8×10^{25} n/m^2. On-line fission gas release measurements indicated significant failures during irradiation. Post-irradiation examination (PIE) of individual fuel compacts revealed the presence of radial cracks in all layers of the TRISO coating. The levels of cracking measured during PIE are shown in Table 2. The particle dimensions, burn-up, end-of-life fluence, irradiation temperature and [235]U enrichment were set to appropriate values for each experiment. Results of these evaluations are presented in Table 2. Included in these results are the percentage of particles predicted to have a cracked IPyC and the percentage of particles predicted to fail due to a cracked SiC. It is seen that the program predicts that the IPyC layer cracks in 100% of

Table 2. Comparison of ceramographic observations to PARFUME calculations for TRISO-coated fissile fuel particles

Irradiation conditions			
Fuel compact ID	Fast fluence (10^{25} n/m^2)	Irradiation temp. (°C)	Burn-up (% FIMA)
NPR-2 A4	3.8	746	79
NPR-1 A5	3.8	987	79
NPR-1 A8	2.4	845	72
NPR-1A A9	1.9	1052	64
IPyC layer[a]				
	Sample size	% failed	95% conf. interval (%)	Calc.
NPR-2 A4	83	65	$54 < p < 76$	100
NPR-1 A5	39	31	$17 < p < 47$	100
NPR-1 A8	53	6	$2 < p < 16$	100
NPR-1A A9	17	18	$5 < p < 42$	100
SiC layer[a]				
	Sample size	% failed	95% conf. interval (%)	Calc.
NPR-2 A4	287	3	$2 < p < 6$	8.2
NPR-1 A5	178	0.6	$0 < p < 3$	1.6
NPR-1 A8	260	0	$0 < p < 2$	4.9
NPR-1A A9	83	1	$0 < p < 5$	0.9

[a] Layer failure is considered as a through wall crack as measured by PIE.

the particles for every compact tested. In reality, the PIE revealed that the actual failure fractions were less than this, as shown in Table 2. It is believed that the creep coefficients currently used in the PARFUME code are too low, which allows the calculated shrinkage stresses to reach too high a value before creep relaxation takes effect.

The failure probabilities predicted by PARFUME for the SiC layer are somewhat high relative to the irradiation test results. However, were the number of cracked IPyCs reduced through the use of a higher creep coefficient, a closer correlation would be observed. The particle samples for examination are typically small (<300 particles), rendering precise correlations with the test results difficult. The predictions are in agreement with the test results in indicating that per cent level particle failures are expected. This type of correlation is generally not achieved if shrinkage cracks in the IPyC are ignored.

The question arises as to how much the creep coefficient would have to be increased before the predicted number of IPyC failures would match the test results. This was determined for all three tests, and the results for the compact NPR-2 A4 are presented in Figure 8. The horizontal dashed line in the graph corresponds to the actual percentage of IPyC failures occurring in the fuel compact. These results together with those of the other tests indicate that the creep coefficient would have to be increased by a factor in the range of 2 to 3 to gain a good correlation with the test results.

The effect of a higher creep coefficient on the SiC failure percentages cannot yet be ascertained because the statistical algorithm used to calculate SiC stresses in a cracked particle was developed on the basis of the lower creep coefficients.

Figure 8. Predicted IPyC failures as a function of the creep coefficient

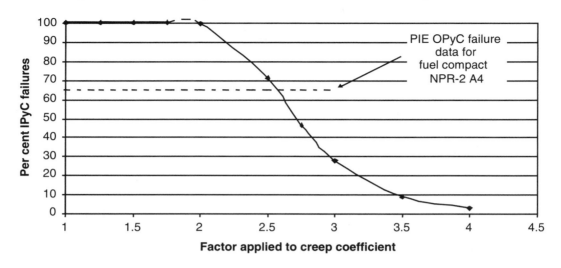

Standard particle model and results for EU high burn-up case

The closed form solution for the standard (un-cracked) particle of the PARFUME code was used to calculate stress levels in EU (German) fuel particles. A major difference between the EU particle and the NPR particle is that the former has a much larger kernel diameter (500 vs. 200 μm), which makes for a significantly larger particle. A calculated time history for the maximum tangential stress in the SiC layer of a nominal EU particle is presented in Figure 9. This calculation was made for a particle having a ^{235}U enrichment of 8%, an end-of-life burn-up of 8.5% FIMA, and an end-of-life fluence of 2.3×10^{25} n/m^2.

Figure 9. Time history for tangential stress in the SiC layer of EU particle

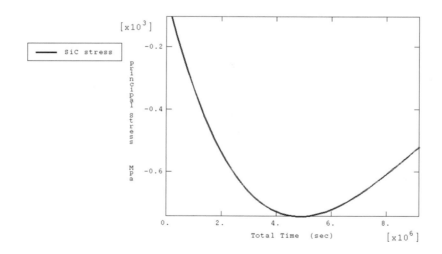

Calculations have also been performed at various levels of burn-up, up to a maximum of 21% FIMA. A range of ^{235}U enrichment from 8 to 20% was considered in these calculations, but the enrichment had no effect on the magnitude of the calculated stress. The maximum tangential stress occurring in the SiC layer at the end of life is plotted in Figure 10 as a function of burn-up, showing

Figure 10. Calculated SiC stress as a function of burn-up for the EU particle at 900°C

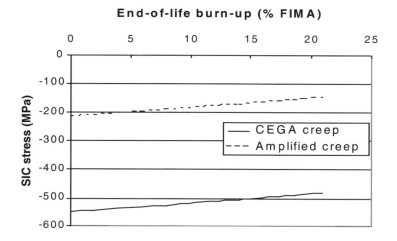

that an increasing burn-up results in an increasing stress. As it is believed that the creep coefficients in PARFUME are too low, the same calculations were performed where the creep coefficient in each case was amplified by a factor of 2.5. These results are also shown in Figure 10, which demonstrate that the higher creep coefficient results in a significantly higher stress in the SiC layer.

The results of Figures 9 and 10 correspond to an irradiation temperature of 900°C. As temperature variations affect the material properties of the coating layers, analyses were also performed at temperatures of 700 and 1 100°C (using the amplification factor of 2.5 on creep). Results for these temperature variations are presented in Figure 11. Because creep is greater at a higher temperature, the compressive stress in the SiC layer is reversed earlier during irradiation. This results in a higher (less compressive) stress at the end of life.

Figure 11. Effect of temperature on the SiC stress in the EU particle

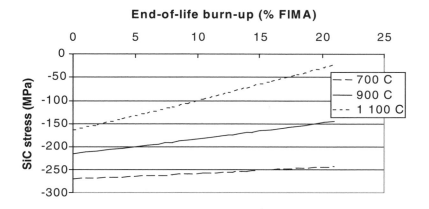

Summary and future work

The INEEL has begun the development of an integrated mechanistic fuel performance model for TRISO-coated gas-reactor particle fuel termed PARFUME (PARticle FUel ModEl). The objective of PARFUME is to physically describe the behaviour of the fuel particle under irradiation. Both the mechanical and physico-chemical behaviour of the particle under irradiation are being considered.

PARFUME is based on multi-dimensional finite element modelling of TRISO-coated gas reactor fuel. The model with modest refinement in some PyC materials properties does a good job of calculating the failures that were observed in the NP-MHTGR fuel irradiation experiments performed in the early 1990s. Future efforts for the code include:

- Completion of the fission-product chemistry module for the code.

- Development of the fission-product transport module.

- Comparison of the code to other older gas reactor irradiations data.

- Exercising the code's predictive capabilities to examine the performance of TRISO-coated fuel at very high burn-up.

REFERENCES

[1] J.L. Kaae, "Effect of Irradiation on the Mechanical Properties of Isotropic Pyrolytic Carbons", *J. Nucl. Mat.*, 46, p. 121 (1973).

[2] J.L. Kaae, J.C. Bokros and D.W. Stevens, "Dimensional Changes and Creep of Poorly Crystalline Isotropic Carbons and Carbon-silicon Alloys During Irradiation", *Carbon*, 10, p. 571 (1972).

[3] J.C. Bokros, R.W. Dunlap and A.S. Schwartz, "Effect of High Neutron Exposure on the Dimensions of Pyrolytic Carbons", *Carbon*, 7, p. 143 (1969).

[4] J.C. Bokros, G.L. Guthrie, R.W. Dunlap and A.S. Schwartz, "Radiation-induced Dimensional Changes and Creep in Carbonaceous Materials", *J. Nucl. Mat.*, 31, p. 25 (1969).

[5] J.L. Kaae and J.C. Bokros, "Irradiation-induced Dimensional Changes and Creep of Isotropic Carbon", *Carbon*, 9, p. 111 (1969).

[6] J.L. Kaae, D.W. Stephens and J.C. Bokros, "Dimensional Changes Induced in Poorly Crystalline Isotropic Carbons by Irradiation", *Carbon*, 10, p. 561 (1972).

[7] J.L. Kaae, "Behavior of Pyrolytic Carbons Irradiated Under Mechanical Restraint", *Carbon*, 12, p. 577 (1974).

[8] N.N. Nemeth, J.M. Mandershield and J.P. Gyekenyesi, "Ceramics Analysis and Reliability Evaluation of Structures (CARES) User's and Programmer's Manual", NASA Technical Paper 2916 (1989).

[9] ABAQUS User's Manual, Version 5.8, Hibbitt, Karlsson and Sorenson, Inc. (1988).

[10] G.K. Miller, D.A. Petti, D.J. Varacalle, J.T. Maki, "Consideration of the Effects on Fuel Particle Behavior from Shrinkage Cracks in the Inner Pyrocarbon Layer", *J. Nucl. Mat.*, 295, p. 205 (2001).

[11] P. Whitcomb, *et al.*, Design-Expert, Version 4.0, Stat-Ease Inc. (1993).

[12] G.K. Miller and R.G. Bennett, "Analytical Solution for Stresses in TRISO-coated Particles", *J. Nucl. Mat.*, 206, p. 35 (1993).

[13] G. Croff, "ORIGEN2 – A Revised and Updated Version of the Oak Ridge Isotope Generation and Depletion Code", ORNL-5621, Oak Ridge National Laboratory, July 1980.

[14] D.R. Olander, "Fundamental Aspects of Nuclear Reactor Fuel Elements", TID-26711-P1, ERDA (1976).

[15] U. Littmark and J.F. Ziegler, "Handbook of Range Distributions for Energetic Ions in All Elements", Pergamon Press, New York (1980).

[16] H. Booth and G. Rymer, "Determination of the Diffusion Constant of Fission Xenon in UO_2 Crystals and Sintered Compacts", CRDC-720, Atomic Energy of Canada Ltd., August 1958.

[17] J.A. Turnbull, *et al.*, "The Diffusion Coefficients of Gaseous and Volatile Species during the Irradiation of Uranium Dioxide", *J. Nucl. Mater.*, 107, 168 (1982).

[18] K. Minato, *et al.*, "Fission Product Behavior in TRISO-coated UO_2 Fuel Particles", *J. Nucl. Mater.*, 208, 266 (1994).

[19] W.J. Kovacs, *et al.*, "High-temperature Gas-cooled Reactor Fuel Pressure Vessel Performance Models", *Nuc. Tech.*, 68, 344 (1985).

[20] C.A. Baldwin, *et al.*, "Interim Post-irradiation Examination Data Report For Fuel Capsule HRB 21", ORNL/M-2850, Oak Ridge National Laboratory, September 1993.

SESSION SUMMARIES

Chair: Y. Sudo

SESSION I

Overviews of High-temperature Engineering Research in Each Country and Organisation

Chair: W. von Lensa

The session was comprised of six oral presentations and one paper which was not presented orally:

- *Research Activities on High-temperature Gas-cooled Reactors (HTRs) in the 5th EURATOM RTD Framework Programme* (presented by J. Martín-Bermejo, European Commission).

  An R&D programme has been started in Europe addressing the main innovation potentials of HTR technology with regard to high burn-up fuel, symbiotic fuel cycles for Pu burning, waste minimisation, fuel disposal, material qualification, licensing issues and component development with a total budget about euros 20 million during the next four years. The European HTR Technology Network is open for international collaboration as is already established with Japan, the United States of America and China.

- *Update on IAEA High-temperature Gas-cooled Reactor Activities* (J. Kendall, IAEA).

  An abstract and full paper were presented, but no oral presentation was made. In this paper, IAEA activities on high-temperature gas-cooled reactors are conducted with the review and support of Member States, primarily through the Technical Working Group on Gas-cooled Reactors (TWG-GCR).

- *Present Status of the Innovative Basic Research on High-temperature Engineering using the HTTR* (presented by Y. Sudo, JAERI, Japan).

  HTR technology is embedded in more general R&D fields making use of high temperatures and radiation effects, e.g. neutron irradiation processing of superconductors, neutron transmutation doping of semiconductors, radiation-induced chemical reactions and production of new materials like SiC fibres, ceramic composites and superplastic ceramics.

- *Present Status in the Netherlands of Research Relevant to High-temperature Gas-cooled Reactor Design* (presented by J.C. Kuijper, NRG Petten, the Netherlands).

  HTR-related R&D in the Netherlands is mainly performed by NRG, JRC-Petten and IRI. Most activities are part of the European HTR Technology Network or contracted for the PBMR project. Concepts for industrial co-generation have been developed for HTR in the power range around 50 MWth. HFR is used for HTR fuel and material irradiation tests.

- *Current Status of High-temperature Engineering Research in France* (presented by J. Rouault, CEA, France).

  A consistent approach on technology development of gas-cooled reactors has been launched in France. This ranges from the support of European suppliers for offering direct cycle HTR till 2010 via medium-term development of specialised HTR for waste transmutation, nuclear process heat applications and robust export versions to long-term realisation of gas-cooled fast reactors using the same technology elements. The budget of the current R&D programme is around 100 million FF per year, allowing for fabrication and qualification of HTR fuel, material tests, helium technology advancement and component development.

- *Overview of High-temperature Reactor Engineering and Research* (presented by K. Kugeler, FZJ, Germany).

  Nuclear technologies have to comply with the general requirements for sustainability of energy supply systems. This will lead to "catastrophe-free" reactor designs with limited radioactive release ($<10^{-5}$ of inventory) in all conceivable accident conditions. Underground siting of NPP can also be a response to cope with terrorist attack. Impacts of nuclear accidents have to be reduced in such a way that they can be handled by normal insurance practice. The ongoing R&D on HTR in Germany indicates the feasibility of these goals.

- *The Contribution of UK Organisations in the Development of New High-temperature Reactors* (presented by A.J. Wickham, Consultant, United Kingdom).

  The UK history of design, construction and operation of GCR also offers a unique technology platform for HTR. Therefore, many UK organisations are involved in the HTR-related European R&D programme as well as in commercial projects like PBMR. Special knowledge is available on the "design" of graphite and the ceramic structures of HTR.

Recommendations with respect to Session I

- Continuation and extension under the 6th EURATOME RTD Framework Programme with a focus on a consistent technology evolution for gas-cooled reactors.

- Optical fibre systems for in-core instrumentation are important for HTR and control of core heat-up accident sequences of all reactor types. HTTR compliments MTR.

- Benefits of nuclear co-generation of electricity and heat should be quantified for different power sizes beyond dedicated electricity production.

- The French R&D strategy on GCR is compatible with the objectives of the EURATOM Framework Programme, the US Generation IV initiative, the IAEA INPRO and commercial HTR activities and should be used as a basis for elaboration of a progressive global load map on GCR R&D.

- The concept of sustainability has to be adopted as generally agreed for nuclear energy too. Requirements for sustainability and for ultimate accident scenarios have to be specified in more detail, including fuel cycle.

- As former resources become scarce, new grades of nuclear graphite have to be developed and qualified as an international effort. Remote techniques and adaptation of the reflector have to be improved to allow replacement of critical parts.

- The presentations did not cover all HTR-programmes as performed in non-OECD countries like Russia and China. They should be invited to future meetings to permit the full scope of global HTR R&D activities.

SESSION II

Improvement in Material Properties by High-temperature Irradiation

Chairs: T. Shikama, B. Marsden

In this session, four papers relevant to improvement of material properties were presented:

- *Improvement in Ductility of Refractory Metals by Neutron Irradiation* (presented by H. Kurishita, Tohoku University, Japan).

  The possibility of using HTR systems as a bed for improving properties of new advanced materials (MOW) was presented.

- *Effects of Superplastic Deformation on Thermal and Mechanical Properties of 3Y-TZP Ceramics (Review)* (presented by T. Hoshiya, JAERI, Japan).

  The application of new advanced materials to HTR systems (superplastic Y3-TZP, CFC/Si (8-10%) was presented.

- *Development of Innovative Carbon-based Ceramic Material: Application in High-temperature, Neutron and Hydrogen Environment* (presented by C.H. Wu, Max-Planck Institut, Germany).

  The subject of advanced ceramic material was presented. This material has very high thermal conductivity, dimensional stability under the neutron irradiation, lower chemical erosion (longer life time), lower tritium retention and lower reactivity with water and oxygen (safety concern).

- *A Carbon Dioxide Partial Condensation Cycle for High-temperature Reactors* (presented by T. Nitawaki, Tokyo Institute of Technology, Japan).

  The re-evaluation of CO_2 as a coolant in HTR systems (CO_2 partial condensation cycle) was presented.

Recommendations with respect to Session II

- Internationally co-operative survey for possibility of application of new advanced materials to HTR systems.

- Re-evaluate usage of CO_2 as a coolant in HTR systems (materials issues should be accessed).

- A deeper understanding of the mechanisms behind the observations made concerning the improvements to the material property changes is requested.

SESSION III

Development of In-core Material Characterisation Methods and Irradiation Facility

Chairs: C. Vitanza, M. Yamawaki

In this session, six papers were presented:

- *Development of the I-I type Irradiation Equipment for the HTTR* (presented by M. Ishihara, JAERI, Japan).

  The main design features of the HTTR were outlined. There are several possible locations for test irradiations in the HTTR, both in the core region and in the reflector. The space available in these locations is large, typically from 100 to 367 mm in diameter. Both thermal and fast flux are moderate, typically less than 2.10^{13} h/cm^{2}s (fast flux) and 3 to 7.10^{13} n/cm^{2}s (thermal flux). Temperature ranges between ~400 and ~900°C. The design characteristics of a first test irradiation rig to be installed in the HTTR were provided. This is intended to investigate the in-pile creep behaviour of stainless steel material in the 550-650°C temperature range.

- *Application of Optical Diagnostics in High-temperature Gas-cooled Systems* (presented by T. Shikama, Tohoku University, Japan).

  The development of optical fibre technology for dose and diagnostic measurements in high-temperature gas-cooled reactors was reviewed. Different materials have been tested in an attempt to enhance the performance in the visible region by means of material doping. The best optical fibre developed so far can be used for about one year in the HTTR core.

- *Measurement Method of In-core Neutron and Gamma-ray Distributions with Scintillator Optical Fibre Detector and Self-powered Detector* (presented by T. Shikama, Tohoku University).

  Novel methods for in-core neutron and gamma-ray distribution, based on scintillator optical fibre detector, were presented. There is currently an important use limitation to rather low temperatures (~100°C). Considerations were also made on the prevailing mechanisms of self-powered neutron detectors at moderate and at high temperatures, i.e. ~500°C.

- *Development of In-core Test Capabilities for Material Characterisation* (presented by C. Vitanza, OECD/NEA).

  The development of irradiation and in-core measuring techniques at the OECD Halden Reactor Project was described. The potential applications to gas reactor investigations were also presented.

- *Application of Work Function Measuring Technique to Monitoring/Characterisation of Material Surfaces under Irradiation* (presented by M. Yamawaki, the University of Tokyo, Japan).

  Development of innovative measuring systems such as the Kelvin probe system introduced above should be further encouraged and promoted in order to expand the capability of testing in HTGR. A new measuring system containing a Kelvin probe has been introduced which may be applicable to monitor/characterise the surface of solid materials under irradiation in gas atmosphere. The results of measurements using such systems have been presented where the microscopic change of the surface of nickel was closely monitored during ion irradiation which could not have been obtained without applying this method.

- *Development of High-temperature Irradiation Techniques Utilising the Japan Materials Testing Reactor* (presented by T. Shikama, Tohoku University, Japan).

  A review of the design features of high-temperature irradiation capsules for use in the JMTR was provided. The good – and when feasible, independent – control of temperature and neutron flux is a key characteristic of this design.

Recommendations with respect to Session III

- The development of irradiation devices for tests in the HTTR is encouraged. The I-I device will be operational in the near future and a test plan for the long-term utilisation of this device should be developed.

- There has been progress in the investigations on fibre optics in reactor properties. One should also concentrate on the application side, possibly beyond the use of fibre optics for dose or diagnostic monitoring.

- Some high-temperature investigations can be done in MTRs. The HTTR can develop synergies with MTRs, but it should also seek areas where the capabilities and characteristics of the HTTR can be utilised in a unique manner For instance, the large space and the high temperature could offer unique opportunities and be used profitably.

SESSION IV

Basic Studies on Behaviour of Irradiated Graphite/Carbon and Ceramic Materials Including their Composites under both Operation and Storage Conditions

Chairs: B. McEnaney, A.J. Wickham

The session was comprised of eight oral presentations:

- *Investigation of High-temperature Reactor (HTR) Materials* (presented by D. Buckthorpe, NNC, United Kingdom).

  A preliminary account of the HTR Materials Test Programme in the HTR-M and HTR-M1 EU programmes was presented in this paper.

- *Radially Keyed Graphite Moderator Cores: An Investigation into the Stability of Finite Element Models* (presented by S.E. Clift, University of Bath, United Kingdom).

  Finite-element analysis of graphite moderator structures was presented. The need to develop finite-element analyses of the mechanical behaviour of graphite cores in new HTRs was shown.

- *Understanding of Mechanical Properties of Graphite on the Basis of Mesoscopic Microstructure* (presented by M. Ishihara, JAERI, Japan).

  Mesoscopic modelling of mechanical properties was presented. The value of mesoscopic analyses was recognised, and the need for further refinement of the models was demonstrated.

- *Advanced Graphite Oxidation Model* (presented by R. Moormann, FZJ, Germany).

  Graphite oxidation under accident conditions was presented. Modelling the kinetics of oxidation of graphite after air ingress is well advanced, but further development is needed to improve the semi-empirical elements of the present model.

- *Irradiation Creep in Graphite – A Review* (presented by B.J. Marsden, the University of Manchester, United Kingdom).

  Irradiation creep of graphite is discussed.

- *An Analysis of Irradiation Creep in Nuclear Graphites* (presented by G.B. Neighbour, University of Hull, United Kingdom).

  Irradiation creep of graphite was shown. This paper and the previous paper highlighted the very complex factors influencing irradiation creep of graphite, and both authors recommended further MTR experiments related to the operating condition of HTR in order to address significant uncertainties in the models.

- *Planning for Disposal of Irradiated Graphite: Issues for the New Generation of HTRs* (presented by A.J. Wickham, Consultant, United Kingdom).

  The entire life cycle of graphite components with respect to planning for decommissioning was presented. The main emphasis of the paper on graphite-reactor decommissioning was the need to plan at the design stage for the eventual disposal of the graphite (and other components) at the outset, thereby potentially avoiding some of the mistakes of the past.

- *Status of IAEA International Database on Irradiated Graphite Properties with Respect to HTR Engineering Issues* (presented by B. McEnaney, University of Bath, United Kingdom).

  The value of archived data, with particular reference to the IAEA International Database on Irradiated Graphite Properties and its application in HTR engineering was presented. The value of preserving knowledge of all forms by whatever means in order to facilitate improvements in future reactor design and to assist in the specification and identification of the most appropriate materials (e.g. graphite type) were principal topics.

Recommendations with respect to Session IV

- Further development is required in the methodology for finite-element analysis of complex interacting structures, and this should be validated by carefully designed experiments. This will then allow much greater confidence in using the methods to investigate more complex assemblies.

- Newly developed basic prediction models on the strength and Young's modulus of graphite in view of microstructure are generally important and are required to understand properties regardless of material grades. These models should be expanded to the irradiation condition to understand more general irradiation data and to make use of the irradiation database for new HTR designs.

- There is a clear need for further experimental data from carefully designed irradiation experiments to resolve the multiple issues arising as a result of irradiation creep. This may perhaps be included in a more general recommendation that efforts should be made to acquire more general irradiation data on the materials which are the most likely candidate graphites for new HTR designs.

- Plans for decommissioning and disposal of the graphite and carbon-based components should form part of the design stage for all new reactors. This factor is relevant at all stages of design development, including the specification of the graphite provided to the manufacturers, since this may well have an important influence on the ease of handling of the components, their method of disposal and their waste categorisation status, for all of which there are potential economic savings.

SESSION V

Basic Studies on HTGR Fuel Fabrication and Performance

Chairs: J. Kuijper, R. Moormann

The session was comprised of two oral presentations and one paper which was not presented orally:

- *Evidence of Dramatic Fission Product Release from Irradiated Nuclear Ceramics* (presented by L. Thomé, CSNSM, France).

  One lecture (collaboration of France and Poland) dealt with fission-product (Cs) transport in ceramic matrices such as ZrO_2. It was shown, through ion implantation and by Rutherford backscattering, that Cs is very stable in ZrO_2 up to concentrations of about 1%. At higher concentrations, diffusion rates increase substantially at temperatures of 650°C.

- *Method of Reactor Tests of HTGR Fuel Elements at an Impulse Loading* (presented by A.S. Chernikov, Ministry of Russian Federation on Atomic Energy, France).

  A second lecture from Russia concerns development and testing of a facility for the investigation of the effect of impulse loading on HTR fuel elements in the IGR facility in Kazakstan. Four spherical fuel elements and one graphite ball were exposed to a maximum power excursion of 33.5 kW. The maximum temperature reached was about 2 000 K. The elements containing 1 g of fuel were intact, whereas the elements containing 1.5 g of fuel were damaged within under 35 s.

- *Gas Reactor TRISO-coated Particle Fuel Modelling Activities at the Idaho National Engineering and Environmental Laboratory* (D. Petti, INEEL, United States of America).

  An abstract and a full paper were presented, but no presentation was made. This paper discusses the INEEL's development of an integrated mechanistic fuel performance model for TRISO-coated gas reactor particle fuel termed PARFUME (PARticle FUel ModEl).

Recommendations with respect to Session V

- The research should also be extended to coatings and structural materials relevant to HTR.

- Extend the measurement to conditions more relevant to actual HTR concepts.

RECOMMENDATIONS

In the latter half of the summary session, possible international activities in the field of basic studies on high-temperature engineering were discussed within the framework of the OECD/NEA Nuclear Science Committee (NSC). It was recommended to include topics relevant to fission-product behaviour and safety issues of HTGR in next meeting, in addition to the topics discussed in this meeting. The chairperson of the last session summarised the recommendations to be presented to the NSC into the following five topics as possible international activities:

- Basic studies on behaviour of irradiated graphite/carbon and ceramic materials including their composites under both operation and storage conditions.

- Development of in-core material characterisation and instrumentation methods.

- Improvement in material properties through high-temperature irradiation.

- Basic studies on HTGR fuel fabrication and performance including fission-product release.

- Basic studies on safety issues of HTGR.

It was also recommended that a further information exchange meeting focused on the organisation of the interactive collaboration activity with regard to the above topics be planned in 2003, tentatively in Oarai, Japan.

LIST OF PARTICIPANTS

FRANCE

BONIN, Bernard
COGEMA
Direction de la recherche et du développement
2, rue Paul Dautier
F-78141 Vélizy-Villacoublay

Tel: +33 1-39-26-38-41
Fax: +33 1-39-26-27-13
Eml: bbonin@cogema.fr

DAMIAN, Frederic
CEA
Saclay DRN/DMT/SERMA
F-91191 Gif-Sur-Yvette

Tel: +33 01 69 08 46 64
Fax: +33 01 69 08 99 35
Eml: damian@soleil.serma.cea.fr

IDE, Hidekazu
Deputy Director
NFI Europe Office
10, rue de Louvois
F-75002 Paris

Tel: +33 1 44 86 05 75
Fax: +33 1 44 86 05 80
Eml: ide@nfi.co.jp

LANGUILLE, Alain
Commissariat à l'Énergie Atomique
Centre d'étude de Cadarache
F-13108 Saint-Paul-lez-Durance

Tel: +33 (0)4 42 25 27 22
Fax: +33 (0)4 42 25 70 42
Eml: Alain.LANGUILLE@cea.fr

MORITA, Takuji
Toyo TANSO Co. France
Z.I du Manet et du Bois de la Couldre
9-10, rue Eugène Hénaff
F-78190 Trappes

Tel: +33 1 30 66 35 35
Fax: +33 1 30 66 31 69
Eml: tmorita6@hotmail.com

ROUAULT, Jacques
Département d'Études des Réacteurs
Service d'études des réacteurs innovants
Bât. 212 CE Cadarache
F-13108 Saint-Paul-lez-Durance

Tel: +33 4 42 25 7265
Fax: +33 4 42 25 48 58
Eml: jrouault@drncad.cea.fr

THOMÉ, Lionel
CSNSM Orsay
Bât. 108
F-91405 Orsay

Tel: +33 1 69 15 52 60
Fax: +33 1 69 15 52 68
Eml: thome@csnsm.in2p3.fr

GERMANY

KUGELER, Kurt
Forschungzentrum Julich GmbH (FZJ)
Institute for Safety Research and Technology
P.O. Box 1913
D-5170 Jülich

Tel: +49 (2461) 61 59 91
Fax: +49 (2461) 61 68 56
Eml: k.kugeler@fz-juelich.de

MOORMANN, Rainer
Forschungszentrum Jülich GmbH – ESS-FZJ
D-52425 Jülich

Tel: +49 2461 614 644
Fax: +49 2461 618 255
Eml: R.Moormann@fz-juelich.de

PSCHOWSKI, Joachim
ESI Energie-Sicherheit-Inspektion GmbH
Besselstrasse 21
D-68219 Mannheim

Tel: +49 621 8764 606
Fax: +49 621 8764 666
Eml: joachim.pschowski@esi-mannheim.de

SARTORI, Rudolf
RWTUEV-AT GmbH
Langemarckstr. 20
D-45141 Essen

Tel: +49 2018252790
Fax: +49 2018252486
Eml: sartori@rwtuev-at.de

VON LENSA, Werner
Research Centre Jülich
Institute for Safety Research and
Nuclear Research Technology (ISR)
D-52425 Jülich

Tel: +49 2461 616629
Fax: +49 2461 615342
Eml: w.von.lensa@fz-juelich.de

WU, Chung
EFDA
Max-Planck-Institut für Plasmaphysik
Boltzmannstrasse 2
D-85748 Garching bei München

Tel: +49 89-32994232
Fax: +49 89-32994198
Eml: wuc@ipp.mpg.de

JAPAN

HOSHIYA, Taiji
High Temperature Irradiation Laboratory
JAERI
Oarai-machi, Ibaraki-ken, 311-1394

Tel: +81 29-264-8605
Fax: +81 29-264-8712
Eml: hoshiya@oarai.jaeri.go.jp

ISHIJIMA, Yasuhiro
Oarai Branch
Institute for Materials Research
Tohoku University
Oarai, Ibaraki, 311-1313

Tel: +81 29 267 3181
Fax: +81 29 267 4947
Eml: yasuishi@imr.tohoku.ac.jp

ISHIHARA, Masahiro
High Temperature Irradiation Laboratory
JAERI
Oarai-machi, Ibaraki-ken, 311-1394

Tel: +81 29 264 8734
Fax: +81 29 264 8712
Eml: ishihara@popsvr.tokai.jaeri.go.jp

ISHINO, Shiori
Dept. of Applied Science
Tokai University
1117, Kitakaname, Hiratsuka-shi
Kanagawa-ken 259-1292

Tel: +81 463 58 1211 (ext.4153)
Fax: +81 463 50 2017
Eml: ishino@keyaki.cc.u-tokai.ac.jp

KURISHITA, Hiroaki
The Oarai Branch
Institute for Materials Research (IMR)
Tohoku University
Narita-machi, Oarai, Higashi-Ibaraki-gun
Ibaraki-ken 311-1313

Tel: +81 29-267-4157
Fax: +81 29-267-4947
Eml: kurishi@imr.tohoku.ac.jp

NITAWAKI, Takeshi
Research Laboratory for Nuclear Reactors
Tokyo Institute of Technology
O-okayama, Meguro-ku, Tokyo 152-8550

Tel: +81 3-5734-3293
Fax: +81 3-5734-2959
Eml: nitawaki@nr.titech.ac.jp

SANOKAWA, Konomo
Executive Director
Japan Marine Science Foundation
4-1-2 Futabadai, Mito-shi
Ibaraki-ken, 311-4143

Tel: +81 29 253 0752
Fax: +81 29 253 0752
Eml:

SHIKAMA, Tatsuo
Institute for Materials Research
Tohoku University
Katahira, Sendai, 980-8577

Tel: +81 29-267-4205(-3181) (Oarai)
Fax: +81 29 267-4947; 022 215 2061
Eml: shikama@imr.tohoku.ac.jp

SUDO, Yukio
JAERI
Oarai-machi, Ibaraki-ken 311-1394

Tel: +81 29-264-8201
Fax: +81 29-264-8471
Eml: sudo@hems.jaeri.go.jp

YAMAWAKI, Michio
Graduate School of Engineering
University of Tokyo
7-3-1 Hongo, Bunkyo-ku
Tokyo 113-8656

Tel: +81 3 5841 7422
Fax: +81 3 5841 8633
Eml: yamawaki@q.t.u-tokyo.ac.jp

YOSHIMUTA, Shigeharu
Nuclear Fuel Industries, Ltd.
3135-41 Muramatsu Tokai-mura
Naka-gun Ibaraki-ken 319-1196

Tel: +81 29-287-8212
Fax: +81 29-287-8223
Eml: yosimuta@nfi.co.jp

THE NETHERLANDS

KUIJPER, Jim C.
Fuels, Actinides and Isotopes
NRG
Postbus 25
NL-1755 ZG Petten

Tel: +31 (224) 56 4506
Fax: +31 (224) 56 3490
Eml: kuijper@nrg-nl.com

VAN DER HAGEN, Tim
IRI – Interfaculty ReactorInstitute
Delft University of Technology
Mekelweg 15
NL-2629 JB Delft

Tel: +31 15 278 2105
Fax: +31 15 278 6422
Eml: hagen@iri.tudelft.nl

RUSSIAN FEDERATION

CHERNIKOV, Albert
Scientific and Industrial Association "Lutch"
Ministry of the Russian Federation on Atomic
Energy
24 Zheleznodorozhnaya
142100 Padolsk Moscow Region

Tel: +7 095 137 9876
Fax: +7 096 7 637097
Eml: chernikov@sialuch.ru

SPAIN

BARRIO, Felix
CIEMAT
Departamento de Fision Nuclear
Avenida Complutense 22
E-28040 Madrid

Tel: +34 91 346 6236
Fax: +34 91 346 6233
Eml: felix.barrio@ciemat.es

UNITED KINGDOM

BUCKTHORPE, Derek
NNC Limited
Booths Hall
Chelford Road
Knutsford
Cheshire WA16 8QZ

Tel: +44 1565 84 3845
Fax: +44 1565 84 3878
Eml: Derek.Buckthorpe@nnc.co.uk

CLIFT, Sally
Department of Mechanical Engineering
University of Bath
Claverton Down
BA2 7AY
Bath

Tel: +44 1225 323870
Fax: +44 1225 826928
Eml: s.e.clift@bath.ac.uk

HOLT, Matt
Department of Engineering
The University of Hull
Cottingham Road
HU6 7RX, Hull

Tel:
Fax:
Eml: m.j.holt@eng.hull.ac.uk

MARSDEN, Barry
Manchester School of Engineering
The University of Manchester
Oxford Road
Manchester M13 9PL

Tel: +44 161 275 4399
Fax: +44 161 275 3844
Eml: barry.marsden@man.ac.uk

MCENANEY, Brian
Director, Nuclear Energy Group
Department of Engineering and Applied
Science
University of Bath
Claverton Down

Tel: +44 1225 826969
Fax: +44 1225 826098
Eml: B.McEnaney@bath.ac.uk

NEIGHBOUR, Gareth
Department of Engineering
University of Hull
Cottingham Road, Hull
HU6 7RX

Tel: +44 1482 466535
Fax: +44 1482 466664
Eml: g.b.neighbour@hull.ac.uk

WICKHAM, Anthony
Consultant
Cwmchwefru
Llanafanfawr
Builth Wells
LD2 3PW Powys

Tel: +44 1597 860 244
Fax: +44 1597 860 244
Eml: tony@tonywickham.co.uk

UNITED STATES OF AMERICA

BALL, David R.
DRB Carbon and Graphite Consulting
28910 Westwood Road
Bay Village, Ohio 44140

Tel: +1 440.835.2859
Fax: +1 440.835.5856
Eml: daiball@aol.com

INTERNATIONAL ORGANISATIONS

MARTIN-BERMEJO, Joaquin
European Commission
DG RTD Unit J-4
Rue de la Loi, 200
Building: MO75/5/30
B-1049 Brussels

Tel: +32 (0)2 295 8332
Fax: +32 (0)2 295 4991
Eml: joaquin.martin-bermejo@cec.eu.int

KESSLER, Carol
Deputy Director General
OECD Nuclear Energy Agency
Le Seine St-Germain
12, boulevard des Îles
F-92130 Issy-les-Moulineaux

Tel: +33 (0) 1 45 24 10 02
Fax: +33 (0) 1 45 24 11 10
Eml: carol.kessler@oecd.org

SHIMOMURA, Kazuo
Deputy Director
OECD Nuclear Energy Agency
Le Seine St-Germain
12, boulevard des Îles
F-92130 Issy-les-Moulineaux

Tel: +33 01 45 24 10 04
Fax: +33 01 45 24 11 06
Eml: kazuo.shimomura@oecd.org

NA, Byung Chan
OECD Nuclear Energy Agency
Le Seine St-Germain
12, boulevard des Îles
F-92130 Issy-les-Moulineaux

Tel: +33 1 4524 1091
Fax: +33 1 4524 1110
Eml: na@nea.fr

ROYEN, Jacques
OECD Nuclear Energy Agency
Le Seine St-Germain
12, boulevard des Îles
F-92130 Issy-les-Moulineaux

Tel: +33 1 45 24 10 52
Fax: +33 1 45 24 11 10
Eml: jacques.royen@oecd.org

SUYAMA, Kenya
OECD Nuclear Energy Agency
Le Seine St-Germain
12, boulevard des Îles
F-92130 Issy-les-Moulineaux

Tel: +33 (0) 1 45 24 11 52
Fax: +33 (0) 1 45 24 11 06
Eml: suyama@nea.fr

VITANZA, Carlo
OECD Nuclear Energy Agency
Le Seine St-Germain
12, boulevard des Îles
F-92130 Issy-les-Moulineaux

Tel: +33 01 45 24 10 62
Fax: +33 01 45 24 11 29
Eml: carlo.vitanza@oecd.org

NEA Publications of General Interest

2000 Annual Report (2001) *Free: paper or Web.*

NEA News
ISSN 1605-9581 Yearly subscription: € 37 US$ 45 GBP 26 ¥ 4 800

Geologic Disposal of Radioactive Waste in Perspective (2000)
ISBN 92-64-18425-2 Price: € 20 US$ 20 GBP 12 ¥ 2 050

Nuclear Science

Fission Gas Behaviour in Water Reactor Fuels (2002)
ISBN 92-64-19715-X Price: € 120 US$ 107 GBP 74 ¥ 12 100
Shielding Aspects of Accelerators, Targets and Irradiation Facilities – SATIF 5 (2001)
ISBN 92-64-18691-3 Price: € 84 US$ 75 GBP 52 ¥ 8 450
Nuclear Production of Hydrogen (2001)
ISBN 92-64-18696-4 Price: € 55 US$ 49 GBP 34 ¥ 5 550
Pyrochemical Separations (2001)
ISBN 92-64-18443-0 Price: € 77 US$ 66 GBP 46 ¥ 7 230
Evaluation of Speciation Technology (2001)
ISBN 92-64-18667-0 Price: € 80 US$ 70 GBP 49 ¥ 7 600
Comparison Calculations for an Accelerator-driven Minor Actinide Burner (2002)
ISBN 92-64-18669-4 *Free: paper or web.*
Forsmark 1 & 2 Boiling Water Reactor Stability Benchmark (2001)
ISBN 92-64-18669-4 *Free: paper or web.*
3-D Radiation Transport Benchmarks for Simple Geometries with Void Regions
(2000)   ISBN 92-64-18274-8 *Free: paper or web.*
Benchmark Calculations of Power Distribution Within Fuel Assemblies
Phase II: Comparison of Data Reduction and Power Reconstruction Methods in Production Codes
(2000)   ISBN 92-64-18275-6 *Free: paper or web.*
Benchmark on the VENUS-2 MOX Core Measurements (2000)   ISBN 92-64-18276-4 *Free: paper or web.*
Calculations of Different Transmutation Concepts: An International Benchmark Exercise
(2000)   ISBN 92-64617638-1 *Free: paper or web.*
Prediction of Neutron Embrittlement in the Reactor Pressure Vessel: VENUS-1 and VENUS-3 Benchmarks
(2000)   ISBN 92-64-17637-3 *Free: paper or web.*
Pressurised Water Reactor Main Steam Line Break (MSLB) Benchmark
(2000)   ISBN 92-64-18280-2 *Free: paper or web.*

International Evaluation Co-operation (*Free on request - paper or CD-ROM*)
Volume 1: *Comparison of Evaluated Data for Chromium-58, Iron-56 and Nickel-58* (1996)
Volume 2: *Generation of Covariance Files for Iron-56 and Natural Iron* (1996)
Volume 3: *Actinide Data in the Thermal Energy Range* (1996)
Volume 4: ^{238}U *Capture and Inelastic Cross-Sections* (1999)
Volume 5: *Plutonium-239 Fission Cross-Section between 1 and 100 keV* (1996)
Volume 8: *Present Status of Minor Actinide Data* (1999)
Volume 10: *Evaluation Method of Inelastic Scattering Cross-sections for Weakly Absorbing Fission-product Nuclides* (2001)
Volume 12: *Nuclear Model to 200 MeV for High-Energy Data Evaluations* (1998)
Volume 13: *Intermediate Energy Data* (1998)
Volume 14: *Processing and Validation of Intermediate Energy Evaluated Data Files* (2000)
Volume 15: *Cross-Section Fluctuations and Shelf-Shielding Effects in the Unresolved Resonance Region* (1996)
Volume16: *Effects of Shape Differences in the Level Densities of Three Formalisms on Calculated Cross-Sections* (1998)
Volume 17: *Status of Pseudo-Fission Product Cross-Sections for Fast Reactors* (1998)
Volume 18: *Epithermal Capture Cross-Section of* ^{235}U (1999)

Order form on reverse side.

ORDER FORM

OECD Nuclear Energy Agency, 12 boulevard des Iles, F-92130 Issy-les-Moulineaux, France
Tel. 33 (0)1 45 24 10 15, Fax 33 (0)1 45 24 11 10, E-mail: nea@nea.fr, Internet: www.nea.fr

Qty	Title	ISBN	Price	Amount
			Total*	

* Including postage fees.

Charge my credit card   ❑ VISA   ❑ Mastercard   ❑ Eurocard   ❑ American Express

Card No.	Expiration date	Signature
Name		
Address	Country	
Telephone	Fax	
E-mail		

OECD PUBLICATION, 2, rue André-Pascal, 75775 PARIS CEDEX 16
PRINTED IN FRANCE
(66 2002 10 1 P) ISBN 92-64-19796-6 – No. 52527 2002